AF327132

# CISM COURSES AND LECTURES

INTERNATIONAL CENTRE FOR MECHANICAL SCIENCES

COURSES AND LECTURES - No. 391

# FREE SURFACE FLOWS

EDITED BY

HENDRIK C. KUHLMANN AND HANS-JOSEF RATH
UNIVERSITY OF BREMEN

SpringerWienNewYork

Le spese di stampa di questo volume sono in parte coperte da
contributi del Consiglio Nazionale delle Ricerche.

This volume contains 122 illustrations



In order to make this volume available as economically and as
rapidly as possible the authors' typescripts have been
reproduced in their original forms. This method unfortunately
has its typographical limitations but it is hoped that they in no
way distract the reader.

ISBN 3-211-83140-1 Springer-Verlag Wien New York

# PREFACE

*Free surface flows are ubiquitous in nature and technology. They are important in many branches of science ranging from physics and oceanography to engineering and materials science.*

*A course entitled Free Surface Flows was held at CISM in Udine during September 1-5, 1997. It was aimed at giving an up-to-date introduction to selected topics in this area and attracted many young scientists working on hydrodynamics or free boundary value problems. The audience's education ranged from graduate to post-doctoral level in the fields of engineering, physics, and applied mathematics. The present lecture notes have been written primarily for advanced students in applied natural sciences. But we hope that they will also be useful to others working, e.g. in industrial research, as an introduction and a reference source on a number of current free surface flow problems.*

*The lectures published in this volume range from introductory texts to reviews on recent advances in selected problems. The fluid physics is emphasized in the majority of the lectures for a thorough understanding of the mechanisms behind the various observable phenomena. Since today most theoretical progress relies heavily on the use of numerical methods, one lecture exclusively devoted to this topic has been included.*

*The fluid dynamics of systems with liquid-liquid interfaces is usually determined by capillary forces, surface stresses, and body forces. These effects can give rise to a variety of phenomena most of which are strongly gravity-dependent. Some attention has been paid, therefore, to those effects, that are relevant under microgravity conditions.*

*The topics covered range from classical capillary phenomena and Marangoni effects to modern numerical techniques. Isothermal phenomena, such as linear and nonlinear gravity and capillary waves are treated and supplemented with a presentation of recent developments in the Faraday problem. Other subjects include the equilibrium shape, stability, and dynamics of capillary surfaces, in particular, the dynamics of liquid bridges and jets. The thermocapillary migration of drops and bubbles is discussed as well as dynamic contact angles in problems of spreading and wetting. Hydrodynamic instabilities and nonlinear flows in liquid layers heated perpendicular to the interface (Rayleigh-Bénard-Marangoni problem) are considered with emphasis on the influence of lateral walls on the pattern formation. Other types of instabilities with applications in crystal growth occur in thermocapillary flows generated by strong temperature gradients tangential to the interface. Recent advances in the analysis of convection in thermocapillary liquid bridges are discussed. Flows in thin liquid films are strongly governed by solutal Marangoni effects. The particular hydrodynamics of quasi-two-dimensional flows and waves in thin liquid films are explained and theoretical predictions are compared with experimental results. These more physics-oriented contributions are supplemented by a review on the state-of-the-art of Eulerian based computational methods for free surface flows. The topics among others include volume tracking methods, particle based methods, and interface dynamics. It is hoped that this volume will serve as a useful starting point for studies in these fields of science which are rapidly developing.*

*The editors are indebted to Wilhelm Schneider for his encouragement to organize this course and for his continued interest and support. All participants have been impressed by the stimulating atmosphere of the center. We are very grateful to Sandor Kaliszky and Giovanni Bianchi for their kind hospitality and thank the CISM staff for their friendly and efficient help in all administrative matters.*

*Hendrik C. Kuhlmann*
*Hans-Josef Rath*

# CONTENTS

Page

# WAVES ON INTERFACES


**S. Fauve**
**Ecole Normale Supérieure, Paris, France**


## Abstract


I review basic results about waves at the interface between a horizontal fluid layer and air at atmospheric pressure or at the interface between two non-miscible fluids. The restoring mechanisms are mostly gravity and surface tension but coupling with relative fluid motion or with an electric or a magnetic field will also be considered. The effect of dissipation on the dispersion relation is discussed. At the nonlinear level, perturbative methods are described in the case of long wavelength approximations or slowly varying wave packets. Finally, parametric amplification of surface waves on a vertically vibrated fluid interface, the Faraday instability, is considered. These lecture notes should not be considered as a review article on the subject. In particular, I have not tried to give a complete bibliography and have only quoted the material used to prepare the notes. Many aspects of the three first parts can be found in well-known books: Non dissipative linear surface waves are presented in details in "Theoretical mechanics of particles and continua" by Fetter and Walecka [1]. A simple discussion of various dissipative effects, both in the bulk and boundary layers can be found in Landau and Lifshitz, "Fluid Mechanics" [2]. Nonlinear aspects are discussed in details by Whitham, "Linear and Nonlinear Waves" [3] and Newell, "Solitons" [4].

# 1 Surface waves

We consider the interface between a horizontal layer of fluid and air at atmospheric pressure (figure 1.a) or between two non-miscible fluids (figure 1.b) and we study the propagation of an interface deformation. The fluid is assumed incompressible of density $\rho$, respectively $\rho_i$ ($i = 1, 2$). $h_0$, respectively, $h_i$, are the heights of the unperturbed fluid layers, and $g$ is the acceleration of gravity. $\tau$ is the surface tension of the liquid-air, respectively the liquid-liquid interface.

In this first section we will study the following linear aspects: dispersion relations of gravity-capillary waves, effect of a relative motion of the fluids in the two-layer case and the Kelvin-Helmholtz instability, effect of electric or magnetic fields and field-coupled surface waves, and viscous damping of surface waves. Dispersion relations of gravity-capillary waves on the surface of inviscid fluids are well known and have been derived in many books (see for instance [1]). We emphasize in this section, the use of dimensional analysis, in particular when both gravity and capillarity are acting, and discuss the effect of dissipation, which, to the best of our knowledge, is not considered in details in textbooks.

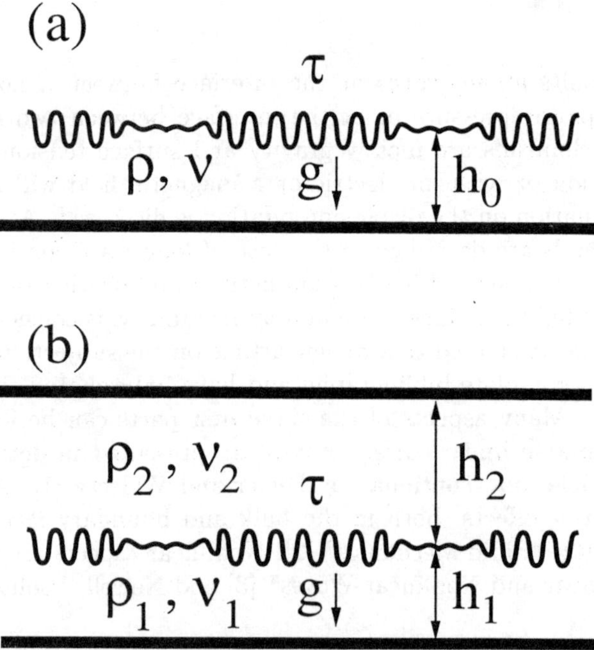


Figure 1. The interface between a fluid layer and air at atmospheric pressure (a), or between two non miscible fluid layers.

## 1.1 Dimensional analysis

First consider the simplest situation: a deep layer of fluid, $h_0 \to \infty$, i.e; $h_0 \ll \lambda$, and assume that the fluid is incompressible and non viscous. Guess the dispersion relations for linear sinusoidal waves of pulsation $\omega$ and wavenumber $k$, in the gravity and capillary regimes.

- Gravity waves : there are four parameters, $\omega$, $k$, $g$, $\rho$, and three units, length, time and mass. There is thus only one non dimensional parameter, $\omega^2/gk$ and the dispersion relation is of the form

$$\omega^2 = C\,gk,\tag{1}$$

where $C$ is a constant ($C = 1$, see 1.2).

- Capillary waves : similarly, we have

$$\omega^2 = C'\,\frac{\tau}{\rho}k^3,\tag{2}$$

where $C'$ is a constant ($C' = 1$, see 1.2).

When the height of the layer is not large compared to the wavelength, $h_0$ should be taken into account and this gives a second non dimensional parameter, $kh_0$. then we have,

- Gravity waves :

$$\omega^2 = gk\,f(kh_0),\tag{3}$$

- Capillary waves :

$$\omega^2 = \frac{\tau}{\rho}k^3\,g(kh_0),\tag{4}$$

where $f$ and $g$ are arbitrary functions of $kh_0$.

*Question 1: could you prove without calculations that the dispersion relation is of the form*

$$\omega^2 = \left(gk + \frac{\tau}{\rho}k^3\right)f(kh_0),\tag{5}$$

*when both gravity and capillary effects have to be taken into account? Two important features should be noted: first, the expressions for $\omega^2$ are additive; why ? Second, although the two restoring mechanisms operate respectively in the bulk (gravity) and at the interface (capillarity), the functional dependence on $kh_0$ is the same; why ? (See 1.3 for help).*

Thus, gravity waves are dominant at large wavelength and capillary waves at short wavelength. The characteristic transition wavelength is the only length scale which can be found with $g, \rho, \tau$, the capillary length

$$l_c = \sqrt{\frac{\tau}{\rho g}}. \tag{6}$$

Both gravity and capillary waves are dispersive. Gravity waves propagate faster for large wavelength whereas their velocity vanishes in the limit of zero wavelength. The opposite is true for capillary waves.

*Question 2: compute the phase velocity $V$ and the group velocity $U$ of gravity-capillary waves. Show that in all realistic situations, capillary-gravity waves propagate much slower than sound waves. By evaluating density perturbations, show that compressibility can be neglected when studying surface waves.*

We now consider the "shallow water" limit, i.e. $kh_0 \to 0$. What is the limit of $f(kh_0)$? We could expect that the wave velocity does not depend any more on $k$ when $k \to 0$ for $h_0$ fixed. then, $V \propto \sqrt{gh_0}$ and we have

$$\omega^2 \approx gh_0 k^2. \tag{7}$$

Thus, dispersion vanishes in the shallow water limit. Note however that the above argument is not rigorous (see "shallow water approximations").

*Question 3: $\sqrt{\tau/\rho h_0}$ is homogeneous to a velocity. Is it the velocity of a corresponding wave ? (See Chomaz' lectures).*

## 1.2  Dispersion relation for gravity-capillary waves

We consider the two-layer case (figure 1.b) and assume that the fluids are incompressible and inviscid. In each layer ($i = 1, 2$), the fluid motion is governed by the incompressibility and the Euler equations

$$\nabla \cdot \mathbf{v}_i = 0 \tag{8}$$

$$\left[ \frac{\partial \mathbf{v}_i}{\partial t} + (\mathbf{v}_i \cdot \nabla)\mathbf{v}_i \right] = -\frac{1}{\rho_i}\nabla p_i + \mathbf{g}. \tag{9}$$

The normal velocity should vanish at the solid boundaries

$$\mathbf{v}_1(z = -h_1) \cdot \hat{\mathbf{z}} = \mathbf{v}_2(z = -h_2) \cdot \hat{\mathbf{z}} = 0. \tag{10}$$

The kinematic boundary condition at the interface, $z = \zeta(x, y, t)$, is

$$\frac{\partial \zeta}{\partial t} + \mathbf{v}_i \cdot \nabla \zeta = \mathbf{v}_i \cdot \hat{\mathbf{z}}. \tag{11}$$

In the absence of viscous stresses, the balance of forces along the normal of the free surface with principal radii of curvature $R_1$ and $R_2$ is given by Laplace formula, and the dynamic boundary condition at the interface, $z = \zeta(x, y, t)$, is

$$p_1 - p_2 = \tau \left( \frac{1}{R_1} + \frac{1}{R_2} \right) = -\tau \nabla \cdot \frac{\nabla \zeta}{[1 + (\nabla \zeta)^2]^{1/2}}. \tag{12}$$

When the interface is flat, the basic state corresponds to the hydrostatic solution,

$$\mathbf{v}_{0i} = 0, \quad p_{0i} = -\rho_i g z. \tag{13}$$

When the interface is deformed, the velocity field in each layer generates a dynamic pressure component, $\tilde{p}_i \equiv p_i + \rho_i g z$. If the velocity field has initially zero vorticity, the flow remains potential, thus $\mathbf{v}_i = \nabla \phi_i$, and from the incompressibility equation, we get

$$\Delta \phi_i = 0. \tag{14}$$

From the Euler equation for a potential flow, we get the Bernoulli formula and the expression for $p_i$ is

$$p_i = -\rho_i g z - \rho_i \frac{\partial \phi_i}{\partial t} - \frac{1}{2} \rho_i v_i^2. \tag{15}$$

The boundary conditions at the free surface $z = \zeta(x, y, t)$ thus become

$$\frac{\partial \zeta}{\partial t} + \nabla \phi_i \cdot \nabla \zeta = \frac{\partial \phi_i}{\partial z}, \tag{16}$$

$$(\rho_2 - \rho_1) g \zeta + \rho_2 \frac{\partial \phi_2}{\partial t} - \rho_1 \frac{\partial \phi_1}{\partial t} + \frac{1}{2} \left[ \rho_2 (\nabla \phi_2)^2 - \rho_1 (\nabla \phi_1)^2 \right] = -\tau \nabla \cdot \frac{\nabla \zeta}{[1 + (\nabla \zeta)^2]^{1/2}}. \tag{17}$$

Equations (10) give

$$\frac{\partial \phi_1}{\partial z}(z = -h_1) = \frac{\partial \phi_2}{\partial z}(z = -h_2) = 0. \tag{18}$$

When the amplitude of the disturbances of the interface is small compared to their wavelength, we can linearize equations (11, 12) and get

$$\frac{\partial \zeta}{\partial t} \approx \frac{\partial \phi_i}{\partial z}(z = 0), \tag{19}$$

$$(\rho_2 - \rho_1) g \zeta + \rho_2 \frac{\partial \phi_2}{\partial t}(z = 0) - \rho_1 \frac{\partial \phi_1}{\partial t}(z = 0) \approx -\tau \Delta \zeta. \tag{20}$$

We next consider a sinusoidal wave of the form

$$\phi_i = f_i(z) \exp i(kx - \omega t) + cc, \tag{21}$$

$$\zeta = C \exp i(kx - \omega t) + cc. \tag{22}$$

Equation (14) together with the boundary conditions (18) gives

$$\phi_1 = A\frac{\operatorname{ch} k(z + h_1)}{\operatorname{ch} kh_1} \exp i(kx - \omega t) + cc \tag{23}$$

$$\phi_2 = B\frac{\operatorname{ch} k(z - h_2)}{\operatorname{ch} kh_2} \exp i(kx - \omega t) + cc \tag{24}$$

Note that for deep layers, i.e. $kh_i \to \infty$, the amplitude of the velocity field decays exponentially away from the interface. We emphasize that the characterisitic "penetration length" of the wave is equal to the wavelength $\lambda$. Equations (19, 20) then give

$$-i\omega C = Ak \operatorname{th} kh_1 \tag{25}$$

$$-i\omega C = -Bk \operatorname{th} kh_2 \tag{26}$$

$$(\rho_2 - \rho_1)gC - i\omega\rho_2 B + i\omega\rho_1 A = \tau k^2 C. \tag{27}$$

Eliminating $A$ and $B$ leads to the dispersion relation

$$\omega^2 = \frac{(\rho_1 - \rho_2)gk + \tau k^3}{\rho_1 \operatorname{coth} kh_1 + \rho_2 \operatorname{coth} kh_2}. \tag{28}$$

The dispersion relation for waves on the surface of a liquid layer of height $h_0$ (figure 1.a) can be obtained with the substitution $\rho_1 = \rho$, $\rho_2 = 0$, $h_1 = h_0$,

$$\omega^2 = \left[gk + \frac{\tau}{\rho}k^3\right] \operatorname{th} kh_0. \tag{29}$$

It is interesting to discuss equation (28) in the case of a liquid layer in contact with its vapor phase in the vicinity of the liquid-vapor critical point $(\rho_c, T_c)$, for which we have

$$\rho_1 - \rho_2 \propto (T_c - T)^\alpha \tag{30}$$

$$\tau \propto (T_c - T)^\beta, \tag{31}$$

with $\alpha \approx 0.3$ and $\beta \approx 1.2$. For a given excitation frequency $\omega$, the wavenumber should become infinite when $T \to T_c$ if viscosity is neglected. Note also that although $\lambda \to 0$, gravity waves are dominant in the vicinity of the critical point! This is because the values of $\alpha$ and $\beta$ strongly deviate from their mean field values 0.5 and 1.5.

When $\rho_1 < \rho_2$, i.e. the heavier fluid is above the lighter one, $\omega^2 < 0$ for modes with wavenumbers smaller than $\sqrt{(\rho_2 - \rho_1)g/\tau}$; thus $\omega$ is pure imaginary and there is a mode with an exponentially growing amplitude. This is the Rayleigh-Taylor instability, discussed in details by Chandrasekhar [5].

*Question 4: find the dispersion relation in the two-layer case when the upper fluid surface is free. Discuss the above comment about the "penetration length" of the wave.*

## 1.3  Integral relations

The propagation of a wave always involves periodic transfers between kinetic energy and potential energy. The dispersion relation could be understood as a mathematical expression of this phenomenon. We will show this for surface waves by considering several integral relations involving the fluid energy, $E$

$$E = \frac{1}{2}\rho \int_V v^2 dV + \rho g \int_V z dV + \tau \int_S dA. \tag{32}$$

The first term is the fluid kinetic energy; the second term is the gravitational potential energy which could be writen in the form

$$E_{pg} = \rho g \int dxdy \int_{-h_0}^{z=\zeta} z dz = \frac{1}{2}\rho g \int \zeta^2(x,y,t)\, dxdy + \text{constant}; \tag{33}$$

the last term is the surface potential energy which could be writen in the form

$$E_{ps} = \tau \int \sqrt{1+(\nabla\zeta)^2}\, dxdy \approx \frac{1}{2}\tau \int (\nabla\zeta)^2\, dxdy + \text{constant}. \tag{34}$$

We now multiply the equation $\Delta\phi = 0$ by $\phi$ and integrate by part on the fluid volume $V$,

$$\int \phi\Delta\phi\, dV = -\int (\nabla\phi)^2 dV + \int \left[\phi\frac{\partial\phi}{\partial z}\right]_{z=\zeta} dxdy = 0. \tag{35}$$

We substitute relation (19) in the above equation and average it on one period of the wave. Integrating by part, we replace $\phi\zeta_t$ by $-\phi_t\zeta$ and use relation (20). We get

$$\frac{1}{2}\rho \overline{\int_V (\nabla\phi)^2 dV} \approx \frac{1}{2}\overline{\int [\rho g \zeta^2 + \tau(\nabla\zeta)^2]\, dxdy}. \tag{36}$$

We thus obtain that in the small amplitude limit we have equipartition between kinetic and potential energy. If we subsitute in this equation expressions (22-24) for $\zeta$ and $\phi$, we find the dispersion relation, $\omega^2$ coming from the kinetic energy and the gravity and surface terms coming from gravitational and surface potential energies.

*Question 5: show formula (35) by multiplying (20) by $\zeta$ and integrating on the fluid surface.*

The above integral relation (36) is useful to find corrections to the dispersion relation due to edge constraints. In a closed container, the fluid meniscus strongly affects the profile of the interface in the vicinity of the lateral boundaries. This generates an additional coupling between the interface and the fluid flow, except with very unrealistic boundary conditions (if one assumes that the fluid interface remains perpendicular to the lateral boundaries). It is possible to eliminate the effect of the meniscus by using a "brimful" configuration [6], i. e. by pinning the fluid interface at the edge of the lateral boundaries. However, this type of edge constraint also generates an additional coupling between the motion of the interface and the flow generated beneath. The corrections to the dispersion relation have been calculated in the case of an inviscid fluid [6, 7].

## 1.4  Coupling with a mean-flow: the Kelvin-Helmholtz instability

We consider now the two-layer situation when the fluids are in relative motion at velocities $U_1$ and $U_2$ along the $x$-axis. The basic state is thus

$$\mathbf{v}_{0i} = U_i\hat{\mathbf{x}}, \quad p_{0i} = -\rho_i g z. \tag{37}$$

The linearized Euler equation for the disturbances $\tilde{v}_i$, $\tilde{p}_i$, gives

$$\frac{\partial \tilde{v}_i}{\partial t} + U_i\frac{\partial \tilde{v}_i}{\partial x} = -\frac{1}{\rho_i}\nabla \tilde{p}_i. \tag{38}$$

Equations (14, 18) are unchanged, thus the expressions for the velocity potentials $\tilde{\phi}_i$ are similar to the ones derived above for $\phi_i$. The boundary conditions at the free surface are modified as follows to leading order

$$\frac{\partial \zeta}{\partial t} + U_i\frac{\partial \zeta}{\partial x} \approx \frac{\partial \tilde{\phi}_i}{\partial z}(z = 0), \tag{39}$$

$$\rho_1 g\zeta + \rho_1\left(\frac{\partial \tilde{\phi}_1}{\partial t} + U_1\frac{\partial \tilde{\phi}_1}{\partial x}\right)(z = 0) - \rho_2 g\zeta - \rho_2\left(\frac{\partial \tilde{\phi}_2}{\partial t} + U_2\frac{\partial \tilde{\phi}_2}{\partial x}\right)(z = 0) \approx \tau\Delta\zeta. \tag{40}$$

Thus, in the deep layer limit ($kh_i \to \infty$),

$$i(-\omega + kU_1)C = Ak, \tag{41}$$

$$i(-\omega + kU_2)C = -Bk, \tag{42}$$

$$(\rho_1 - \rho_2)gC + i\rho_1(-\omega + kU_1)A - i\rho_2(-\omega + kU_2)B = -\tau k^2 C, \tag{43}$$

and the dispersion relation is

$$\rho_1(\omega - kU_1)^2 + \rho_2(\omega - kU_2)^2 = (\rho_1 - \rho_2)gk + \tau k^3. \tag{44}$$

Solving for $\omega$ gives

$$\omega = k\frac{\rho_1 U_1 + \rho_2 U_2}{\rho_1 + \rho_2}$$

$$\pm \left[\frac{\rho_1 - \rho_2}{\rho_1 + \rho_2}gk + \frac{\tau}{\rho_1 + \rho_2}k^3 - \frac{\rho_1\rho_2}{(\rho_1 + \rho_2)^2}(U_1 - U_2)^2 k^2\right]^{1/2}. \tag{45}$$

An instability occurs when the expression under the square-root vanishes and becomes negative. This is the Kelvin-Helmholtz instability. This occurs for a critical value of the velocity difference $\Delta U \equiv |U_1 - U_2|$ such that

$$(\rho_1 - \rho_2)\frac{g}{k} + \tau k = \frac{\rho_1 \rho_2}{(\rho_1 + \rho_2)^2}\Delta U_c^2(k). \tag{46}$$

The minimum $\Delta U_c$ of $\Delta U_c^2(k)$ is reached for $k = k_c$ such that

$$k_c = \sqrt{\frac{g(\rho_1 - \rho_2)}{\tau}}, \tag{47}$$

and

$$\Delta U_c = 2\frac{\rho_1 + \rho_2}{\rho_1 \rho_2}\sqrt{g\tau((\rho_1 - \rho_2)}. \tag{48}$$

We refer to Chandrasekhar [5] for a more detailed presentation of the Kelvin-Helmoltz instability. The nonlinear regime has been discussed by Weissman [8].

## 1.5  Field-coupled surface waves

We consider two layers of electrically conducting or dielectric (respectively magnetic) liquids, submitted to an externally applied electric (respectively magnetic) field. Waves at the interface are obviously affected by this external field because the stress tensor is modified. For instance, we have for an inviscid dielectric liquid of permitivity $\epsilon$,

$$\sigma_{ik} = -p\delta_{ik} + \epsilon E_i E_k - \frac{E^2}{2}\left[\epsilon - \rho\left(\frac{\partial \epsilon}{\partial \rho}\right)_T\right]\delta_{ik}, \tag{49}$$

and the dynamic boundary condition at the interface becomes

$$\left(\sigma_{ik}^{(2)} - \sigma_{ik}^{(1)}\right)n_k = -\tau\nabla \cdot \frac{\nabla\zeta}{[1 + (\nabla\zeta)^2]^{1/2}}. \tag{50}$$

In turn, the electric or magnetic field is affected by the motion of the interface and does not remain uniform. The field lines are indeed distorted in order to satisfy the boundary conditions at the free surface. Thus, the dispersion relation is modified by the addition of a new term, the form of which can be guessed by dimensional analysis. For deep layers, we have

$$\omega^2 = \omega^2(E = 0) + C(\epsilon_1, \epsilon_2)\frac{\epsilon_0 E^2}{\rho}k^2, \tag{51}$$

respectively

$$\omega^2 = \omega^2(H = 0) + C'(\mu_1, \mu_2)\frac{\mu_0 H^2}{\rho}k^2, \tag{52}$$

where $C$ (respectively $C'$) is a dimensionless constant that vanishes if the two fluids have the same electric permitivity (respectively the same magnetic permeability $\mu$).

We have similar couplings when free charges or free currents exist within the fluid layers. Consider for instance a liquid metal with a surface density $\sigma_0$ of electric charges. When the interface is flat, the electric field is $\mathbf{E_0} = \sigma_0/\epsilon_0\hat{\mathbf{z}}$, so the electric potential is $\psi = -\sigma_0 z/\epsilon_0$. When the interface is deformed, we have

$$\psi = -\frac{\sigma_0}{\epsilon_0}z + \tilde{\psi}(x, y, z, t),\tag{53}$$

with

$$\Delta\tilde{\psi} = 0, \quad \tilde{\psi} \to 0 \quad \text{for} \quad z \to \infty.\tag{54}$$

Thus for a sinusoidal surface wave along the $x$ direction

$$\tilde{\psi} = D\exp(-kz)\,\exp i(kx - \omega t).\tag{55}$$

The electric potential at the surface of the liquid metal should be a constant, which we take to be zero. Thus, at leading order

$$\tilde{\psi} \approx \frac{\sigma_0}{\epsilon_0}\zeta\exp(-kz).\tag{56}$$

The electrostatic pressure is to leading order

$$p_e =\approx \frac{\sigma_0^2}{2\epsilon_0}(1 + 2k\zeta) = \text{constant} + \frac{\sigma_0^2}{\epsilon_0}k\zeta,\tag{57}$$

and is directed outward from the liquid metal, thus destabilizing. The dynamic boundary condition at the interface for an uncharged surface (20) becomes at leading order

$$\rho g\zeta + \rho\frac{\partial\phi}{\partial t}(z = 0) + \tau k^2\zeta - \frac{\sigma_0^2}{\epsilon_0}k\zeta \approx 0,\tag{58}$$

whereas the kinematic boundary condition and the bulk equation for the velocity potential are unchanged. Thus, the dispersion relation becomes

$$\omega^2 = gk - \frac{\sigma_0^2}{\epsilon_0\rho}k^2 + \frac{\tau}{\rho}k^3,\tag{59}$$

which is of the form guessed by dimensional analysis.

If the charge surface density is large enough, $\omega$ vanishes and an instability occurs for a mode of wavenumber

$$k_c = \sqrt{\frac{\rho g}{\tau}},\tag{60}$$

when

$$\sigma_0^2 = \sigma_{0C}^2 = 2\epsilon_0\sqrt{\rho g\tau}.\tag{61}$$

Note also that for large enough $\sigma_0$ but less than $\sigma_{0C}$, there exists a range of wavenumbers for which the group and phase velocities are opposite, $UV < 0$.

*Question 6 : consider two dielectric (respectively magnetic) liquids without free charges or currents submitted to an electric (respectively magnetic) field. Sketch the field lines perturbed by a wavy interface in the case of an applied field perpendicular and then parallel to the unperturbed interface. Could you show that the perpendicular field configuration leads to an instability for large enough fields whereas the parallel field configuration is stable?*

The dispersion relations for various types of field coupled surface waves are derived in details in [9].

## 1.6   Viscous damping of surface waves

The flow generated by surface waves cannot be potential everywhere when the fluid viscosity is taken into account. Even if $\nu$ is small, vorticity is generated in thin boundary layers. For small wave and thus velocity amplitude, the typical size of the boundary layer is obtained by balancing $\partial v/\partial t$ and $\nu\Delta v$ in the Navier-Stokes equation; we thus get the "viscous penetration length", $\delta$,

$$\delta = \sqrt{\frac{\nu}{\omega}}. \tag{62}$$

Consequently, in the limit of small viscosity, the flow can be approxiated by a potential flow in the bulk, but vorticity should be taken into account in boundary layers of size $\delta$ in the vicinity of solid boundaries and of the free surface.

The power dissipated by viscosity is given by

$$P_d = \frac{1}{2}\rho\nu \int_V \left( \frac{\partial v_i}{\partial x_k} + \frac{\partial v_k}{\partial x_i} \right)^2 dV. \tag{63}$$

This leads to the decay of the fluid total energy $E$ according to

$$\frac{dE}{dt} = -P_d. \tag{64}$$

We define the dissipation $\gamma$ by

$$\gamma = \frac{\overline{P_d}}{\overline{E}}, \tag{65}$$

where the overbar denotes averaging on one period of the wave.

If $\delta \ll h_0$, the bulk dissipation can be evaluated by assuming that the flow is potential [2]. We get

$$\overline{P_d} = \frac{1}{2}\rho\nu \int_V 4\overline{\left( \frac{\partial^2\phi}{\partial x_i \partial x_k} \right)^2} dV = 8\rho\nu k^4 \int_V \overline{\phi^2} dV. \tag{66}$$

Because of equipartition of energy, we have

$$\overline{E} = 2\overline{E}_c = \rho \int_V \overline{v^2} dV = 2\rho k^2 \int_V \overline{\phi^2} dV. \tag{67}$$

Consequently, the bulk dissipation, $\gamma_B$, is

$$\gamma_B = 2\nu k^2. \tag{68}$$

We could have easily guessed this form from the $\nu \Delta v$ term of the Navier-Stokes equation or from dimensional arguments assuming that the dissipation is proportional to the viscosity. However, this is not true in general as we will see now by evaluating the dissipation due to the bottom boundary layer. At a solid boundary, the velocity should vanish and the velocity gradient thus scales like $v_0/\delta$ where $v_0$ is a typical velocity amplitude. The dissipated power is the power of the friction force; its value per unit surface scales like

$$\frac{P_d}{S} \propto \rho \nu \frac{v_0}{\delta} v_0 = \rho v_0^2 \sqrt{\nu \omega}, \tag{69}$$

which gives a dissipation

$$\gamma_{BL} \propto \frac{\sqrt{\nu \omega}}{l}, \tag{70}$$

where $l$ is an appropriate lengthscale. We are in the small viscosity case, i.e. $\delta \ll \lambda$ and $\delta \ll h_0$. If $\lambda \ll h_0$, the flow which penetrates under the free surface on a typical length $\lambda$, does not "feel" the bottom. Thus, the bottom boundary layer dissipation is negligible. On the contrary, if $h_0 \ll \lambda$, we expect $h_0$ to be the relevant length and

$$\gamma_{BL} \propto \frac{\sqrt{\nu \omega}}{h_0}. \tag{71}$$

Dissipation at the lateral boundaries could be taken into account similarly [2].

It can be shown that the dissipation due to the surface boundary layer is of higher order for small $\nu$ (see below)

$$\gamma_{SL} \propto (\nu k^2)^{\frac{3}{2}} \, \omega^{-\frac{1}{2}}. \tag{72}$$

In addition, there exist dissipative mechanisms due to the meniscus and to the contamination of the fluid free surface [10]. Since we can suppress the meniscus by pinning the fluid surface at the edge of a solid boundary (brimful conditions) and work with clean fluids, we will discard these physico-chemical aspects.

In conclusion, we have the following limit regimes: weak dissipation means that $\delta \ll \lambda$ and $\delta \ll h_0$ whatever the relation between $\lambda$ and $h_0$.

- $\delta \ll \lambda \ll h_0 : \gamma \approx 2\nu k^2$,
- $\delta \ll h_0 \ll \lambda : \gamma \approx \sqrt{\nu \omega}/h_0$.

Strong dissipation corresponds to a situation for which the viscous penetration length $\delta$ is larger than the size of the layer in which the flow is important, i.e. $\delta \gg \lambda$ for $\lambda \ll h_0$ or $\delta \gg h_0$ for $h_0 \ll \lambda$.

We now consider the effect of dissipation on the dispersion relation in the small viscosity limit. To wit, we first use the equipartition of energy, $\overline{E}_c = \overline{E}_p$. If we assume that this relation is roughly correct in the small viscosity limit, we can find the frequency shift due to damping. Indeed, for a wave of given amplitude and wavenumber, the average potential energy is unchanged. Thus the average kinetic energy is unchanged. Since the velocity is reduced in the boundary layers, this means that the velocity amplitude should be increased in the bulk. The wave amplitude and wavenumber being kept constant, this implies that the pulsation $\omega$ is increased.

*Question 7: on the contrary, the frequency of a simple pendulum is decreased by Stokes damping. Do you understand this using the same simple argument ?*

In order to go beyond the rough arguments and estimations given above, we should consider the Navier-Stokes equation and the corresponding boundary conditions for a viscous fluid,

$$\nabla \cdot \mathbf{v} = 0 \tag{73}$$

$$\left[ \frac{\partial \mathbf{v}}{\partial t} + (\mathbf{v} \cdot \nabla)\mathbf{v} \right] = -\frac{1}{\rho}\nabla p + \nu \Delta \mathbf{v} + \mathbf{g}. \tag{74}$$

$$\mathbf{v}(z = -h_0) = 0. \tag{75}$$

The kinematic boundary condition at the free surface is unchanged

$$\frac{\partial \zeta}{\partial t} + \mathbf{v}_i(z = \zeta) \cdot \nabla \zeta = \mathbf{v}_i(z = \zeta) \cdot \hat{\mathbf{z}}. \tag{76}$$

The dynamic boundary condition at the free surface, $z = \zeta(x, y, t)$, becomes

$$-\sigma_{ik} n_k \, (z = \zeta) = p_{atm} + \tau \left( \frac{1}{R_1} + \frac{1}{R_2} \right) n_i \, (z = \zeta), \tag{77}$$

where

$$\sigma_{ik} = -p\delta_{ik} + \rho\nu \left( \frac{\partial v_i}{\partial x_k} + \frac{\partial v_k}{\partial x_i} \right). \tag{78}$$

The solution corresponding to the flat interface is ,

$$\mathbf{v}_0 = 0, \quad p_0 = p_{atm} - \rho g z. \tag{79}$$

We look for small perturbations, $\tilde{v}$, $\tilde{p} = p - p_0$, for which the Navier-Stokes equation (74) becomes

$$\frac{\partial \tilde{v}}{\partial t} = -\frac{1}{\rho}\nabla \tilde{p} + \nu \Delta \tilde{v}. \tag{80}$$

Taking the divergence of (80) and using the incompressibilty condition (73), we get

$$\Delta \tilde{p} = 0, \tag{81}$$

that we can solve. We then look for a solution of (80) in the form $\tilde{v}_{part} + \tilde{u}$ where $\tilde{u}$ is the general solution of the diffusion equation

$$\frac{\partial \mathbf{u}}{\partial t} = \nu \Delta \mathbf{u}. \tag{82}$$

and $\tilde{v}_{part}$ is a particular solution of (80). We note that $\nabla \phi$ is a particular solution provided that

$$\Delta \phi = 0, \tag{83}$$

$$\frac{\partial \phi}{\partial t} = -\frac{\tilde{p}}{\rho}. \tag{84}$$

Thus we have solved (80) in the form

$$\tilde{v} = \nabla \phi + \tilde{u}, \tag{85}$$

where $\nabla \phi$ represents the potential flow obtained in the inviscid case and $\tilde{u}$ is a "viscous correction", provided that the boundary conditions are satisfied. Using the incompressibility condition (73), we can write all the boundary conditions in a form involving only $\zeta$, $\tilde{p}$, $\phi$ and $\tilde{w} = \tilde{u} \cdot \hat{\mathbf{z}}$, the vertical component of $\tilde{u}$:

$$\tilde{w}(z = -h_0) + \frac{\partial \phi}{\partial z}(z = -h_0) = 0, \tag{86}$$

$$\frac{\partial \tilde{w}}{\partial z}(z = -h_0) + \frac{\partial^2 \phi}{\partial z^2}(z = -h_0) = 0, \tag{87}$$

$$\left( \frac{\tilde{p}}{\rho} - 2\nu \frac{\partial \tilde{w}}{\partial z} - 2\nu \frac{\partial^2 \phi}{\partial z^2} \right)(z = -h_0) \approx g\zeta - \frac{\tau}{\rho}\Delta\zeta, \tag{88}$$

$$\left( \Delta_\perp - \frac{\partial}{\partial z^2} \right)\left( \tilde{w} + \frac{\partial \phi}{\partial z} \right)(z = 0) = 0, \tag{89}$$

$$\frac{\partial \zeta}{\partial t} \approx \left( \tilde{w} + \frac{\partial \phi}{\partial z} \right)(z = 0), \tag{90}$$

where $\Delta_\perp$ is the horizontal Laplacian. Using (82,83) and (90), the boundary condition (89) gives after integration in time

$$w(z = 0) \approx 2\nu \Delta_\perp \zeta. \tag{91}$$

We next look for a solution of the Fourier component $\hat{\phi}(z, t)$ at wavenumber $k$ of $\phi$, in the form

$$\hat{\phi}(z, t) = A(t)\,\mathrm{ch}\,k(z + h_0) + B(t)\,\mathrm{sh}\,k(z + h_0). \tag{92}$$

Using the boundary conditions (86, 90, 91) we obtain

$$\hat{\phi}(z, t) = \frac{\left( \frac{\partial}{\partial t} + 2\nu k^2 \right)\hat{\zeta}}{k\,\mathrm{sh}\,k h_0}\,\mathrm{ch}\,k(z + h_0) + \frac{\hat{w}(z = -h_0)}{k\,\mathrm{sh}\,k h_0}\,\mathrm{sh}\,k(z + h_0). \tag{93}$$

Finally, using (84) the boundary condition (88) gives

$$\left(\frac{\partial}{\partial t} + 2\nu k^2\right)^2 \hat{\zeta} + \omega_0^2(k)\hat{\zeta} + \frac{\left(\frac{\partial}{\partial t} + 2\nu k^2\right)\hat{w}}{\operatorname{ch} kh_0}(z = -h_0) + 2\nu k \operatorname{th} kh_0 \frac{\partial \hat{w}}{\partial z}(z = 0) = 0, \quad (94)$$

where $\omega = \omega_0(k)$ is the dispersion relation for the zero viscosity case. In the first term, $2\nu k^2$ corresponds to the bulk dissipation. The term involving $\hat{w}(z = -h_0)$ vanishes in the deep layer limit, $kh_0 \to \infty$; it thus traces back to the bottom boundary layer contribution, whereas the last term corresponds to the surface boundary layer contribution.

We next need to solve the equation for the "viscous field" $\tilde{u}$. We have for the Fourier component $\hat{w}(z, t)$

$$\left[\frac{\partial}{\partial t} - \nu\left(\frac{\partial}{\partial z^2} - k^2\right)\right]\hat{w}(z, t) = 0, \quad (95)$$

$$\hat{w}(z = 0) = -2\nu k^2 \hat{\zeta}, \quad (96)$$

$$\left(\operatorname{sh} kh_0 \frac{\partial \hat{w}}{\partial z} + k \operatorname{ch} kh_0\right)(z = 0) = -k\left(\frac{\partial}{\partial t} + 2\nu k^2\right)\hat{\zeta}. \quad (97)$$

At this stage we can write the exact dispersion relation. It is however not very illuminating even if we then solve it with a computer. It is more interesting to solve the "Stokes problem", i.e. to find the flow $\tilde{w}$ generated by the moving boundary i.e. the free surface, in the small viscosity case ($\delta \ll \lambda$). We get

$$\hat{w}(z, t) = \left\{-2\nu k^2 \exp\left[(1 + i)\frac{z}{\delta}\right]\right.$$
$$\left. + \frac{(1 + i)k\delta\,\omega_0(k)}{2\operatorname{sh} kh_0} \exp\left[-(1 + i)\frac{z + h_0}{\delta}\right]\right\}\zeta_0 \exp i\omega_0 t. \quad (98)$$

This shows that in the limit $\delta \ll \lambda$, the viscous correction is noticeable in surface and bottom boundary layers of characteristic size $\delta$, the later one being negligible in the deep layer limit, $kh_0 \to \infty$. The bulk flow is thus potential.

The dispersion relation becomes much simpler in the deep layer limit because we have not to describe the bottom boundary layer. We get from (95, 96)

$$\hat{w}(z, t) = -2\nu k^2 \zeta_0 \exp qz \, \exp -i\omega t, \quad (99)$$

with

$$q = \sqrt{k^2 - i\frac{\omega}{\nu}}. \quad (100)$$

In the limit $kh_0 \to \infty$, (94) then gives the dispersion relation

$$(-i\omega + 2\nu k^2)^2 + \omega_0^2(k) - 4\nu^2 k^3 q = 0. \quad (101)$$

In the small viscosity case, $\nu k^2/\omega = (\delta/\lambda)^2 \ll 1$, we expand $\omega$ in powers of $\delta/\lambda$. We get

$$\omega(k) = \omega_0(k) - i2\nu k^2 + \sqrt{2}(1-i)(\nu k^2)^{\frac{3}{2}}\omega_0^{-\frac{1}{2}} + \mathcal{O}\left(\frac{\nu^2 k^4}{\omega_0}\right). \tag{102}$$

We thus find, in agreement with the estimations given above:
- bulk dissipation: $\gamma_B = 2\nu k^2$,
- surface dissipation: $\gamma_{SL} = \sqrt{2}(\nu k^2)^{\frac{3}{2}}\,\omega_0^{-\frac{1}{2}}$,
- frequency shift: $\Delta\omega = \sqrt{2}(\nu k^2)^{\frac{3}{2}}\,\omega_0^{-\frac{1}{2}}$.

# 2 Long wavelength approximations

The aim of this chapter is to find simple evolution equations governing surface waves in the limit of "shallow water" i.e. when $\lambda \gg h_0$. For simplicity, we will only consider gravity waves and inviscid fluid. The presence of different scales, here variations in $x$ very slow compared to variations in $z$, usually greatly simplifies a problem. We already noticed that if $kh_0 \to 0$, the dispersion relation for gravity waves is to leading order

$$\omega^2 \approx gh_0 k^2 - \frac{1}{3}gh_0^3 k^4 + \cdots, \tag{103}$$

so that dispersion is small and could be taken into account perturbatively. We will do that in section 2.2. But first, we will show that the structure of the problem and in particular the nonlinear terms, are strongly simplified in the long wavelength approximation even for order one wave amplitude.

## 2.1 Shallow water equations

We define $\mathbf{v} = (u, 0, w)$ and write the incompressibility and the Euler equations

$$\frac{\partial u}{\partial x} + \frac{\partial w}{\partial z} = 0, \tag{104}$$

$$\frac{\partial u}{\partial t} + u\frac{\partial u}{\partial x} + w\frac{\partial u}{\partial z} = -\frac{1}{\rho}\frac{\partial p}{\partial x}, \tag{105}$$

$$\frac{\partial w}{\partial t} + u\frac{\partial w}{\partial x} + w\frac{\partial w}{\partial z} = -\frac{1}{\rho}\frac{\partial p}{\partial z} - g. \tag{106}$$

If $\lambda \gg h_0$, the incompressibility equation implies $w \simeq h_0 u/\lambda \ll u$. Thus, the left hand side of (106) is small compared to the right hand side. Therefore, we have hydrostatic balance at leading order, and

$$p \approx p_{atm} + \rho g\left[\zeta(x,t) - z\right]. \tag{107}$$

Then, by substituting (107) in (105), we obtain

$$\frac{Du}{Dt} = \frac{\partial u}{\partial t} + u\frac{\partial u}{\partial x} + w\frac{\partial u}{\partial z} \approx -g\frac{\partial \zeta}{\partial x}(x,t). \tag{108}$$

The left hand side of (108) is a function of $x$ and $t$, thus if $u(x,z,t=0) = u(x,t=0)$, $u$ remains independent of $z$ at leading order, and

$$\frac{\partial u}{\partial t} + u\frac{\partial u}{\partial x} \approx -g\frac{\partial \zeta}{\partial x}. \tag{109}$$

We should now find an evolution equation for $\zeta$. To wit, we first derive an exact relation which traces back to mass conservation. We have

$$\int_{-h_0}^{z=\zeta(x,t)} \left(\frac{\partial u}{\partial x} + \frac{\partial w}{\partial z}\right) dz = 0, \tag{110}$$

and thus

$$\int_{-h_0}^{z=\zeta(x,t)} u(x,z,t)\, dz - \frac{\partial \zeta}{\partial x} u(x,z=\zeta,t) + w(x,z=\zeta,t) = 0. \tag{111}$$

Using the kinematic boundary condition (11), we get the required conservation equation

$$\frac{\partial \zeta}{\partial t} + \frac{\partial}{\partial x}\left[\int_{-h_0}^{z=\zeta(x,t)} u(x,z,t)\, dz\right] = 0. \tag{112}$$

At leading order, $u$ does not depend on $z$, and

$$\frac{\partial \zeta}{\partial t} + \frac{\partial}{\partial x}\left[(h_0 + \zeta(x,t))\, u\right] \approx 0. \tag{113}$$

Equations (109, 113) for $u(x,t)$ and $\zeta(x,t)$ are the "shallow water" equations. These equations neglect dispersive terms that are smaller by a factor $(h_0/\lambda)^2$, but involve finite nonlinear terms. At the linear level, we get the usual wave equation

$$\frac{\partial^2 u}{\partial t^2} - gh_0\frac{\partial^2 u}{\partial x^2} = g\frac{\partial^2}{\partial x^2}(u\zeta) - \frac{1}{2}\frac{\partial^2}{\partial x \partial t}\left(u^2\right). \tag{114}$$

Note also that the ratio of the nonlinear terms to the linear ones is of order $\zeta_0/h_0$, where $\zeta_0$ is the typical wave amplitude.

## 2.2   Dispersive terms in the long wavelength limit

We next have to find the leading order dispersive terms in the shallow water limit. We first consider waves of small amplitude ($\zeta_0 \ll h_0$) and forget nonlinear terms. Using $\lambda$

as scale for $x$, $x = \lambda X$, and $h_0$ as scale for $z$, $z = h_0 Z$, the Laplace equation for the velocity potential $\phi$ becomes

$$\left(\frac{h_0}{\lambda}\right)^2 \frac{\partial^2 \phi}{\partial X^2} + \frac{\partial^2 \phi}{\partial Z^2} = 0. \tag{115}$$

Using the boundary condition (18), we get

$$\phi(X, Z, t) = F(X, t) + \mathcal{O}\left(\frac{h_0}{\lambda}\right)^2. \tag{116}$$

Thus, at leading order, the velocity potential does not depend on $z$. We define $\tilde{F}(K, \omega)$

$$F(X, t) = \frac{1}{(2\pi)^2} \int \tilde{F}(K, \omega) \exp i(KX - \omega t)\, dK\, d\omega, \tag{117}$$

and multiply the dispersion relation (103) by $\tilde{F}$. Taking the inverse Fourier transform immediately gives

$$\frac{\partial^2 F}{\partial t^2} = g h_0 \frac{\partial^2 F}{\partial X^2} + \frac{1}{3} g h_0^3 \frac{\partial^4 F}{\partial X^4} + \cdots \tag{118}$$

Equation (118) is a linear evolution equation involving the leading order dispersive term $\partial^4 F / \partial X^4$.

It may be interesting to see that they are contributions to this dispersive term from both the boundary conditions (11, 12). The general solution of the Laplace equation (14) with boundary conditions (18) is

$$\phi(x, z, t) = \frac{1}{2\pi} \int_\infty^\infty \mathrm{ch} k(z + h_0) \hat{A}(k, t) \exp ikx\, dk. \tag{119}$$

For $k h_0 \to 0$, we expand the hyperbolic function in $\phi(x, z = 0, t)$ and $\partial\phi/\partial z((x, z = 0, t)$, and then take the inverse Fourier transform to write these quantities with respect to

$$A(x, t) = \frac{1}{(2\pi)} \int_\infty^\infty \hat{A}(k, t) \exp ikx\, dk. \tag{120}$$

We obtain

$$\phi(x, z = 0, t) \approx \left(1 - \frac{h_0^2}{2}\frac{\partial^2}{\partial x^2} + \cdots\right) A(x, t), \tag{121}$$

$$\frac{\partial\phi}{\partial z}(x, z = 0, t) \approx \left(-h_0\frac{\partial^2}{\partial x^2} + \frac{h_0^3}{6}\frac{\partial^4}{\partial x^4} + \cdots\right) A(x, t). \tag{122}$$

Then, from boundary conditions (11, 12) at $z = 0$

$$\frac{\partial\zeta}{\partial t} = -h_0\frac{\partial^2 A}{\partial x^2} + \frac{h_0^3}{6}\frac{\partial^4 A}{\partial x^4} + \cdots, \tag{123}$$

$$g\zeta = -\frac{\partial A}{\partial t} + \frac{h_0^2}{2}\frac{\partial^3 A}{\partial t \partial x^2} + \cdots. \tag{124}$$

Eliminating $\zeta$ gives

$$\frac{\partial^2 A}{\partial t^2} = gh_0\frac{\partial^2 A}{\partial x^2} - \frac{gh_0^3}{6}\frac{\partial^4 A}{\partial x^4} + \frac{h_0^2}{2}\frac{\partial^4 A}{\partial t^2 \partial x^2} + \cdots. \tag{125}$$

At this order we can substitute

$$\frac{\partial^2 A}{\partial t^2} \approx gh_0\frac{\partial^2 A}{\partial x^2}, \tag{126}$$

in the second dispersive term; we get an equation similar to (118).

## 2.3   Boussinesq and Korteweg-de Vries equations

The final step is to find an evolution equation for the leading order velocity potential $F(X,T)$, containing both dispersive and nonlinear effects. As we already noticed, the dimensionless parameter measuring the strength of the nonlinearities is $\zeta_0/h_0$. Thus, in order to describe dispersion and nonlinearities perturbatively, we should have two small parameters,

$$\epsilon \equiv \left(\frac{h_0}{\lambda}\right)^2 \ll 1 \quad \text{(small dispersion)}, \tag{127}$$

$$\mu \equiv \frac{\zeta_0}{h_0} \ll 1 \quad \text{(small nonlinearities)}, \tag{128}$$

We want to find the expression of the leading order nonlinear term in (114) as a function of $u$ alone, or correspondingly, to find the leading order quadratic nonlinearity in (118) as a function of $F(X,T)$ and its derivatives with respect to time and space. One simple way to proceed is to consider all possible quadratic nonlinearities and to keep the lowest order ones allowed by symmetry arguments (i.e. the less differentiated ones). The quadratic nonlinearities are, $F^2$, $FF_X$, $FF_t$, $FF_{XX}$, $F_XF_X$, $FF_{tt}$, $F_tF_t$, $F_XF_t$, $FF_{Xt}$, $FF_{XXX}$, $F_XF_{XX}$, $\cdots$

We first note that $F$ being a velocity potential, the addition of a constant to $F$ should not change the problem, i.e. should not change the evolution equation for $F$. Thus, the equation should be invariant under the transformation

$$F \to F + \text{constant}. \tag{129}$$

Therefore, all terms with an explicit dependence on $F$ should be discarded.

Then, the full set of equations (8-12) is invariant under time reversal symmetry, $t \to -t$, $\phi \to -\phi$, reflection symmetry, $x \to -x$, $\phi \to \phi$, and both applied simultaneaously, $t \to -t$, $x \to -x$, $\phi \to -\phi$. Therefore, the evolution equation for $F$ should be invariant under the following transformations:

$$t \to -t, \quad F \to -F, \tag{130}$$

$$x \to -x, \quad F \to F, \tag{131}$$

$$t \to -t, \quad x \to -x, \quad F \to -F. \tag{132}$$

Considering the transformation of the linear part (118) under the above operations, we conclude that the quadratic nonlinearities should involve
- an odd number of t-derivatives,
- an even number of X-derivatives,
- an odd number of total derivatives.

The leading order terms obeying the above requirements are

$$\frac{\partial^2 F}{\partial X^2}\frac{\partial F}{\partial t}, \quad \frac{\partial^2 F}{\partial X \partial t}\frac{\partial F}{\partial X}, \quad \frac{\partial^2 F}{\partial t^2}\frac{\partial F}{\partial t}. \tag{133}$$

This rather simplifies the problem. Moreover, if is clear from (114) that it is unlikely to find a quadratic nonlinearity involving only time-derivatives. We thus have,

$$\frac{\partial^2 F}{\partial t^2} = gh_0\frac{\partial^2 F}{\partial X^2} + \frac{1}{3}gh_0^3\frac{\partial^4 F}{\partial X^4} + \alpha\frac{\partial^2 F}{\partial X^2}\frac{\partial F}{\partial t} + \beta\frac{\partial^2 F}{\partial X \partial t}\frac{\partial F}{\partial X}. \tag{134}$$

It is tempting to use at this stage

$$\frac{\partial F}{\partial t} \approx \pm\sqrt{gh_0}\frac{\partial F}{\partial X}, \tag{135}$$

in order to simplify the nonlinear terms of (134) keeping the same order of approximation. However, we note that this breaks space-reflection invariance; it is then meaningless to keep the second order wave equation to leading order. But, if we do so, we get

$$\frac{\partial^2 F}{\partial t^2} = \frac{\partial^2 F}{\partial X^2} + \frac{\partial^4 F}{\partial X^4} + \frac{\partial^2 F}{\partial X^2}\frac{\partial F}{\partial X}, \tag{136}$$

which is the Boussinesq equation, where the coefficients have been simplified by appropriate scalings of amplitude, time and space, and where the sign of the nonlinear term does not matter (write the equation for $-F$).

*Question 8 : equations (8-12) are galilean invariant, i.e. invariant under the transformations $x \to x - ct$, $t \to t$, $u(x,t) \to u(x - ct, t) + c$, $w(x,t) \to w(x - ct, t)$, $p(x,t) \to p(x - ct, t)$. Equation (134) as well as the leading order linear wave equation obtained in the long wavelenth limit, are not galilean invariant. Why ? Could you find additional constraints on the coefficients $\alpha$ and $\beta$ by requiring galilean invariance ?*

We now want to find equation (134) with a perturbative expansion of equations (8-12) when both parameters $\epsilon$ and $\mu$ are small. We first recall these equations with a slight change of notation (the origin of the $z$-axis):

$$\Delta\phi = 0, \tag{137}$$

$$\frac{\partial \phi}{\partial z}(z = 0) = 0 \tag{138}$$

$$\frac{\partial \zeta}{\partial t} + \frac{\partial \zeta}{\partial x}\frac{\partial \phi}{\partial x}(z = h_0 + \zeta) = \frac{\partial \phi}{\partial z}(z = h_0 + \zeta), \tag{139}$$

$$g\zeta + \frac{\partial \phi}{\partial t}(z = h_0 + \zeta) + \frac{1}{2}\left[\left(\frac{\partial \phi}{\partial x}\right)^2 + \left(\frac{\partial \phi}{\partial z}\right)^2\right](z = h_0 + \zeta) = 0. \tag{140}$$

We next take $\lambda$, $h_0$, $\lambda/\sqrt{gh_0}$, $\zeta_0$ and $\zeta_0\lambda\sqrt{g/h_0}$, as scales for $x$, $z$, $t$, $\zeta$ and $\phi$. We obtain for the dimensionless fields and variables

$$\epsilon\frac{\partial^2 \Phi}{\partial X^2} + \frac{\partial^2 \Phi}{\partial Z^2} = 0, \tag{141}$$

$$\frac{\partial \Phi}{\partial Z}(Z = 0) = 0 \tag{142}$$

$$\frac{\partial \Xi}{\partial T} + \mu\frac{\partial \Xi}{\partial X}\frac{\partial \Phi}{\partial X}(Z = 1 + \mu\Xi) = \frac{1}{\epsilon}\frac{\partial \Phi}{\partial Z}(Z = 1 + \mu\Xi), \tag{143}$$

$$\Xi + \left[\frac{\partial \Phi}{\partial T} + \frac{1}{2}\mu\left(\frac{\partial \Phi}{\partial X}\right)^2 + \frac{1}{2}\frac{\mu}{\epsilon}\left(\frac{\partial \Phi}{\partial Z}\right)^2\right](Z = 1 + \mu\Xi) = 0. \tag{144}$$

We start by solving (141) iteratively with the boundary condition (142). We get

$$\Phi(X, Z, T) = F(X, T) - \frac{\epsilon}{2}Z^2\frac{\partial^2 F}{\partial X^2} + \frac{\epsilon^2}{24}Z^4\frac{\partial^4 F}{\partial X^4} + \cdots \tag{145}$$

Note that when computing terms of order $\epsilon^2$, there is an integration constant $F_2(X, T)$, but we can absorb it in $F(X, T) = F_0 + \epsilon^2 F_2 + \cdots$ Substituting (145) in (144), we easily obtain

$$\Xi(X, T) = -\frac{\partial F}{\partial T} + \frac{\epsilon}{2}\frac{\partial^3 F}{\partial X^2 \partial T} - \frac{\mu}{2}\left(\frac{\partial F}{\partial X}\right)^2 + \mathcal{O}(\mu\epsilon, \epsilon^2) \tag{146}$$

We compute at the same order

$$\frac{1}{\epsilon}\frac{\partial \Phi}{\partial Z}(X, Z = 0, T) \approx -\frac{\partial^2 F}{\partial X^2} + \mu\frac{\partial F}{\partial T}\frac{\partial^2 F}{\partial X^2} + \frac{\epsilon}{6}\frac{\partial^4 F}{\partial X^4}, \tag{147}$$

$$\frac{\partial \Xi}{\partial X}\frac{\partial \Phi}{\partial X}(X, Z = 0, T) \approx -\frac{\partial F}{\partial X}\frac{\partial^2 F}{\partial X \partial T}. \tag{148}$$

Substituting in (143) gives the evolution equation for $F$

$$\frac{\partial^2 F}{\partial T^2} - \frac{\partial^2 F}{\partial X^2} = \epsilon\left(\frac{1}{2}\frac{\partial^4 F}{\partial X^2 \partial T^2} - \frac{1}{6}\frac{\partial^4 F}{\partial X^4}\right) - \mu\left(\frac{\partial^2 F}{\partial X^2}\frac{\partial F}{\partial T} + 2\frac{\partial F}{\partial X}\frac{\partial^2 F}{\partial X \partial T}\right), \tag{149}$$

which is similar to equation (134) with $\alpha = 1$ and $\beta = 2$. We emphasize that we should take the scaling $\mu \simeq \epsilon$ in order to bring nonlinear and dispersive terms at the same order. Other scalings lead to different equations. We take $\mu = \epsilon$ and combine the two dispersive terms using the leading order wave equation to get

$$\frac{\partial^2 F}{\partial T^2} - \frac{\partial^2 F}{\partial X^2} = \epsilon \left( \frac{1}{3} \frac{\partial^4 F}{\partial X^4} - \frac{\partial^2 F}{\partial X^2} \frac{\partial F}{\partial T} - 2 \frac{\partial F}{\partial X} \frac{\partial^2 F}{\partial X \partial T} \right), \tag{150}$$

and look for the progressive wave solution in the $x$-direction

$$F(X, T) = f(X - T, \tau) + \epsilon f_1 + \epsilon^2 f_2 + \cdots, \tag{151}$$

where $\tau$ is a "slow time", $\tau = \epsilon T$, that describes the slow deformation of the shape of the linear non dispersive solution $f(X - T)$ under the combined action of small dispersive and small nonlinear terms. Defining $\xi \equiv X - T$, $\eta \equiv X + T$, we have at leading order

$$-4 \frac{\partial^2 f}{\partial \xi \partial \eta} = 0, \tag{152}$$

where we have chosed the solution $f = f(\xi, \tau)$. At the next order

$$-4 \frac{\partial^2 f_1}{\partial \xi \partial \eta} = 2 \frac{\partial^2 f}{\partial \tau \partial \xi} + \frac{1}{3} \frac{\partial^4 f}{\partial \xi^4} + 3 \frac{\partial^2 f}{\partial \xi^2} \frac{\partial f}{\partial \xi}, \tag{153}$$

gives

$$-4 f_1 = \left[ 2 \frac{\partial f}{\partial \tau} + \frac{1}{3} \frac{\partial^3 f}{\partial \xi^3} + \frac{3}{2} \left( \frac{\partial f}{\partial \xi} \right)^2 \right] \eta. \tag{154}$$

Thus, the correction $f_1$ diverges with $\eta$ unless we apply the "solvability condition"

$$\frac{\partial f}{\partial \tau} + \frac{3}{4} \left( \frac{\partial f}{\partial \xi} \right)^2 + \frac{1}{6} \frac{\partial^3 f}{\partial \xi^3} = 0. \tag{155}$$

We define $\Theta \equiv \partial f / \partial \xi$, and differentiating (155), we obtain the Korteweg-de Vries equation

$$\frac{\partial \Theta}{\partial \tau} + + \frac{3}{2} \Theta \frac{\partial \Theta}{\partial \xi} + \frac{1}{6} \frac{\partial^3 \Theta}{\partial \xi^3} = 0, \tag{156}$$

which describe the deformation of the shape of a progressive wave, due to the combined action of nonlinearity and dispersion, in the co-moving reference frame.

## 3   Amplitude equations

We want to understand the behavior of a surface wave-packet in the case of finite dispersion ($\lambda$ not too large compared to $h_0$) and small nonlinearities.

The relevant length scales are, the wavelength, $\lambda$, which is no longer assumed large compared to the height of the layer $h_0$, and the typical length, $l$, of the wave-packet, which is assumed large compared to $\lambda$ in order to have a slow modulation of the carrier wave, and thus a meaningful envelope concept. Although we may be in the deep layer limit ($\lambda \ll h_0$, it is expected that the whole wave train will "feel" the bottom plate if $l$ is not small compared to $h_0$. As in the previous section, for simplicity, we consider only the case of gravity waves and neglect viscous dissipation.

Before looking at the specific case of surface waves, we will consider generic predictions for the behavior of a wave-packet in a nonlinear and dispersive medium.

## 3.1   The nonlinear Schrödinger equation

Consider a wave-packet

$$\zeta(x,t) = \int_{-\infty}^{\infty} \hat{F}(k) \, \exp i\,[kx - \omega(k)t] \, dk + c.c. \tag{157}$$

If $l \gg \lambda$, then $\hat{F}(k)$ is peaked around $k = k_0$ with a typical width $\delta k \simeq k_0 \lambda / l$, and we have

$$\omega(k) \approx \omega(k_0) + (k - k_0)\left(\frac{d\omega}{dk}\right)_0 + \frac{1}{2}(k - k_0)^2\left(\frac{d^2\omega}{dk^2}\right)_0 + \cdots \tag{158}$$

Defining $\Omega(K) \equiv \omega - \omega_0$, $K = k - k_0$, we get

$$\Omega(k) \approx KU + \alpha K^2 + \cdots, \tag{159}$$

where $U$ is the group velocity and $\alpha \equiv (d^2\omega/dk^2)_0 \neq 0$ if the medium is dispersive.

Moreover, the largest contribution to the integral in (157) comes from $k$ in the neighbourhood of $k_0$

$$
\begin{aligned}
\zeta(x,t) &\approx \int_{-\infty}^{\infty} \hat{F}(k_0 + K) \, \exp i\,[Kx - \Omega(K)t] \, dK \,\, \exp i\,[k_0 x - \omega(k_0)t] + c.c. \\
&\approx \int_{-\infty}^{\infty} \hat{A}(K,t) \exp iKx \, dK \,\, \exp i\,[k_0 x - \omega(k_0)t] + c.c. \\
&\approx A(x,t) \exp i\,[k_0 x - \omega(k_0)t] + c.c.,
\end{aligned}
\tag{160}
$$

where the Fourier component $\hat{A}$ of the slowly modulated wave envelope $A(x,t)$ is

$$\hat{A}(K,t) = \hat{F}(k_0 + K) \exp -i\Omega t. \tag{161}$$

Thus, using (159) we have

$$\frac{\partial \hat{A}}{\partial t} = -i\Omega\hat{A} = -iKU\hat{A} - i\alpha K^2 \hat{A} + \cdots, \tag{162}$$

and taking the inverse Fourier transform, we obtain the linear evolution equation for the wave envelope

$$\frac{\partial A}{\partial t} = -U\frac{\partial A}{\partial x} + i\alpha\frac{\partial^2 A}{\partial x^2} + \cdots \tag{163}$$

If $l \gg \lambda$, the scale separation is described by a small parameter $\epsilon \equiv \lambda/l$. The two spatial scales are thus related by $X = \epsilon x$. Note that there exist three temporal scales: the period of the carrier wave, $2\pi/\omega(k_0)$ is the fast one. The first slow time scale traces back to the propagation of the wave packet at the group velocity. The corresponding time coordinate is $T = \epsilon t$. Looking at solutions of (163) in the form $A(\xi \equiv X - UT, \tau)$, i.e; in the reference frame moving at the group velocity, we obtain an equation analogous to the Schrödinger equation for a free particle,

$$\frac{\partial A}{\partial \tau} = +i\alpha \frac{\partial^2 A}{\partial \xi^2}, \tag{164}$$

which describes the dispersion of the wave packet on a slower time scale $\tau$.

*Question 9: we are considering waves without dissipation, i.e. systems with the $t \to -t$ symmetry. Does (164) respect this time reversal symmetry ? The answer should be yes; then, why do we have always dispersion and never contraction of the wave packet ?*

We now have to find the leading order nonlinear terms in (163) or (164). The simplest way to proceed is again to use symmetry arguments.

Equations (8-12) governing the surface waves are invariant by translations in both time and space,

$$t \to t + t_0, \quad x \to x + x_0,$$

and from (160) we see that this corresponds to

$$A \to A \exp i\chi,$$

where $\chi$ can vary through all reals. Considering all possible nonlinear terms, we find that the lowest order term with the right transformation property is $|A|^2 A$. So we can write

$$\frac{\partial A}{\partial \tau} = i\alpha \frac{\partial^2 A}{\partial \xi^2} + \beta'|A|^2 A. \tag{165}$$

There are two further symmetries: time reversal and space reflection. In the general case these can be applied separately, but we have taken the particular form of $\zeta$ given by (160) that consists only of waves propagating to the right, and this constrains us to applying both transformations together. Applying the symmetries together implies the invariance of the amplitude equation under the transformation

$$\tau \to -\tau, \quad \xi \to -\xi \quad A \to \bar{A},$$

and applying this to (165) gives

$$-\frac{\partial \bar{A}}{\partial \tau} = i\alpha \frac{\partial^2 \bar{A}}{\partial \xi^2} + \beta'|A|^2 \bar{A}$$

However the complex conjugate of (165) is

$$\frac{\partial \bar{A}}{\partial \tau} = -i\alpha \frac{\partial^2 \bar{A}}{\partial \xi^2} + \bar{\beta}'|A|^2 \bar{A}$$

Hence $\bar{\beta}' = -\beta'$ , and $\beta'$ is pure imaginary, so $\beta' = -i\beta$.

$$\frac{\partial A}{\partial \tau} = i\alpha\frac{\partial^2 A}{\partial \xi^2} - i\beta|A|^2 A \tag{166}$$

This is the nonlinear Schrödinger equation. It shows that the dynamics of the wave-packet consists of a balance between dispersion, $i\alpha A_{\xi\xi}$, and nonlinearity, $-i\beta|A|^2 A$, that traces back in this problem to the amplitude dependence of the wave frequency.

## 3.2   The Benjamin-Feir instability

We now use the nonlinear Schrödinger equation to study the stability of a quasi-monochromatic wave. The original motivation was to understand the instability of Stokes waves. When a wave train of surface gravity waves is generated with a paddle oscillating at constant frequency, one observes that if the fluid layer is deep enough compared to the wavelength, the quasi-monochromatic wave is unstable and breaks into a series of pulses[11, 12].

Return to the nonlinear Schrödinger equation (166), and consider the particular solution

$$\begin{aligned} A_0 &= Q \exp i\Omega\tau \tag{167} \\ \Omega &= -\beta Q^2, \tag{168} \end{aligned}$$

that represents a quasi-monochromatic wave of amplitude $Q$, wavenumber $k_0$ and frequency $\omega_0 + \Omega$. If we perturb $A_0$ slightly, so that

$$A = [Q + r(x,t)] \exp i[\Omega\tau + \chi(\xi,\tau)], \tag{169}$$

we obtain after separating real and imaginary parts,

$$\frac{\partial r}{\partial \tau} = -\alpha(Q+r)\frac{\partial^2 \chi}{\partial \xi^2} - 2\alpha\frac{\partial r}{\partial \xi}\frac{\partial \chi}{\partial \xi} \tag{170}$$

$$Q\frac{\partial \chi}{\partial \tau} = -\beta r(2Q^2 + 3Qr + r^2) + \alpha\frac{\partial^2 r}{\partial \xi^2} - \alpha(Q+r)\left(\frac{\partial \chi}{\partial \xi}\right)^2 - r\frac{\partial \chi}{\partial \tau}. \tag{171}$$

Linearising and taking $\begin{pmatrix} r \\ \chi \end{pmatrix} \propto \exp(\eta t - iK\xi)$, one finds the dispersion relation

$$\eta^2 = -[2\alpha\beta Q^2 K^2 + \alpha^2 K^4]. \tag{172}$$

$\eta^2$ is always negative for $\alpha\beta > 0$, but has a positive region in $K$ for $\alpha\beta < 0$. Thus if $\alpha\beta > 0$ then $\eta$ is pure imaginary and the quasi-monochromatic wave (167) is a stable solution. On the other hand if $\alpha\beta < 0$ then in the long wavelength region, $\eta$ has both a negative and a positive root. When $\eta$ is positive, there is an instability, the Benjamin-Feir or side-band instability [11]. It has the name "side-band" because if one takes a band of frequencies centred on $\omega_0$, the interaction of one side-mode with the second harmonic is resonant with the other side-mode, causing it to be amplified, i.e. $2\omega_0 - (\omega_0 - \Omega) = \omega_0 + \Omega$ [13].

Another way to understand this instability is to linearize (169, 170) in the form

$$\frac{\partial^2 \chi}{\partial \tau^2} = 2\alpha\beta Q^2 \frac{\partial^2 \chi}{\partial \xi^2} - \alpha^2 \frac{\partial^4 \chi}{\partial \xi^4}. \tag{173}$$

So if $\alpha\beta < 0$, the phase of the wave obeys an unstable propagation equation (the propagation velocity is imaginary).

In the stable case it might be interesting to find the higher order terms of this propagation equation. This is simple only if one considers slowly varying propagative solutions of (170, 171) in the form

$$r = r(X, T) \tag{174}$$
$$\chi = \chi(X, T), \tag{175}$$

with $X = \epsilon(\xi - c\tau)$, $c = Q\sqrt{2\alpha\beta}$, $T = \epsilon^3 \tau$. Expanding $r$ and $\chi$

$$r = \epsilon^2 r_0 + \epsilon^4 r_1 + \cdots \tag{176}$$
$$\chi = \epsilon\chi_0 + \epsilon^3\chi_1 + \cdots, \tag{177}$$

we obtain

$$\frac{\partial r_0}{\partial T} + 6\sqrt{2\alpha\beta}\, r_0 \frac{\partial r_0}{\partial X} - \frac{\alpha}{Q}\sqrt{\frac{\alpha}{2\beta}}\, \frac{\partial^3 r_0}{\partial X^3} = 0, \tag{178}$$

which is the Korteweg-de Vries equation. It has well-known solitary wave solutions consisting of region with a non zero $r$, or correspondingly a region with a non-zero phase-gradient, thus a localized region with a different local wavenumber for the wave train.

*Question 10 : if we were now to consider a slowly modulated wave solution of (178), we would find that its slowly varying amplitude $A_1$ satisfies the nonlinear Schrödinger equation with coefficients depending on the ones of the nonlinear Schrödinger equation we start from at the beginning of this section. Derive the mapping between the old and new coefficients. Is there a fixed point? If yes, what would this mean? Find other similar examples using symmetry arguments to guess the form of the successive equations.*

## 3.3   The Davey-Stewartson equation

As we mentioned above, the nonlinear Schrödinger equation is a generic evolution equation for the complex amplitude of a nonlinear wave-packet in a dispersive medium, provided that the small amplitude wave envelope varies slowly enough such that dispersion and nonlinearity come in perturbatively at the same order. For the particular case of surface waves, the nonlinear Schrödinger equation has been found from the primitive equations (8-12) with the method of multiple scales [14, 15].

However, in the general case of three-dimensional wave-packets of surface waves, it has been found by Davey and Stewartson [16] that the wave envelope $A(x, y, t)$ cannot be decoupled in general, from a large scale mean flow which is generated by the gradients of $A$. More precisely, we have for the envelope of a carrier gravity wave along the $x$-axis

$$\frac{\partial A}{\partial \tau} + i\nu AB = i\alpha \frac{\partial^2 A}{\partial \xi^2} + i\alpha' \frac{\partial^2 A}{\partial \eta^2} - i\beta |A|^2 A \qquad (179)$$

$$(gh_0 - U^2)\frac{\partial^2 B}{\partial \xi^2} + gh_0 \frac{\partial^2 B}{\partial \eta^2} = \gamma \frac{\partial^2 |A|^2}{\partial \xi^2}, \qquad (180)$$

wher $B$ is the real amplitude of a large scale velocity field along the $x$-axis and the slow scales are defined as follows

$$\xi = \epsilon(x - Ut), \quad \eta = \epsilon y, \quad \tau = \epsilon^2 t.$$

The term $i\nu AB$ in (179) describes the advection of the wave by the mean flow which in turn, is generated by the variations of the wave amplitude (180). These equations reduce to the nonlinear Schrödinger equation for two-dimensional waves; note however, that the coefficient of the cubic nonlinearity is affected by the mean flow.

This type of coupling between the complex amplitude of a wave and a mean flow, is not restricted to the problem of surface waves. It also occurs for Tolmien-Schlichting waves in the plane Poiseuille flow [17] and for Rayleigh-Bénard convection [18]. In all cases, the mean flow is another neutral mode that is coupled with the wave envelope. This additional neutral mode can be understood as a Goldstone mode due to broken Galilean invariance [19]. In a different context, $B$ looks like the scalar electrostatic potential that appears when the Schrödinger equation for the complex wave function $A$ is made gauge invariant. Note that a term analogous to the one associated with the vector potential is missing in the Davey-Stewartson equations as well as in the other fluid mechanical examples quoted above.

*Question 11: could you find a fluid dynamical example for which such a term in involved?*

# 4 Parametric amplification of surface waves

## 4.1 Parametric resonance

Parametric resonance was first observed by Faraday in 1831 [20]. He experimentally studied the crispations upon the surface of a fluid layer which oscillates vertically, and observed that the period of the surface waves was twice the one of the container.

Melde made a similar observation with a string maintained in transverse vibrations by connecting one of its extremities to a vibrating fork, the direction of motion of the point of attachment being parallel to the length of the string [21]. These experiments were analyzed by Rayleigh who concluded that the energy of an oscillating system may be increased by supplying energy at a frequency which differs from the fundamental frequency of the oscillator. The simplest illustration of this principle is provided by a pendulum, the length of which is varying in time; a child on a swing for example, knows that he can amplify the oscillatory motion of the swing by lowering his center of gravity on the down swing and raising it on the up swing. He thus pumps at twice the frequency of the swing. In this example, the work made by the child during each cycle is converted into mechanical energy of the pendulum. This mechanism is phase sensitive; if the child shifts the phase of his pumping motion with respect to the swing, thus raises his center of gravity on the down swing and lowers it on the up swing, he strongly damps the oscillatory motion of the swing by taking its mechanical energy (the work made by the child during each cycle being negative). Conversely, for a given pumping motion, there exist two oscillatory responses of the swing which are amplified whereas the two others in quadrature are damped.

The motion of the child on the swing can be modeled by a simple pendulum, the length of which is varying in time. The corresponding equation for the angle $\theta$ of the pendulum with the vertical, is

$$\ddot{\theta} + 2\lambda\dot{\theta} + \omega_0^2(1 + f\sin\Omega t)\sin\theta = 0, \tag{181}$$

where $\lambda$ is the damping, $\omega_0$ is the pendulum eigenfrequency, $f$ is the amplitude of the forcing and $\Omega$ is its frequency. The ratio $2:1$ is not the only possibility for parametric resonance. Any oscillator with an eigenfrequency $\omega_0$ can be parametrically pumped by frequency modulation at $\Omega$ with

$$n\Omega = 2\omega_0, \tag{182}$$

where $n$ is an integer. This corresponds to the resonance condition of one of the sidebands $n\Omega - \omega_0$ with $\omega_0$.

Parametric amplification in continuous media occurs widely in physical situations. Examples include Langmuir waves in plasmas, spin waves in ferromagnets, surface waves on a ferrofluid in a time-dependent magnetic field, or on a liquid dielectric in an alternting electric field, and as mentioned above, parametric amplification of surface waves on a vertically vibrated horizontal layer of fluid. Since the forcing is homogeneous in space, the parametric resonance conditions read

$$\begin{aligned} n\Omega &= \omega_1 + \omega_2 \tag{183} \\ \vec{k}_1 + \vec{k}_2 &= 0, \tag{184} \end{aligned}$$

where $\Omega$ is the pulsation of the external forcing. Note that two propagative waves with wavevectors $\vec{k}_{1,2}$ should be considered in order to satisfy conservation of momentum. The pulsation and wavenumber of each wave should also satisfy the dispersion relation

$\omega_{1,2} = \omega(k_{1,2})$, thus we recover the resonance condition (182) where $\omega_0$ is replaced by $\omega_0(k)$ with $k \equiv |\vec{k}_1| = |\vec{k}_2|$. A review of parametric instabilites is given in reference [22].

## 4.2  Linear stability analysis in the inviscid limit

The parametric excitation of surface waves on a horizontal layer of fluid subjected to vertical oscillations, provides the simplest experimental model of parametric amplification in spatially extended systems. It is also one of the most flexible systems to study pattern forming instabilities, primarily because it involves two experimental control parameters, the oscillation ampplitude, i. e. the magnitude of the external constraint which drives the instability, and the the vibration frequency that controls the instability wavenumber through the dispersion relation of the surface waves. After the early observations of Faraday, the parametric nature of the instability was understood by Rayleigh and its linear stability analysis was performed by Benjamin and Ursell in the case of an inviscid fluid [23]. The analysis was first extended to the weakly nonlinear regime by Miles using the averaged Lagrangian formalism [24].

We consider a horizontal fluid layer of depth $h_0$, open to the air at atmospheric pressure. Neglecting dissipation, we recall the dispersion relation of capillary-gravity surface waves

$$\omega_0^2(k) = \left( gk + \frac{\tau k^3}{\rho} \right) \operatorname{th} kh_0. \tag{185}$$

If the container is vertically vibrated, the fluid is subjected to a time-dependent effective gravity, i.e. the frequencies (185) are time-dependent, and surface waves of the form $\exp i(\omega_1 t - k_1 x)$ or $\exp i(\omega_2 t + k_2 x)$ in any direction $x$, can be parametrically amplified. We neglect here the effect of lateral boundaries. As mentioned above, momentum and energy conservation imply

$$\omega_1 = \omega_2 = \omega_0(k) = n\frac{\Omega}{2}, \tag{186}$$

where $\Omega$ is the pulsation of the external forcing, $f(t)$, and $n$ is an integer. Thus a standing wave, or a superposition of standing waves with different directions, are generated at the instability onset. Their wavenumber $k$ is fixed by the excitation frequency $\Omega$. In this inviscid limit, one writes the surface deformation in the form

$$\xi(\vec{r}, t) = a_{\vec{k}}(t)\, S_{\vec{k}}(\vec{r}) + \cdots, \tag{187}$$

where $S_{\vec{k}}(\vec{r})$ is any linear mode of the fluid surface; one gets from the linearized Euler equations [23]

$$\ddot{a}_{\vec{k}} + \omega_0^2(k)\, (1 + \Gamma_k \sin \Omega t)\, a_{\vec{k}} = 0, \tag{188}$$

with

$$\Gamma_k \equiv \frac{\Gamma}{1 + \tau k^2/\rho g}, \tag{189}$$

where $\Gamma = f/g$ is the amplitude of the modulation of the acceleration relative to $g$. This shows that the amplitude of each linear surface mode obeys a Mathieu equation.

It should be noted that a rather unrealistic boundary condition is needed in order to get (188). As mentioned in paragraph 1.3, it should be assumed that the fluid interface remain perpendicular to the lateral boundaries. Any more realistic boundary condition leads to a much more heavy analysis and modifies both the dispersion relation (185) and the form of the normal modes $S_{\vec{k}}(\vec{r})$. However, a more involved analysis is not necessary at this stage because the inviscid limit is not the appropriate one anyway for the study of pattern formation (see below).

## 4.3   Effect of viscosity

It is possible at this stage to take into account a small dissipation in a phenomenological way by adding a damping term to (188)

$$\ddot{a}_{\vec{k}} + 2\lambda_{\vec{k}}\dot{a}_{\vec{k}} + \omega_0^2(k)\left(1 + \Gamma_k \sin \Omega t\right) a_{\vec{k}} = 0. \tag{190}$$

In the deep layer limit, $kh_0 \gg 1$, this damping term is mainly due to viscous dissipation in the bulk, and we have

$$\lambda_{\vec{k}} \approx 2\nu k^2, \tag{191}$$

where $\nu$ is the fluid kinematic viscosity. The effect of non-zero damping in (190) is to shift all the resonances given by equation (186) to finite thresholds $f_{c,\vec{k}}$. Thus, in extended containers, and for not too high dissipation, the $n = 1$ resonance is selected because it has the lowest threshold

$$f_0 \approx 4\nu k_0 \omega_0. \tag{192}$$

Consequently, linear analysis predicts the parametric amplification of a standing-wave with wavenumber $k_0$ such that

$$\omega(k_0) \approx \frac{\Omega}{2}, \tag{193}$$

where $\Omega$ is the external forcing pulsation, when the modulation of the acceleration is increased above $f_0$. This parametric-resonance condition is satisfied only approximately in a finite size container because of the wavenumber quantization due to lateral boundary conditions.

The reduction to (190) is generally not possible for a fluid with a finite viscosity. The exact linear equation for $a_{\vec{k}}(t)$ has been found by Nam Hong U [25] and rediscovered recently by many authors. It involves an integral term, due to the presence of the oscillating interface, and (190) should be replaced by an integrodifferential equation. The integral term traces back to viscous boundary layers, wich generate an additional damping of free surface waves. In the case of the deep layer limit, it scales like $(\nu k^2)^{3/2}\omega_0^{-1/2}$. For a shallow layer of depth $h_0$, the leading order additional damping is proportional to $\sqrt{\omega\nu}/h_0$. As a technical curiousity, note that these integral terms take the form of fractional derivatives of $a_{\vec{k}}(t)$.

Equation (192) for the critical acceleration $f_0$ at instability onset can be understood with dimensional arguments. In the case of low viscosity (i. e. $\delta \ll \lambda$, $\delta \ll h_0$), viscous damping depends on $\nu$, $k$, $\omega$ and $h_0$, and should be overcomed at instability onset $f_0$. We thus have three dimensionless parameters and

$$f_0 = \frac{\omega_0^2}{k_0} \, F\left(\frac{\nu k_0^2}{\omega_0}, k_0 h_0\right). \tag{194}$$

In the deep layer limit, $k_0 h_0 \to \infty$, we expect $f_0$ to become independant of $h_0$. It is then tempting to recover (192) by assuming that $F$ is linear in its remaining argument. However, this limit should be checked carefully because the bottom friction term scales like $\sqrt{\nu}$ and becomes important for small viscosity. We also know from section 1.6 that the next order correction for $F$ in the deep-layer limit involves $\nu^{3/2}$. Thus the next order correction to the critical acceleration is proportional to $\nu^{3/2} k_0^2 \omega_0^{1/2}$. This term has been found analytically; note that its coefficient is negative [25].

New qualitative phenomena occur when bottom friction becomes important. Performing the exact Floquet analysis for arbitrary viscosity and height, Kumar found that the subharmonic response ($n = 1$) may be preempted by the harmonic one ($n = 2$) [26]. The mechanism is as follows: long wavelengths are strongly inhibited by bottom friction when they become comparable to the height of the layer. The threshold of the first harmonic resonance, which corresponds to a smaller wavelength than the first subharmonic one, then becomes lower. As the frequency is decreased further, the lowest threshold corresponds successively to higher and higher resonances. The harmonic response has been observed experimentally by Müller et al.[27].

The low viscosity scaling for the instability onset (192) is also modified in the high viscosity case or for shallow water with finite viscosity [28]. In the case of shallow water, the damping does not depend on $k$ any more and $h_0$ becomes the relevant lengthscale. It is thus appropriate to write for the instability threshold

$$f_0 = \omega_0^2 h_0 \, G\left(\frac{\nu}{\omega_0 h_0^2}, k_0 h_0\right), \tag{195}$$

and to expect that $G$ becomes independent of $k_0 h_0$ in the shallow water regime. It is clear that a lubrication approximation can be used in this limit when $\delta$ becomes larger than $h_0$. A more complex analysis is required when $\delta \gg \lambda$ but with $\delta \ll h_0$ [29].

In the high viscosity case, the instability threshold $f_0$ can easily reach values much larger than the acceleration of gravity [30]. Thus, it seems that $g$ can be neglected, i. e. the fluid container can be turned up side down without modifying the parametric stability threshold of the flat interface! This is true but one has to consider the Rayleigh-Taylor instability, i. e. the instability of the interface between a heavy fluid on a lighter one; the later can be stabilized by parametric forcing in containers of finite lateral extent [31], just as the vertical position of a simple pendulum submitted to vertical vibrations.

Another very important effect of damping is that it controls the width $\Delta\omega$ of the parametric resonance above the instability onset. Using Equation (190), we get

$$\Delta\omega = 2\sqrt{2}\,\lambda\sqrt{\frac{f - f_0}{f_0}}. \tag{196}$$

Using (185) in the deep layer ($kh_0 \gg 1$) and capillary wave ($\tau k^2/\rho g \gg 1$) limits, one obtains the width of the band $\Delta k$ of unstable wavenumbers above instability onset

$$\Delta k \approx \frac{8\sqrt{2}}{3}\frac{\eta\omega}{\tau}\sqrt{\frac{f - f_0}{f_0}}, \tag{197}$$

where $\eta = \rho\nu$ is the fluid dynamic viscosity. $\Delta k$ is to be compared with the quantization imposed by the box, i. e. with $\pi/L$ where $L$ is the horizontal size of the fluid container. When dissipation is small enough, the band is too small to contain many modes and the system is monomode above the instability onset. Consequently, one of the linear mode of the container is selected so that the effect of the lateral boundaries is dominant. Many experimental studies in large aspect ratio containers have been performed with fluids of a few cP viscosity and 20 to 30 dyne/cm surface tension, driven at 100 to 500 Hz [32-34], thus giving $\Delta k$ smaller than $\pi/10\,\mathrm{cm}^{-1}$ for a 2% supercriticality, whereas $\pi/L \approx \pi/10\,\mathrm{cm}^{-1}$. Although generated in large aspect ratio containers, the patterns were thus strongly constrained by the lateral boundaries. In reference [35] $\Delta k$ is even smaller because of the large surface tension of mercury. On the contrary, when the dissipation is large, the parametric resonance width is large and several unstable modes can fit in the container even in the vicinity of the instability threshold. Moreover, secondary time-dependent instabilities are delayed and we can get a stationary pattern with several linearly unstable modes; the boundary effects are then considerably reduced. This situation is achieved in the experiments of reference [30] by using a glycerol-water mixture with about 100 times the dynamic viscosity of water. Although the above simple evaluation of $\Delta k$ in the capillary limit no longer applies, $\Delta k$ is also large compared to $\pi/L$ for the experiments of reference [36].

## 4.4 Amplitude equations in different limits

The evolution equations for the complex amplitude of weakly nonlinear surface waves generated by the Faraday instability, have been mostly derived in the limit of small viscosity. The nature of the problem, even at the linear level, is qualitatively modified when viscosity is neglected at leading order. The normal modes are propagative sinusoidal waves for zero viscosity whereas they are standing anharmonic waves for finite viscosity. They involve a larger and larger number of frequency harmonics when the viscosity in increased [37]. This makes the nonlinear regime very difficult to handle analytically for finite viscosity.

We begin with the case of zero viscosity and consider one space dimension for simplicity. We write the surface deformation

$$\zeta(x,t) = A(X,T) \exp i(kx - \frac{\Omega}{2}t) + B(X,T) \exp i(kx + \frac{\Omega}{2}t) + cc + \cdots \qquad (198)$$

The evolution equations for the complex amplitudes $A(X,T)$ and $B(X,T)$ of the two counter-propagating waves are to leading nonlinear order

$$\frac{\partial A}{\partial T} + U\frac{\partial A}{\partial X} - i\alpha_i\frac{\partial^2 A}{\partial X^2} + i(\beta_i|A|^2 + \beta_i'|B|^2)A \;=\; -i\nu A + \mu B \qquad (199)$$

$$\frac{\partial B}{\partial T} - U\frac{\partial B}{\partial X} + i\alpha_i\frac{\partial^2 B}{\partial X^2} + i(\beta_i'|A|^2 + i\beta_i|B|^2)B \;=\; i\nu B + \mu A. \qquad (200)$$

The left hand sides are just coupled nonlinear Schrödinger equations for the two counterpropagating free waves and the right hand sides represent the effect of parametric forcing: $\nu$ is the detuning of the free wave frequency $\omega_0(k)$ from subharmonic parametric resonance. As said above, the wave selects its wavenumber in order to be as close as possible to parametric resonance, but because of quantization due to the boundary conditions, detuning should be taken into account for waves amplified at very small values of the forcing. Moreover, if one changes the forcing frequency from a regime where finite amplitude waves are developed, one will be able to study situations with large detunings. The terms proportional to the forcing trace back to the parametric resonance conditions (183, 184). Note that progressive waves ($A = 0$ or $B = 0$) cannot be parametrically amplified. The reason is that the phase of a propagative wave continuously drifts with respect to the phase of the forcing and one cannot transfer energy from the forcing to the waves if they are not phase-locked. This phase locking betweeen the waves and the forcing is possible with a standing wave.

If dissipation is taken into account perturbatively at this stage, both the coefficients of the linear and nonlinear terms are modified. We get

$$\frac{\partial A}{\partial T} + U\frac{\partial A}{\partial X} - \alpha\frac{\partial^2 A}{\partial X^2} + (\beta|A|^2 + \beta'|B|^2)A \;=\; (-\lambda - i\nu)A + \mu B \qquad (201)$$

$$\frac{\partial B}{\partial T} - U\frac{\partial B}{\partial X} - \overline{\alpha}\frac{\partial^2 B}{\partial X^2} + (\beta'|A|^2 + \beta|B|^2)B \;=\; (-\lambda + i\nu)B + \mu A, \qquad (202)$$

where $\alpha = \alpha_r + i\alpha_i$, $\beta = \beta_r + i\beta_i$, $\beta' = \beta_r' + i\beta_i'$. In the deep layer limit, $\lambda$ is to leading order bulk dissipation, i. e. is proportional to viscosity. The status of the coefficients of nonlinear damping is more subtle. They involve contributions from both the potential flow (bulk dissipation) and from the vortical surface layer. The second one is ignored in most small viscosity approximations. It is claimed that it is of the same order as the first, i. e. proportional to the viscosity, and that this is the reason for which amplitude equations derived in this limit do not predict the patterns that are experimentally observed [38]. The situation may be even worse: the vertical component of the surface vortical flow being proportional to viscosity (98), the continuity relation implies that the horizontal component is proportional to $\sqrt{\nu}$. The nonlinear interaction of this component with the potential flow may generate nonlinear damping coefficients of order $\sqrt{\nu}$, i. e. larger than the linear one in the small viscosity limit.

As said above, the expansion (198) is valid only in the limit of vanishing dissipation. In the case of finite dissipation, propagative waves are not neutral modes, and one has from Floquet theory,

$$\zeta(x, t) = \epsilon \left[ W(X, T) \exp ikx + c.c. \right] \left[ \pi(t) \exp i\sigma t + c.c. \right] + \cdots, \tag{203}$$

where $\pi(t)$ is a time-periodic function with period $2\pi/\Omega$, and where $i\sigma$ is the Floquet exponent ($\sigma = \Omega/2$ for the subharmonic response). The temporal part of the above neutral mode has a fixed phase with respect to the one of the forcing, and could involve a large number of higher harmonics when the viscosity is high. At instability onset, the spatial part is a sum of modes of the form $\exp \pm ikx$ with wavenumbers centered around the one selected by parametric resonance, i.e. contrary to the temporal part, the space-dependence is nearly sinusoidal. Symmetry requirements on $W(X, T)$ follow from

- continuous translational invariance in space, which implies the invariance under the transformation

$$W \to W \exp(-i\phi),$$

- space-reflection symmetry, $x \to -x$, which implies

$$X \to -X, \quad W \to \overline{W}.$$

The first constraint implies that the evolution equation for $W$ should have the form

$$\frac{\partial W}{\partial T} = (\mu - \mu_c)W + a\frac{\partial W}{\partial X} + \sigma\frac{\partial^2 W}{\partial X^2} - \gamma|W|^2 W. \tag{204}$$

Taking the complex conjugate of (204) and using the space-reflection symmetry, we get that $\sigma$ and $\gamma$ should be real and $a$ should be pure imaginary. However, the term proportional to $\partial W/\partial X$ in (204) can be easily eliminated with the change of variable

$$W = \tilde{W} \exp(iqX)$$

that gives

$$\frac{\partial \tilde{W}}{\partial T} = (\mu - \tilde{\mu}_c)\tilde{W} + \sigma\frac{\partial^2 \tilde{W}}{\partial X^2} - \gamma|\tilde{W}|^2 \tilde{W},$$

with $\tilde{\mu}_c = \mu_c + \frac{a^2}{4\sigma}$ if one takes $q = -\frac{a}{2i\sigma}$ in order to eliminate the term proportional to $\partial W/\partial X$. This means that this term is present as soon as the field $\zeta(x, t)$ is not expanded at the critical wavenumber $k$ which corresponds to the smallest value of $\mu$ for instability onset. If the expansion (203) is performed at criticality, the evolution equation for $W$ thus reads (we drop the tildes)

$$\frac{\partial W}{\partial T} = (\mu - \mu_c)W + \sigma\frac{\partial^2 W}{\partial X^2} - \gamma|W|^2 W, \tag{205}$$

where $\sigma$ and $\gamma$ are real numbers. This is a Ginzburg-Landau equation, analogous to the one obtained for the amplitude of the flow velocity at the onset of Rayleigh-Bénard convection.

The two-wave amplitude equations (201, 202) can be reduced to a Ginzburg-Landau equation (205) very close to the marginal stability curve

$$\mu_c{}^2(q) = (\lambda + \alpha_r)^2 + (\nu + qU_0 - \alpha_i q^2)^2, \tag{206}$$

which describes the onset of parametric instability as a stationary bifurcation of the system (201, 202). Note that the most unstable wavenumber $q_c$ for which $\mu_c(q)$ is minimum, say $\mu_c$, is in general different from zero, except if $\nu = 0$. We write $W(\xi, \tau)$ the complex amplitude of the marginal mode of (201, 202), i.e. we look for an expansion in the form

$$\begin{pmatrix} A \\ B \end{pmatrix} = \epsilon \begin{pmatrix} 1 \\ v_c \end{pmatrix} W(\xi, \tau) \exp iqX + \cdots,$$

where $v_c = \frac{1}{\mu_c}(\lambda + i\nu + iq_c U + \overline{\alpha}q_c{}^2)$. We scale space and time as $\xi = \epsilon X$, $\tau = \epsilon^2 T$. The solvability condition then shows that $W(\xi, \tau)$ obeys a Ginzburg-Landau equation [39]. As soon as a finite dissipation is present, counter-propagating waves are not neutral modes of the parametric instability when they are taken separately, and only the standing wave part is a neutral mode; adiabatic elimination of the other part, which is damped in the vicinity of the instability onset, leads to the above reduction. In the case of zero dissipation, $\lambda = 0$, $\gamma$ can have both signs leading to a super or subcritical onset of parametric waves, depending on the sign of the detuning, and vanishes if $\nu = 0$. For zero detuning, parametric instability is thus saturated by a quintic nonlinearity. If a small viscosity is taken into account, the coefficient of the cubic nonlinearity remains non zero even for zero detuning, and as we mentioned, may be large compared to the one of the linear damping term. Because of this mixing of orders when both the distance to the threshold and viscosity are small, one understands why the small viscosity approximation is unable to give correct predictions just above the instability onset. The only advantage of the analysis in the zero dissipation limit is that (201, 202) can provide a qualitative description of secondary instabilites of parametrically generated waves as the forcing is increased [40, 41]. For a review of secondary instabilities of hydrodynamic patterns, see reference [42].

## 4.5   Patterns selection

The problem of pattern selection above instability onset was adressed first by Faraday himself [20]: he remarked that, "the hexagon, the square, and the equilateral triangle are the only regular figures that can fill an area perfectly. The square and triangle are the only figures that can allow of one half alternating symmetrically with the other, ...; and of these two the boundary lines between squares are of shorter extent than those between equilateral triangles of equal area. It is evident therefore that one of these two will be finally assumed, and that that will be the square arrangement; because then

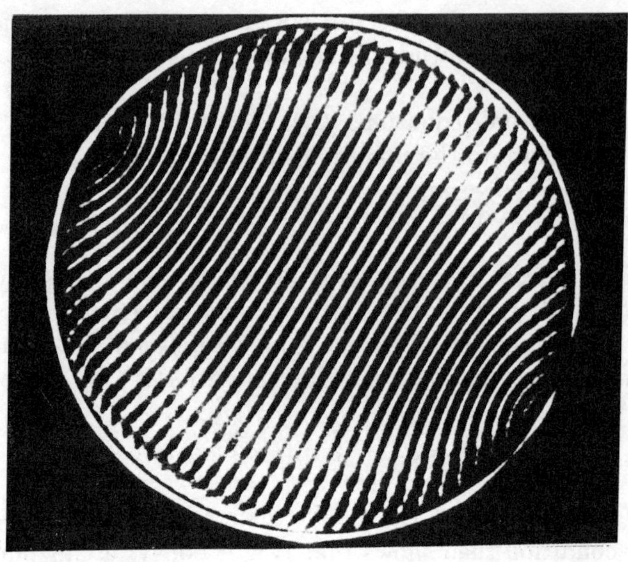

Figure 2. One-dimensional standing-wave, $\Omega/2\pi = 60$ Hz. All the container of diameter $D = 12$ cm is displayed. The fluid is a mixture of glycerol and water, which has a kinematic viscosity $\nu = 1\,cm^2/s$, and a surface tension $\tau = 65dyn/cm$.

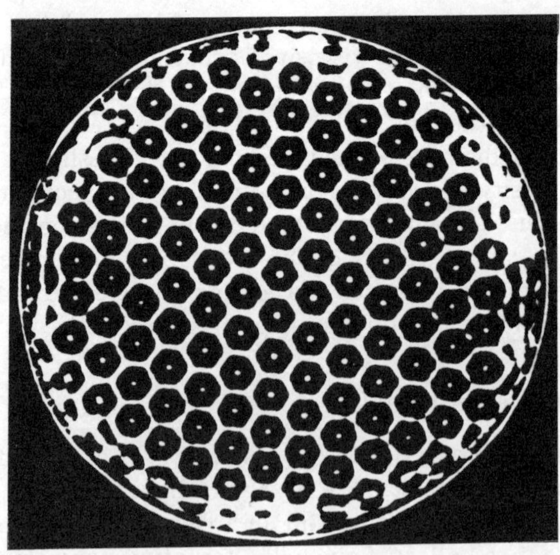

Figure 3. Hexagonal pattern, $\Omega/2\pi = 30$ Hz, $m = 2$, $n = 3$. Same fluid properties as in figure 2.

the fluid will offer the least resistance in its undulations to the motion of the plate, or will pass most readily to those positions into which the forces it receives from the plate conspire to impel it". Until recently most experiments performed in large aspect ratio containers displayed square patterns [32, 33, 35, 43], and it came to be believed that, in the absense of strong boundary effects, nonlinear interactions between the growing waves in the bulk of the container select squares. This was even showed analytically for capillary waves in the limit of a nearly inviscid fluid [44]. However, as said above, viscosity was not taken into account properly in the nonlinear interactions.

It has been shown experimentally that when the dissipation is increased, one observes a transition from squares to a one-dimensional standing wave at instability onset [36, 45] (see figure 2). When viscosity is small enough such that squares are observed at high frequency, it has been shown that hexagons are observed when the frequency is decreased [46]. Note that hexagons were also observed earlier but not at instability onset [34, 43]. More complex patterns such as targets or spirals, have been also observed at instability onset [30] and also quasiperiodic patterns [34, 47], but these later are much easier to control with a two-frequency forcing [30, 45, 48]. (see section 4.6). The Faraday instability thus provides one of the richest pattern-forming systems.

Above the instability onset, we write the surface deformation, $\zeta(\vec{r}, t)$, in the form

$$\zeta(\vec{r}, t) = \sum_{p=1}^{N} \left[ W_p \exp\left(i\vec{k}_p \cdot \vec{r}\right) + c.c. \right] \left[ \pi(t) \exp i\sigma t + c.c. \right] \cdots, \tag{207}$$

where $\pi(t)$ is a time-periodic function with period $2\pi/\Omega$, and where $i\sigma$ is the Floquet exponent.

We first consider the case of a sinusoidal external forcing of pulsation $\Omega$, and a subharmonic response i. e. an oscillation with pulsation $\Omega/2$. In other words, the Floquet multiplier, $\exp\left(2\pi i\sigma/\Omega\right)$, is -1. This is the usual case in the deep layer limit. As said above, when the viscosity is large enough, the Faraday instability generates a plane wave ($N = 1$). At smaller viscosity, squares ($N = 2$, $\vec{k}_1.\vec{k}_2 = 0$) at high frequency, and hexagons or triangles ($N = 3$, $\vec{k}_1 + \vec{k}_2 + \vec{k}_3 = 0$) at low frequency, are observed.

The subharmonic response implies that the invariance $t \to t + 2\pi/\Omega$ is broken at instability onset. If $\zeta(\vec{r}, t)$ is a non-zero solution, this invariance implies that $\zeta(\vec{r}, t + 2\pi/\Omega) = -\zeta(\vec{r}, t)$ is also a solution. Therefore, (207) implies that the evolution equations for the amplitudes $W_p$ should be invariant under the transformation $W_p \to -W_p$. Consequently they cannot involve quadratic nonlinearities. This means that, even when hexagons or triangles are observed, they are not generated by resonant triad interactions. Amplitude equations up to cubic order have thus the form (ignoring spatial derivatives)

$$\frac{\partial W_p}{\partial T} = (\mu - \mu_c)W_p - \sum_{q=1}^{N} \gamma(\theta_{pq})|W_q|^2 W_p, \tag{208}$$

where $\theta_{pq}$ is the angle between $\vec{k}_p$ and $\vec{k}_q$. This system of equations has a Lyapunov funtional and stable symmetric $2N$-fold patterns of any $N$ are possible, depending on the cubic coupling function $\gamma(\theta)$ [49, 50]. $\gamma(\theta)$ has been calculated in the limit of small viscosity by several authors [44, 51, 52]. As said above, they do not give the correct predictions. A recent calculation, made for finite viscosity, seems to be in agreement with the experimental results quoted above [38].

A very important aspect of the pattern-forming problem in the Faraday instability is that $\gamma(\theta)$ is strongly influenced by nonlinear interactions that involve wavenumbers other than $k_c$. Second harmonics of wavevector $\vec{k}_m + \vec{k}_n$ and frequency $2\omega$ are neutral in the small viscosity limit if

$$\omega_0(\vec{k}_m + \vec{k}_n) = 2\omega = 2\omega_0(k_c). \tag{209}$$

A simple energy argument shows that when the dispersion relation is such that this condition can be fulfilled for an angle $\theta_{mn}$, the system tries to avoid pairs of wave vectors separated by $\theta_{mn}$ [30]. For capillary waves in the inviscid limit, $\theta_{mn} = 74.9^o$ [44] whereas for a viscous type dispersion relation, $\omega_0(k) \propto k^2$, $\theta_{mn} = 90^o$; this may explain qualitatively why squares are no longer observed when the fluid viscosity is increased.

## 4.6  "Quasipatterns" generated by a two-frequency periodic forcing

The idea is to modify the forcing $f(t)$ in order to break the invariance of the amplitude equations under the transformation $W_p \to -W_p$. To wit, we consider a two-frequency forcing,

$$f(t) = f \left[ \cos\theta \, \cos m\Omega t \, + \, \sin\theta \, \cos(n\Omega t + \phi) \right], \tag{210}$$

where $f \cos\theta$ (respectively $f \sin\theta$) is the amplitude at pulsation $m\Omega$ (respectively $n\Omega$). $m$ and $n$ are relatively prime, i. e. the vibration is periodic with period $2\pi/\Omega$. The phase $\phi$ can be taken within $[0, 2\pi/m]$, since using Bezout's theorem, there exists an integer $p$, such that the transformation, $\phi \to \phi + 2\pi/m$, $t \to t + 2p\pi/m\Omega$, keeps (210) invariant. If $m$ and $n$ are of different parity, the invariance $t \to t + 2\pi/\Omega$ is not broken at the instability onset. Therefore the invariance $W_p \to -W_p$ is no longer true, and the amplitude equations may involve quadratic nonlinearities corresponding to triad interactions. Their effect is to generate hexagonal patterns [30]. We do observe hexagons with a two-frequency forcing when $m$ and $n$ are of different parity, as for instance in the case of figure 3 where $m = 2$, $n = 3$ and $\Omega/2\pi = 30$ Hz.

The experimental results are plotted on the two-parameter diagram of figure 4 for $m = 4$ and $n = 5$ with $\Omega/2\pi = 14.60$ Hz and $\phi = 75^o$. The axes $\theta = 0$ and $\theta = \pi/2$ correspond to sinusoidal forcings with respective pulsations $4\Omega$ and $5\Omega$. Two neutral curves depart from these instability thresholds. Along the upper branch, a pattern of lines of wavenumber $8.8 \pm 0.3$ cm$^{-1}$ bifurcates supercritically (L1). Stroboscopic observation shows that its response pulsation is $\Omega/2$. The corresponding Floquet multiplier

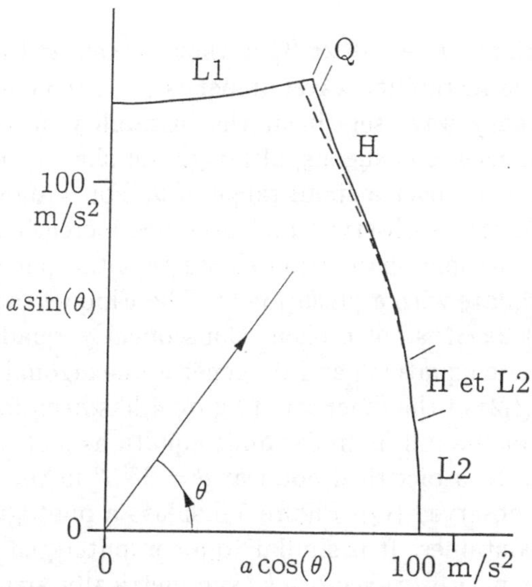


Figure 4. Phase diagram obtained with the external forcing (210), where $\Omega/2\pi = 14.60$ Hz, $m = 4$, $n = 5$, $\phi = 75^o$. The full lines are the stability limit of the flat surface when the excitation amplitude $a$ is increased. The different bifurcated states are, (L1), (L2): one-dimensional standing waves, (H): hexagons, (Q): quasicrystalline pattern. The dashed line shows the subcriticality of hexagons.

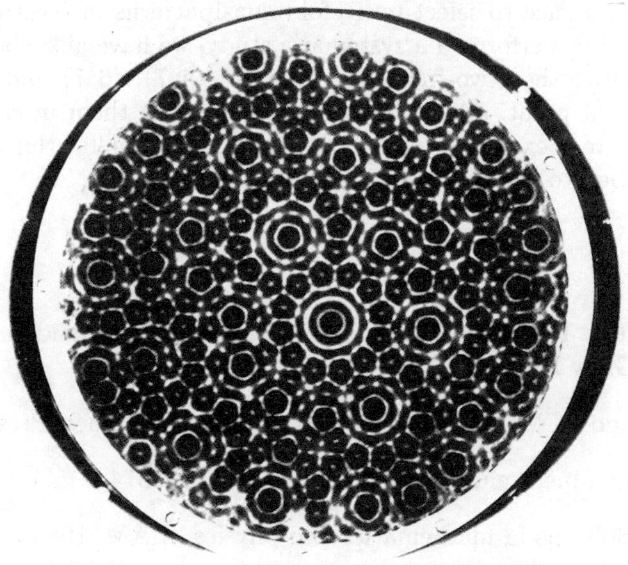

Figure 5. Quasicrystalline pattern. Same fluid properties as in figure 2.

in (207) is $-1$. The invariance $t \to t + 2\pi/\Omega$ is thus broken at the instability onset. Along the lower branch, the instability wavenumber is $7.5 \pm 0.3$ cm$^{-1}$. For $\theta$ less than about $20^o$, a plane stationary wave sets in at the instability onset (L2). For larger values of $\theta$, the pattern changes to hexagons, although not sharply, and hexagons, lines and disordered patterns coexist over a small range of $\theta$. For $\theta$ above $30^o$, a transition to a perfect hexagonal pattern is observed and becomes increasingly subcritical as $\theta$ is increased further. Stroboscopic observation shows that the patterns that bifurcate along the lower branch oscillate with a pulsation $\Omega$. The Floquet multiplier is thus $+1$, and the invariance $t \to t + 2\pi/\Omega$ is not broken. Consequently, quadratic nonlinearities are possible in the amplitude equations and do generate hexagonal patterns [30]. The instability lines (L1) and (L2) of the diagram of figure 4 has been found by performing a linear stability analysis of the full hydrodynamic equations [53].

The two branches meet at a bicritical point at $\theta = 65.5^o$ in the vicinity of which a twelvefold quasipattern is observed (Q). Figure 5 displays a photograph of this "quasipattern" in a cylindrical container. It is similar to a computer-generated image of the sum of twelve plane waves with vave-vectors $\vec{k}_p$ symmetrically arranged on a circle of radius $k_0$ [45]. The transition from (L1) to this quasipattern is subcritical whereas it is supercritical from the hexagonal pattern.

Our observations show that the quasicrystalline pattern can be understood as the superposition of two hexagonal modes, the orientations of which differ from $30^o$. The quadratic nonlinearities do not couple the two hexagonal modes. The selection between the hexagonal patterns and the quasicrystalline one, thus depends on cubic nonlinearities. We speculate that a sufficient change in those, due to the proximity of the bicritical point, would suffice to select twelvefold quasipatterns in favor of hexagons.

Although we have not performed a systematic study, we have also observed twelvefold quasipatterns with other two-frequency forcings, $(4,7)$, $(6,7)$ and $(8,9)$, in the vicinity of the bicritical point. Moreover, we have observed them in containers with different shapes (square, hexagonal, octogonal and irregular) [30], thus showing that they cannot be ascribed to sidewall boundary effects.

# References

[1] Fetter, A. L. and J. D. Walecka: Theoretical mechanics of particles and continua, chap. 10, Mac Graw Hill, 1980.

[2] Landau, L. D. and E. M. Lifshitz: Fluid Mechanics, Pergamon Press, 1959.

[3] Whitham, G. B.: Linear and nonlinear waves, Wiley, 1974.

[4] Newell, A. C.: Solitons in mathematics and physics, SIAM, 1985.

[5] Chandrasekhar, S.: Hydrodynamic and hydromagnetic stability, Clarendon Press, 1961.

[6] Benjamin, T. B. and J. C. Scott: Gravity-capillary waves with edge constraints, J. Fluid Mech. 92 (1970), 241-267.

[7] Graham-Eagle, J.: A new method for calculating eigenvalues with applications to gravity-capillary waves with edge constraints, Math. Proc. Camb. Phil. Soc. 94 (1983), 553-564.

[8] Weissman, M. A.: Nonlinear wave packets in the Kelvin-Helmholtz instability, Phil. Trans. R. Soc. Lond. 290 (1979), 639-681.

[9] Melcher, : Field coupled surface waves, MIT Press, Cambridge 1963.

[10] Miles, J.: Surface-wave damping in closed basins, Proc. Roy. Soc. Lond. A 297 (1967), 459-475.

[11] Benjamin, T. B. and J. E. Feir: The disintegration of wave trains on deep water, J. Fluid Mech. 27 (1967), 417-430.

[12] Lake, B. M., M. C. Yuen, M. Rungaldier and W. E. Ferguson: Nonlinear deep-water waves: theory and experiment. Part 2. Evolution of a continuous wave train, J. Fluid Mech. 83 (1977), 49-74.

[13] Stuart, J. T. and R. C. DiPrima: The Eckhaus and Benjamin-Feir resonance mechanisms, Proc. Roy. Soc. Lond. A 362 (1978), 27-41.

[14] Benney, D. J. and G. J. Roskes: Wave instabilities, Stud. Applied Math. (USA), 48 (1969), 337-385.

[15] Hasimoto, H. and H. Ono: Nonlinear modulation of gravity waves, J. Phy. Soc. Japan, 33 (1972), 805-811.

[16] Davey, A. and K. Stewartson: On three-dimensional packets of surface waves, Proc. R. Soc. Lond. A, 338 (1974), 101-110.

[17] Davey, A., L. M. Hocking and K. Stewartson: On the nonlinear evolution of three-dimensional disturbances in plane Poiseuille flow , J. Fluid Mech. 63 (1974), 529-536.

[18] Zippelius, A. and E. Siggia: Stability of finite amplitude convection, Phys. Fluids 26 (1982), 2905-2915

[19] Coullet, P. and S. Fauve: Propagative phase dynamics for systems with Galilean invariance, Phys. Rev. Lett. 55 (1985), 2857-2860.

[20] Faraday, M.: On the form and states of fluids on vibrating elastic surfaces, Phil. Trans. R. Soc. Lond. 52 (1831), 319-340.

[21] Melde, F.: Uber erregung stehender Wellen eines fadenformigen Körpers, Ann. Phys. Chem., Ser. 2, 109 (1859) 193-215.

[22] Fauve, S.: Parametric instabilities, in: Dynamics of nonlinear and disordered systems (Ed. G. Martinez-Mekler and T. H. Seligman), World Scientific, 1995, Chap. 3.

[23] Benjamin, T. B. and F. Ursell: The stability of the plane free surface of a liquid in vertical periodic motion, Proc. Roy. Soc. Lond. A 225 (1954), 505-515.

[24] Miles, J.: Nonlinear Faraday resonance, J. Fluid Mech. 146 (1984), 285-302.

[25] Nam Hong, U.: A new approach to parametric excitation of stationary surface waves in a viscous liquid, Bull. Russian Acad. Sci., Phys./Supplement 57 (1993) 131-135.

[26] Kumar, K.: Linear theory of Faraday instability in viscous liquids, Proc. Roy. Soc. Lond. A 452 (1996) 1113-1126.

[27] Müller, H. W, H. Wittmer, C. Wagner, J. Albers and K. Knorr: Analytic stability theory for Faraday waves and the observation fo the harmonic surface response, Phys. Rev. Lett. 78 (1997), 2357-2360.

[28] Lioubashevski, O., J. Fineberg and L. S. Tuckerman: Scaling of the transition to parametrically driven surface waves in highly dissipative systems, Phys. Rev. E 55 (1997), 3832-3835.

[29] Cerda, E. A. and E. L. Tirapegui: Faraday instability in viscous fluids, (preprint 1998).

[30] Edwards, W.S. and S. Fauve: Patterns and quasipatterns in the Faraday experiment, J. Fluid Mech. 278 (1994), 123-148.

[31] Wolf, G. H.: Dynamic stabilization of the interchange instability of a liquid-gas interface, Phys. Rev. Lett. 24 (1970), 444-446.

[32] Ezerskii, A. B., M. I. Rabinovich, V. P. Reutov and I. M. Starobinets: Spatiotemporal chaos in the parametric excitation of a capillary ripple, Sov. Phys. JETP 64 (1986), 1228-1236.

[33] Tufillaro, N., R. Ramshankar and J. P. Gollub: Order disorder transition in capillary ripples, Phys. Rev. Lett. 62 (1989) 422-425.

[34] Christiansen, B., P. Alstrom and M. T. Levinsen: Ordered capillary waves states: quasicrystals, hexagons and radial waves, Phys. Rev. Lett. 68 (1992), 2157-2160.

[35] Ciliberto, S., S. Douady and S. Fauve: Investigating space-time chaos in Faraday instability by means of the fluctuations of the driving acceleration, Europhysics Lett. 15 (1991) 23-28.

[36] Fauve, S. K. Kumar, C. Laroche, D. Beysens and Y. Garrabos: Parametric instability of a liquid-vapor interface close to the critical point, Phys. Rev. Lett. 68 (1992) 3160-3163.

[37] Edwards, W. S, S. Fauve, K. Kumar and C. Laroche: Pattern formation in parametrically excited surface waves, in: Turbulence in spatially extended systems, (Ed. R. Benzi, C. Basdevant and S. Ciliberto), Nova Science Publishers, 1993, 259-273.

[38] Chen P. and J. Vinals: Amplitude equations and pattern selection in Faraday waves, (preprint 1998).

[39] Riecke, H.: Stable wavenumber kinks in parametrically excited surface waves, Europhysics Lett. 11 (1990), 213-218.

[40] Douady, S., S. Fauve and O. Thual: Oscillatory phase modulation of parametrically forced surface waves, Europhysics Lett. 10 (1989), 309-315.

[41] Fauve, S., S. Douady and O. Thual: Drift instabilites of cellular patterns, J. Phys. France II 1 (1991), 311-322.

[42] Fauve, S.: Pattern forming instabilities, in: Hydrodynamics and nonlinear instabilities (Ed. C. Godrèche and P. Manneville), Cambridge University Press, 1998, Chap. 4.

[43] Douady, S. and S. Fauve: Pattern selection in Faraday instability, Europhysics Lett. 6 (1988), 221-226.

[44] Milner, S. T.: Square patern and secondary instabilities in driven capillary waves, J. Fluid Mech. 225 (1991) 81-100.

[45] Edwards, W.S. and S. Fauve: Parametrically excited quasicrystalline surface waves, Phys. Rev. E 47 (1993), R 788-791.

[46] Kumar, K. and K. M. S. Bajaj, Competing patterns in the Faraday experiment, Phys. Rev. E 52 (1995), R 4606-4609.

[47] Binks, D. and W. van de Water: Nonlinear pattern formation of Faraday waves, Phys. Rev. Lett. 78 (1997) 4043-4046.

[48] Edwards, W.S. and S. Fauve: Structure quasicristalline engendrée par instabilité paramétrique, C. R. Acad. Sci. Paris, 315-II (1992), 417-420.

[49] Malomed, B. A., A. A. Nepomnyashchii and M. I. Tribelskii: Two-dimensional quasiperiodic structures in non equilibrium systems, Sov. Phys. JETP 69 (1989), 388-396.

[50] Newell, A. C. and Y. Pomeau: Turbulent crystals in macroscopic systems, J. Phys. A 26 (1993), L429-434.

[51] Miles, J.: Faraday waves: rolls versus squares, J. Fluid Mech. 269 (1994), 353-371.

[52] P. L. Hansen and P. Alstrom: Perturbation theory of parametrically driven capillary waves at low viscosity, J. Fluid Mech. 351 (1997), 301-344.

[53] Besson, T., W. S. Edwards and L. S. Tuckerman: Two-frequency parametric excitation of surface waves, Phys. Rev. E 54 (1996), 507-513.

# THIN FILM DYNAMICS


**J.M. Chomaz**
**Ecole Polytechnique, Palaiseau, France**

**M. Costa**
**University of Naples "Federico II", Naples, Italy**


## Abstract


Thin films[1] possess two radically distinct typical scales associated with their transverse and their longitudinal dimensions. Two distinct dynamics are thus associated to these length scales: transverse or longitudinal dispersive waves linked to the film thickness, and longitudinal quasi-two-dimensional (2D) motion scaling on the film length. The physics of both waves and 2D motion are studied here. The response of a film to a localized impulse is computed, and the behaviour is interpreted in the light of group-velocity notions. When air is blown on the film, the waves turn into instability modes, as demonstrated by a simple pressure argument in the limit of small density ratios. The different behavior observed in the case of a water jet and in the case of air blowing on a film is explained by introducing the equivalent of group velocity for instability waves, which naturally leads to discriminate between the absolute and the convective type of instability. In the long-wave limit, waves become similar to the elastic waves propagating on a stretched membrane. In recent experiments, Couder [7] and Gharib [13] use soap films as a two-dimensional fluid. In the present paper, we show that the necessary condition for the film to comply to Navier–Stokes equations is that the typical flow velocity be small compared to the Marangoni elastic wave velocity.


---

[1]even if traditionally the term film refers to soap film and, for pure water, the term sheet is more generally used, we are going to use indifferently both, since soap films are now produced by a nozzle through an expansion as water sheets and since both are subject to similar instabilities and wavy motions.

# 1 Introduction

Water films are a matter of study since the prior papers of Squire [27] and Taylor [28] due to their natural beauty, their theoretical interest and their variety of applications ranging from atomization and sprays in combustion to curtain coating processes. Water films sustain waves originating from the interaction of the capillary waves developing on each of its interfaces. A large part of thin film dynamics can be understood from this dispersive-wave point of view and from the over-simplified stretched solid membrane model. Inertial effects in the surrounding fluid (taken into account by the Bernoulli equation) turn to be destabilizing when the fluid moves faster than the wave. This subtle destabilizing effect of the surrounding fluid (which may also be demonstrated on the solid membrane model) eventually leads to the breaking of the film. The above traditional applications of films refer to transverse motions in the film. When soap is added to water, the dependence of surface tension with the superficial soap concentration makes the film elastic and therefore reduce its tendency to break. In this case, large scale motions in the film plane have been observed by Y. Couder [7] (1981 and following papers with co-workers), who introduced the use of soap films as a thin fluid layer in which two-dimensional hydrodynamics can be studied. The technique has been adapted and improved by Gharib (1989) [13] (and, later on, by Goldburg's group [18]), who has developed an actual 2D soap tunnel.

# 2 The elastic membrane

The case relative to the motion of an infinitely thin elastic membrane under tension is analyzed since, in the asymptotic limit for large wavelength, it represents the anti-symmetric wave dynamics on a thin film, for which internal motion may be neglected. With the aim to illustrate basic reasoning concerning wave propagation phenomena, a simple mathematical description is derived which allows straightforward individuation of wave properties. The effect of wind blowing on the membrane is then considered. We show that, depending on the phase of the pressure variation induced in the air by the potential flow imposed by the elasic wave propagation, the air will just increase the film inertia (when the wind is slower than the wave), or it may destabilize the wave when the wind speed is greater than the phase velocity of the wave.

## 2.1 Wave propagation on elastic membrane

We consider a membrane whose initial mass per unit length is $\rho_m$, and on the ends of which constant tension forces are applied. The membrane is elastic with a tension linearly proportional to the elongation. It is assumed to be flat in the $xz-$plane. For simplicity, only two-dimensional waves will be considered on the membrane, with their crests parallel to the $z-$axis. End effects will also be neglected, as the membrane is assumed to be infinite in the $x-$direction.

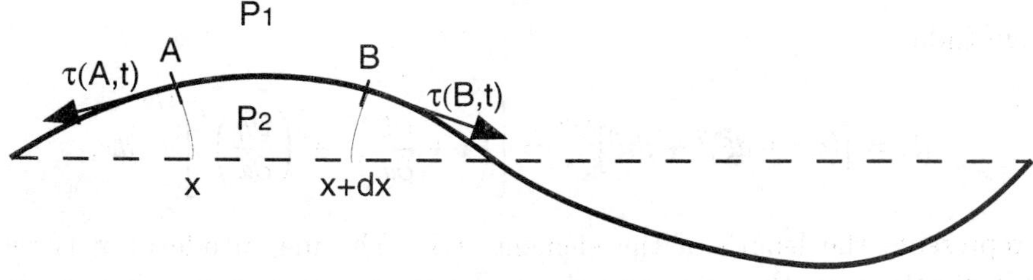


Figure 1: Forces acting on a displaced and stretched membrane element.

A perturbation to the initial configuration, namely a local displacement of a small amount, leads to wave motion. The tension in the membrane gives rise to forces which tends to restore the initial configuration (evident on figure 1). Tension opposes membrane bending, whereas elasticity opposes membrane stretching. On the other hand, an immediate return is delayed by the inertia of the displaced portion. The momentum acquired by this portion causes it to overshoot the initial position. Moreover the local perturbation will spread along the whole membrane as time increases.

In order to formally describe the phenomenon, let us concentrate on a Lagrangian point of the membrane $A$ which, in the unperturbed state, is localized at $(x, 0)$, and is perturbatively displaced to some location specified by the vector

$$\mathbf{r} = \mathbf{i}\xi + \mathbf{j}\eta, \tag{1}$$

where $\mathbf{i}$ and $\mathbf{j}$ repectively designate the unit vectors in the $x-$ and $y-$directions. A neighboring point $B$ situated at $(x + dx, 0)$ is moved to a location specified by the displacement vector

$$\mathbf{r} + \mathbf{dr} = \mathbf{i}(\xi + d\xi) + \mathbf{j}(\eta + d\eta). \tag{2}$$

We apply Newton's second law of dynamics considering that, in the perturbed configuration, the net vector force acting on a unitary length of the element $AB$ is the sum of the tension force $-\boldsymbol{\tau}$ acting at $A$ and of the force $\boldsymbol{\tau} + (\partial\boldsymbol{\tau}/\partial x)\, dx$ acting at $B$ ($x$ is the Lagrangian coordinate), and that the mass of the membrane element $AB$ is the same as at rest, namely $\rho_m dx$:

$$\frac{\partial\boldsymbol{\tau}}{\partial x} = \rho_m \frac{\partial^2 \mathbf{r}}{\partial t^2}. \tag{3}$$

The direction of $\boldsymbol{\tau}$ is given by the unit vector tangent to the membrane $\hat{\mathbf{ds}} = \mathbf{ds}/ds$, where the vector $\mathbf{ds}$ given by

$$\mathbf{ds} = \mathbf{i}(dx + d\xi) + \mathbf{j}d\eta \tag{4}$$

has magnitude

$$ds = \left[(dx + d\xi)^2 + d\eta^2\right]^{1/2} = \left[\left(1 + \frac{\partial\xi}{\partial x}\right)^2 + \left(\frac{\partial\eta}{\partial x}\right)^2\right]^{1/2} dx, \tag{5}$$

which represents the length of the element $AB$. The magnitude of $\tau$ is related to the tension acting on the undisplaced membrane $\sigma_0$: at any position, $\sigma_0$ is indeed incremented by the fractional amount $(ds - dx)/dx$ owing to elasticity. The relation:

$$\tau = \sigma_0 + \sigma' \frac{ds - dx}{dx} \tag{6}$$

is assumed to hold. It follows that

$$\boldsymbol{\tau} = \left[\sigma_0 + \sigma'\left(\frac{ds}{dx} - 1\right)\right]\hat{\mathbf{ds}}. \tag{7}$$

If small values of $\partial\xi/\partial x$ and $\partial\eta/\partial x$ are considered, the expression for $\boldsymbol{\tau}$ can be linearized as:

$$\boldsymbol{\tau} = \mathbf{i}\left(\sigma_0 + \sigma'\frac{\partial\xi}{\partial x}\right) + \mathbf{j}\sigma_0\frac{\partial\eta}{\partial x}. \tag{8}$$

Consequently, equation (3) separates into the following scalar equations:

$$\frac{\partial^2\xi}{\partial x^2} = \frac{\rho_m}{\sigma'}\frac{\partial^2\xi}{\partial t^2} \tag{9}$$

$$\frac{\partial^2\eta}{\partial x^2} = \frac{\rho_m}{\sigma_0}\frac{\partial^2\eta}{\partial t^2}. \tag{10}$$

Two decoupled families of waves prevail in the linearized regime. Equation (9) governs the propagation of small-amplitude longitudinal waves (*elastic* waves) due to membrane elasticity; equation (10) describes small amplitude transverse waves, which correspond to membrane *flapping* due to transverse restoring tension forces. The elastic mode of vibration leads to waves of the form:

$$(\xi, \eta) = (Ae^{i(\omega t - kx)}, 0), \tag{11}$$

where $A$ is a constant arbitrary amplitude. Substitution of equation (11) into equation (9) leads to the phase velocity:

$$c_e = \frac{\omega}{k} = \sqrt{\frac{\sigma'}{\rho_m}}. \tag{12}$$

For the flapping mode, the waves have the form:

$$(\xi, \eta) = (0, Ae^{i(\omega t - kx)}) \tag{13}$$

and the phase velocity:

$$c_f = \frac{\omega}{k} = \sqrt{\frac{\sigma_0}{\rho_m}}. \tag{14}$$

Relations (14) and (12) for phase velocity of both longitudinal and transverse waves may be directly derived using a dimensional analysis similar to the one presented in Fauve's contribution to this volume. The particularity of both equations is that the phase velocity is independent of the wavelength, i.e. that these waves are not dispersive, like usual sound waves, and propagate information at a constant speed. This remark is crucial when transverse instability of jet or compressible effects in soap films are considered.

## 2.2 Instability of an elastic membrane in the wind

The presence of air (on the upper side only, for simplicity) is taken into account, assuming that the air is inviscid and the flow is potential. The air motion is then forced by the deformation of the interface and the uniform mean air velocity $U_a$. It may be described in the two-dimensional framework by a potential function $\phi_a$ such that:

$$\Delta\phi_a = 0 , \tag{15}$$

and complying with the linearized boundary condition:

$$\frac{\partial\eta}{\partial t} + U_a\frac{\partial\eta}{\partial x} - \frac{\partial\phi_a}{\partial y} = 0 , \quad \text{at } y = 0 . \tag{16}$$

Symmetrically, the effect of air blowing is to add a pressure force acting normal to the membrane, the magnitude of which is given by applying the linearized **unstationary** Bernoulli equation. The Newton's law for the membrane reads:

$$\rho_m\frac{\partial^2\eta}{\partial t^2} = \sigma_0\frac{\partial^2\eta}{\partial x^2} + \rho_a\left(\frac{\partial\phi_a}{\partial t} + U_a\frac{\partial\phi_a}{\partial x}\right) , \tag{17}$$

where the $\phi_a$ terms are evaluated at $y = 0$ and where $\rho_a$ is the density of air. The pressure term $\left(\frac{\partial\phi_a}{\partial t} + U_a\frac{\partial\phi_a}{\partial x}\right)$ changes sign depending if the phase velocity of the wave considered is larger or smaller than the air velocity $U_a$. When the wave goes faster than the air the pressure term adds to the restoring force (added mass) whereas when it goes slower (as in the standard Kelvin Helmoltz instability) the pressure term is destabilizing.

We look for instability-wave solutions where $k$ and $\omega$ may be complex:

$$(\phi_a, \eta) = (\hat{\phi}_a(y), A)e^{i(\omega t - kx)} . \tag{18}$$

Equation (15) together with the boundary condition according which $\hat{\phi}_a(y)$ vanishes at $+\infty$ gives:

$$\hat{\phi}_a(y) = Be^{-k\,\text{sign}(\Re(k))y} . \tag{19}$$

The presence of the "absolute value" $k\mathrm{sign}(\Re(k))$ corresponds to the fact that the potential flow is induced by the membrane, and not by any other source at infinity which would displace the membrane. This absolute value is crucial (see ref. [15]) for the analysis of flow instability and the resulting singularity may be avoided by imposing $\Re(k)$ to be positive.

The two conditions at $y = 0$, given by equations (16) and (17), lead to the modified dispersion relation for the *flapping* mode in the presence of air stream:

$$\omega^2 = \frac{\sigma_0}{\rho_m}k^2 - \frac{\rho_a}{\rho_m k}(\omega - U_a k)^2 = c_f{}^2 k^2 - \frac{\rho_a}{\rho_m k}(\omega - U_a k)^2 \ . \tag{20}$$

This equation may be written as :

$$\omega^2(1 + \frac{\rho_a}{\rho_m k}) - 2\omega\frac{\rho_a}{\rho_m}U_a - c_f{}^2 k^2 + \frac{\rho_a U_a^2}{\rho_m}k \ . \tag{21}$$

It admits complex solutions when the discriminant is negative, i.e. when $0 < k\rho_m/\rho_a < -1 + (U_a/c_f)^2$. This is possible only if the Weber number $We = U_a/c_f$ is greater than unity. This means that the air just increases the film inertia (see factor $(1 + \frac{\rho_a}{\rho_m k})$ in the first term of equation (21)) when the wind velocity in smaller than the wave velocity. When the Weber number $We$ is greater than 1, equation (21) admits complex solutions, indicating that the wave may be destabilized when the wind speed is faster than the phase velocity of the wave (14). When the linear mass ratio $\rho_a/\rho_m k$ is small (say of order $\epsilon$), the wave is destabilized only at order $\epsilon$ whereas, at leading order, it keeps propagating at the constant phase and group velocities given by equation (14).

# 3    General theories

This section presents the general results and physical ideas pertaining to wave and instability with a particular emphasis on the idea of the propagation of information. In dealing with waves a change in space of the sign of the group velocity measured in the laboratory frame is well known to induce a stationary shock wave (equivalent for thin film which are dominated by surface tension to the hydraulic jump for gravity waves on shallow water). For instability the equivalence of the group velocity may be found in the absolute and convective instability concept and the shock wave is replaced by the global instability of non parallel unstable flow.

The present section is somehow formal and may be skipped in the first reading and refereed to when reading the next section where the theory is actually applied to the thin film problem.

## 3.1    Dispersive wave propagation

Waves occurring at an interface between a liquid and a gas are probably the most familiar example of wave propagation phenomena. In fluids, contrary to the elastic membrane case, the waves are *dispersive* since different wavelengths move at different speed (see next section for details). Waves generated by a locolized impulse are found at a different location in space when time evolves.

Usually the study of this kind of problem is performed by referring to elementary solutions of the governing equations in the form of sinusoidal wave:

$$\phi(\mathbf{r},\, t) = \Re\{Ae^{i(\mathbf{k}\cdot\mathbf{r}-\omega t)}\}, \tag{22}$$

with vavenumber vector $\mathbf{k} = (k_1, k_2)$, frequency $\omega$ and amplitude $A$, deriving the link existing between wavenumber and frequency, namely the *dispersion relation*

$$D(\mathbf{k},\, \omega) = 0. \tag{23}$$

Within the study of ordinary waves, equation (23) is solved by searching for real roots

$$\omega = \Omega(\mathbf{k}). \tag{24}$$

Complex solutions allow for growth in time of the perturbations introduced into the basic state and are thus taken under consideration in the study of the system stability. When several roots of equation (23) are found, each of them is called a *mode*.

General solution of the linearized problem does not have to follow the wave form (22) but if it does locally then the wave train may be represented using the local amplitude $A$ and the local phase $\theta$ :

$$\phi(\mathbf{r},\, t) = \Re\{A(\mathbf{r},\, t)e^{i\theta(\mathbf{r},\, t)}\}. \tag{25}$$

The gradient of the phase function $\theta$ represents the local wavevector $\mathbf{k}_l$ which is a function of time and space, whose magnitude $k_l$ represents the average number of crests over a distance of $2\pi$. The derivative in time of $\theta$, taken with the opposite sign, represents the local frequency $\omega_l$ which is a function of time and space, hence the average number of wave crests per $2\pi$ units of time. The local wavelength and the period are $\lambda = 2\pi/k_l$ and $T = 2\pi/\omega_l$ respectively. Any phase surface $\theta = const$ moves in the direction of $\mathbf{k}_l$ with a velocity equal to $\omega/k_l$. Therefore the phase velocity is defined as

$$\mathbf{c} = \frac{\omega_l}{k_l}\hat{\mathbf{k}}_l, \tag{26}$$

where $\hat{\mathbf{k}}_l$ is the unit vector in the direction of $\mathbf{k}_l$.

We want to emphasize that the locally defined quantities $(\mathbf{k}_l, \omega_l)$ do not have to follow the dispersion relation for plane wave (equation (23)) except if the amplitude is small and the variation of $\mathbf{k}_l$ and $\omega_l$ are slow enough in time and space for the WKBJ theory to apply. Furthermore formulation 25 is not limited to small amplitude and may be used to describe a non linear stage [29].

Planar waves have a major interest for the linear theory since the way of understanding how a fluid interface may react to a small perturbation is by Fourier analysis.

The system response to the perturbation is regarded as the superposition (linear combination) of different sinusoidal disturbances. The superposition principle holds if the equations governing the wave motion are linear and the boundary conditions are required to be satisfied on the unperturbed boundaries instead on the perturbed one (no non linearities introduced through the boundary conditions).

Without being entangled in the complexity of the multidimensional Fourier analysis, we examine one dimensional plane waves of arbitrary shape, for which $\omega = \Omega(k)$, whose description comes directly from the Fourier transform pair definition in terms of the coordinate $x$ and the wavenumber $k$ :

$$\phi(x) = \frac{1}{2\pi} \int_{-\infty}^{+\infty} F(k)e^{ikx}dk \tag{27}$$

$$F(k) = \int_{-\infty}^{+\infty} \phi(x)e^{-ikx}dx. \tag{28}$$

The complex spectral density function $F(k)$, multiplied by $e^{ikx}$ and integrated over the whole wavenumbers range, gives a particular function of space $\phi(x)$.

In a nondispersive medium where the wave velocity is a positive constant $c = \omega/k$, the representation of a wave traveling in the positive $x$ direction is obtained by replacing $x$ by $x - ct$ in equation (27):

$$\phi(x,\,t) = \frac{1}{2\pi} \int_{-\infty}^{+\infty} F(k)e^{ik(x-ct)}dk = \frac{1}{2\pi} \int_{-\infty}^{+\infty} F(k)e^{i(kx-\Omega(k)t)}dk. \tag{29}$$

$\phi(x,\,t)$ is thus a continuum of sinusoidal wave components of the kind defined by equation (22), where a complex amplitude has been taken. Since

$$\phi(x,\,0) = \frac{1}{2\pi} \int_{-\infty}^{+\infty} F(k)e^{ikx}dk, \tag{30}$$

the function $F(k)$ can be determined as inverse Fourier transform of the wave shape at $t = 0$.

When a dispersive medium is considered and the property :

$$\frac{d^2\Omega}{dk^2}(k) \neq 0, \tag{31}$$

is assumed, subscripts denoting derivative with respect to $k$, the superposition of different Fourier components leads to the important concept of *group velocity*.

For simplicity, we first consider the combination of only two waves with the same amplitude and whose wavenumbers $k_a$ and $k_b$ differ slightly. The corresponding frequencies are $\omega_a = \Omega(k_a)$ and $\omega_b = \Omega(k_b)$. The sum of these two components

$$\phi(x,\,t) = \Re\{Ae^{i(k_a x - \omega_a t)} + Ae^{i(k_b x - \omega_b t)}\} = A\cos(k_a x - \omega_a t) + A\cos(k_b x - \omega_b t) =$$
$$2A\cos\left\{\frac{1}{2}[(k_b - k_a)x - (\omega_b - \omega_a)t]\right\}\cos\left\{\frac{1}{2}[(k_b + k_a)x - (\omega_b + \omega_a)t]\right\} \tag{32}$$

is an oscillating wave of frequency $(\omega_b + \omega_a)/2$ with wavenumber $(k_b + k_a)/2$, whose amplitude is a slowly modulated in space and time. A point where the *local* amplitude (i.e. the envelop smoothing the fast oscillation) reaches, say, the value $2A$, moves at velocity

$$\frac{\omega_b - \omega_a}{k_b - k_a}. \tag{33}$$

This last, in the limit $k_b \to k_a$, represents the derivative of $\omega$ with respect to $k$, defined as the group velocity

$$c_g(k) = \frac{d\omega}{dk}. \tag{34}$$

In the previous nondispersive case since $\omega$ is a linear function of $k$, the group velocity is independent of $k$ and equales the phase velocity.

We consider now the more general situation where $\phi(x, t)$ is the combination of continuum of waves:

$$\phi(x, t) = \frac{1}{2\pi} \int_{-\infty}^{+\infty} F(k)e^{i[kx - \Omega(k)t]}dk. \tag{35}$$

The factor $1/(2\pi)$ has been introduced in the sum in analogy with equation (29) so as to obtain also in this case $F(k)$ as inverse Fourier transform of the wave shape at $t = 0$. Obviously in general the existence of several modes should be kept into account and the theory that follows would be applicable to each of the mode $\omega = \Omega(k)$ separatly (see [29]). Assuming $\phi(x, t)$ to have the shape of a single wave packet of almost constant wavenumber $k_0$ and slowing slowly varying amplitude, so that $|F(k)|$ is very small except where $k$ is very close to $k_0$, it is possible to replace $\Omega(k)$ by its Taylor's series expansion

$$\Omega(k) \simeq \Omega(k_0) + c_g(k - k_0) \tag{36}$$

in equation (35), although this is strictly valid only near to $k_0$. Thus $\phi(x, t)$ rewrites as a pure harmonic wave of wavenumber $k_0$ multiplied by a function of $x$ and $t$ only through the particular combination $x - c_g t$:

$$\phi(x, t) = \frac{1}{2\pi}e^{i[k_0 x - \Omega(k_0)t]} \int_{-\infty}^{+\infty} F(k)e^{i(k-k_0)(x-c_g t)}dk. \tag{37}$$

The integral in the above formula represents an amplitude modulation for the harmonic wave of wavenumber $k_0$ in the form of a packet which moves as a whole at a velocity equal to $c_g$. Since wave motion makes for energy to be propagated, and the packet is where all energy is, group velocity is, on an intuitive ground, also the velocity of propagation of energy.

Once a wave packet is constructed to represent how a given system reacts to a perturbation introduced at $t = 0$, it is generally of major interest to determine the behavior of the system response as the time and space tend to infinity, namely what is usually called the *far field analysis* of waves. A convenient way to perform it, is by evaluating the limit as $t \to \infty$ of equation (35) along a ray $x/t$ fixed in the $(x,\,t)$ plane. The variable

$$\chi(k) = \Omega(k) - k\frac{x}{t} \tag{38}$$

is introduced so as to rewrite equation (35) in the form:

$$\phi(x,\,t) = \frac{1}{2\pi} \int_{-\infty}^{+\infty} F(k)e^{-i\chi t}dk. \tag{39}$$

At this stage, the method of stationary phase [1] suggests that the main contribution at large $t$ to this integral comes from waves whose wavenumber is in the vicinity of the stationary points $k^*$ for which:

$$\frac{d\chi}{dk}(k^*) = \frac{d\Omega}{dk}(k^*) - \frac{x}{t} = 0. \tag{40}$$

The contribution from other waves are rapidly oscillating and hence of negligible net value. Near to $k^*$ one can write:

$$F(k) \simeq F(k^*) \tag{41}$$

and

$$\chi(k) \simeq \chi(k^*) + \frac{1}{2}(k - k^*)^2\frac{d^2\chi}{dk^2}(k^*). \tag{42}$$

It follows that

$$\phi(x,\,t) \simeq \frac{1}{2\pi}F(k^*)e^{-i\chi(k^*)t} \int_{-\infty}^{+\infty} e^{-\frac{i}{2}(k-k^*)^2\frac{d^2\chi}{dk^2}(k^*)t}dk. \tag{43}$$

Rotation of $\pm\frac{\pi}{4}$, depending on the sign of $\frac{d^2\chi}{dk^2}(k^*)$, of the integration contour transforms the integral into the following real one:

$$\int_{-\infty}^{+\infty} e^{-a\zeta^2}d\zeta = \left(\frac{\pi}{a}\right)^{(1/2)} \tag{44}$$

and

$$\phi(x,\,t) \simeq F(k^*)\left(\frac{1}{2\pi t\left|\frac{d^2\chi}{dk^2}(k^*)\right|}\right)^{1/2} e^{-i\chi(k^*)t - \frac{i\pi}{4}sgn\frac{d^2\chi}{dk^2}}, \tag{45}$$

is the searched result.

In situations where $\frac{d^2\Omega}{dk^2}(k) = 0$ for some point $k^*$ with $\frac{d^2\chi}{dk^2}(k^*) = 0$, but $\frac{d^3\chi}{dk^3}(k^*) \neq 0$, the Taylor's series modifies into:

$$\chi(k) \simeq \chi(k^*) + \frac{1}{3!}(k - k^*)^3\frac{d^3\chi}{dk^3}(k^*) \tag{46}$$

and equation (43) becomes:

$$\phi(x,\,t) \simeq \frac{1}{2\pi}F(k^*)e^{-i\chi(k^*)t} \int_{-\infty}^{+\infty} e^{-\frac{i}{3!}(k-k^*)^3\frac{d^3\chi}{dk^3}(k^*)t}dk. \tag{47}$$

This gives a different amplitude modulation of the wave:

$$\phi(x,\,t) \simeq \frac{3^{-1/6}2^{-1/3}}{2\pi} F(k^*) \left( t \left| \frac{d^3\chi}{dk^3}(k^*) \right| \right)^{-1/3} e^{-i\chi(k^*)t}. \tag{48}$$

A direct interpretation of group velocity comes by observing that in each of the case considered $\phi(x,\,t)$ has been represented as in equation (25), namely as an oscillatory quantity of phase $\theta$ whose amplitude is modulated in space and time. If purely sinusoidal waves are considered, the phase is a linear function of $x$ and $t$, and

$$k = \frac{\partial\theta}{\partial x} \qquad \omega = -\frac{\partial\theta}{\partial t} \tag{49}$$

are constant quantities.

In more general situations (even in the nonlinear case see [29] where they both depend on $x$ and $t$, $k$ and $\omega$ are related, since they are defined as partial derivatives, by the relation

$$\frac{\partial k}{\partial t} + \frac{\partial\omega}{\partial x} = 0, \tag{50}$$

which can be rewritten as:

$$\frac{\partial k}{\partial t} + \frac{d\omega}{dk}\frac{\partial k}{\partial x} = 0. \tag{51}$$

This equation tells us that a particular wavenumber is propagated at the group velocity! Equation (51) is a transport equation for $k$ and $c_g = \frac{d\omega}{dk}$ is the velocity an observer sould move at in order to see a wave of constant wavelength. This is obviously a consequence of the hyperbolic character of equation (51) whose solution $k = f(x - c_g t)$ remains constant if $x - c_g t$ does. An equally intuitive consideration concerning energy propagation, as the one previously made, is that the path along which any wavenumber $k$ is found is the one along which the energy of the wave of that wavenumber is propagated. Indeed, a formal proof would show that, in general, the rate of energy transport by a sinusoidal wave is given by the group velocity times the average total energy density. In the case the group velocity is greater than the phase velocity, energy propagates faster than crests. In the opposite situation wave crests may be observed to continually appear at the back of a packet and disappear at the front, their motion being not sustained by an energy moving fast enough.

*Extension to three dimensional wave and lee waves*

Extension to three dimensional wave propagation ($\mathbf{k} = (k_1,\,k_2)$) of the above considerations gives the following definition of a wave of arbitrary shape:

$$\phi(\mathbf{r},\,t) = \frac{1}{(2\pi)^2} \int F(\mathbf{k}) e^{i[\mathbf{k}\cdot\mathbf{r} - \Omega(\mathbf{k})t]} d\mathbf{k}. \tag{52}$$

whose asymptotic time behavior is

$$\phi(\mathbf{r},\,t) \simeq F(\mathbf{k}^*) \left(\frac{1}{2\pi t}\right) \left( t \cdot det \left| \frac{\partial\Omega}{\partial k_l \partial k_j} \right| \right)^{-1/2} e^{i[\mathbf{k}^*\cdot\mathbf{r} - \Omega(\mathbf{k}^*)t + iv]} \tag{53}$$

with $v$ related to the number of factors $\pi i/4$ due to path rotation and being $\mathbf{k}^*$ the wavenumber vector for which

$$\frac{x_l}{t} = \frac{\partial \Omega}{\partial k_l}(\mathbf{k}^*) \qquad l = 1,\, 2. \tag{54}$$

Definition of phase (equation (25)) leads to the following kinematic description of waves:

$$\frac{\partial k_l}{\partial t} + \frac{\partial \omega}{\partial x_l} = 0 \tag{55}$$

$$\frac{\partial k_l}{\partial x_j} - \frac{\partial k_j}{\partial x_l} = 0, \tag{56}$$

Equation (55) can be written as

$$\frac{\partial k_l}{\partial t} + \frac{\partial \Omega}{\partial k_j}\frac{\partial k_j}{\partial x_l} = 0. \tag{57}$$

or, being

$$c_j(\mathbf{k}) = \frac{\partial \Omega}{\partial k_j}(\mathbf{k}), \tag{58}$$

as

$$\frac{\partial k_l}{\partial t} + c_j(\mathbf{k})\frac{\partial k_l}{\partial x_j}. \tag{59}$$

When propagation in more than two dimensions is considered, attention is to be devoted to the possible existence of preferred directions of wave motion. Dispersion relations in which the dependence on the wavenumber vector concerns only its magnitude are relative to isotropic media. An example is given by waves originating from a point source in still water, as a stone thrown into a pond. The same argumentation leading to equation (45) applies with $k$ as the modulus of the wave vector and $x$ replaced by the distance from the source of waves.

On the contrary, the introduction of a moving obstacle into still water produces non isotropic waves. These last may be regarded as originating from a fixed obstacle introduced into a current moving at the same speed but in the opposite direction. Further important property of the dispersion relation is that it constitutes the only tool needed to draw the shape the crest lines assume in clashing with the obstacle (wave patterns). In searching for these, it is necessary to establish the reference frame with respect to which their shape remains steady. Dispersion relation for waves propagating into still water can be transferred to a reference frame moving with a relative velocity $-\mathbf{U}$, writing the frequency $\omega'$ in the moving frame as:

$$\omega' = \mathbf{U} \cdot \mathbf{k} + \omega(\mathbf{k}), \tag{60}$$

where $\omega$ is the frequency in the fixed frame, according to the Doppler effect. In analogous way any dispersion relation derived for waves on a current can be written in a reference frame moving with the current itself.

From equation (60), stationary waves are found for wavenumber vector $\mathbf{k} = (k_1,\ k_2)$ satisfying:

$$H(k_1,\ k_2) = \mathbf{U} \cdot \mathbf{k} + \omega(\mathbf{k}) = 0. \tag{61}$$

Equations (55) and (56) reduce to the single equation:

$$\frac{\partial k_2}{\partial x_1} - \frac{\partial k_1}{\partial x_2} = 0. \tag{62}$$

Since equation (61) holds,

$$k_1 = f(k_2) \tag{63}$$

and

$$\frac{\partial k_2}{\partial x_1} - \frac{df}{dk_2}(k_2)\frac{\partial k_2}{\partial x_2} = 0. \tag{64}$$

$k_1$ and $k_2$ are constant on characteristics which are assumed to pass all through the point where the source is situated and are obtained in the $x_1 x_2$ plane by solving:

$$\frac{dx_2}{dx_1} = -\frac{df}{dk_2}(k_2). \tag{65}$$

Regarding $H(k_1,\ k_2)$ as an implicit function, the following relation holds:

$$\frac{df}{dk_2}(k_2)\frac{\partial H}{\partial k_1} + \frac{\partial H}{\partial k_2} = 0 \tag{66}$$

and equation (65) gives:

$$\frac{x_2}{x_1} = \frac{df}{dk_2}(k_2) = \frac{\partial H}{\partial k_2}\bigg/\frac{\partial H}{\partial k_1} \tag{67}$$

which, together with equation (63), yields $k_1$ and $k_2$ on each characteristic. The distribution of $\mathbf{k}$ gives the patterns; the lines of constant phases give the shape of the crest lines.

The above consideration holds without any limitation about the character of the current. In the case the current is uniform a simple computation gives the physics behind the previous equations defining the stationary waves pattern. Figure 2 shows a current encountering an obstacle which is in the origin of a reference frame $(x_1,\ x_2)$. At each point $P$ the wavenumber vector $\mathbf{k}$ characterising the lee wave of the obstacle $O$ should be such that its frequency should be zero in reference frame of the obstacle and its wave group velocity in the reference frame of the obstacle should be parallel and in the same direction than the vector $OP$ since information (or energy) should have travelled from the source $O$ to the observation point $P$. From this geometric optic approximation we immediately deduce that characteristics are ray coming from $O$. This condition, which is analogous to equation (64), together with the one corresponding to take waves of null frequency, is analogous to solve the more complicate problem above addressed. The angles $\xi$ and $\gamma$ emphasised that phase and group velocity have nothing in common.

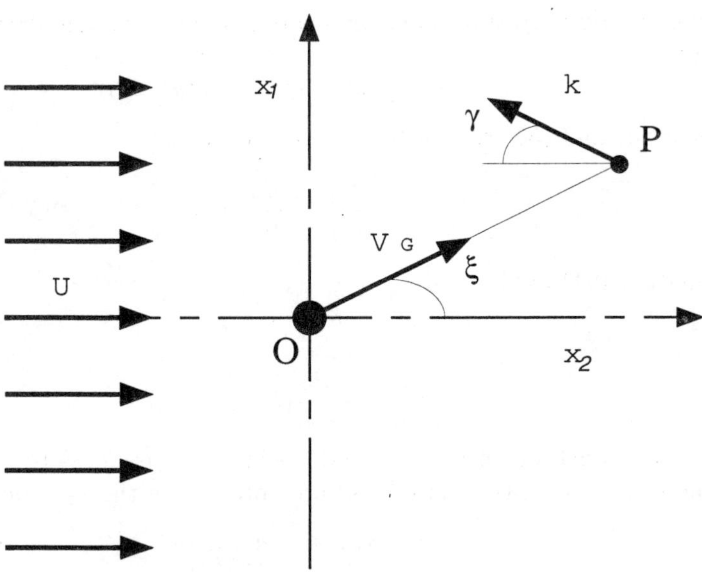


Figure 2: Schematic representation of the geometric optic approximation for waves generated by an obstacle.

*Remark*

No dissipation has been until now considered in the wave propagation phenomena. Indeed, wave crests, once generated, are observed only over finite distances of space. The reason for this is attenuation, namely dissipation of energy. It can occur through three main ways: *bottom friction*, where propagation happens over a finite depth, *internal dissipation*, due to the viscosity of the interested liquid, and *surface dissipation*. The first mechanism is related to the presence of the boundary layer over the rigid surface, but is indeed negligible if the wavelength is small enough to not produce motion in proximity of the boundary layer itself. Internal dissipation is always present and found to be inversely proportional to wavelength and frequency of waves. Surface dissipation comes from possible gradients of surface tension due to the presence of surfactants or temperature gradients. It can be thought to as an extra viscous dissipation due to enhanced shearing stresses within the surface boundary layer [20].

## 3.2   Propagation of instability waves

Stability analysis of fluid systems is of fundamental importance both from a theoretical and a technological point of view. Engineering relevant phenomena related to loss of stability and transition from laminar flows to turbulence are drastic changes in global properties such as drag, aeroacoustic noise generation, heat transfer and mixing. The study of the stability of a given configuration, the so-called *basic state*, concerns the break up of this configuration when externally imposed constrains, such as shear, temperature or pressure gradients, reach critical values. These last are usually individuated in terms of dimensionless control parameters. On a mathematical basis, the question is that any solution to the equations governing a given flow must be stable to perturbations. A perturbation, in fact, introduced at time $t = 0$, may decay, remain into the flow with an almost constant amplitude, or grow so to make the initial configuration to cease to be observed. Within the linear approximation, considering disturbances of small amplitude superimposed to the basic flow, it is possible to analyze the initial stages of growth and to achieve a certain comprehension of the physical mechanisms responsible of break up. This comprehension involves *instability mode* which are the natural extension of dispersive wave except that $\omega$ and $k$ may be complex. Propagation of information may be analyzed using wave packet as for dispersive wave and an extension of the stationary phase argument give a precise analogous of the classical group velocity (which has a meaning only when the complex derivative $\partial \omega / \partial k$ is real).

As previously observed, small amplitude wave propagation and instability are examined within the same theoretical context until the dispersion relation between wavenumber and frequency is obtained. The existence of at least one complex solution for the frequency with positive imaginary part implies the presence of a perturbation growing with time, hence an unstable behavior of the flow. Any disturbance excites all the modes, and only if each of them decays with time, the system can be said to be stable. This kind of approach, called *temporal modes analysis*, leads to the identification, in the control parameters space, of two domains corresponding to stability and instability separated by a *neutral surface*. As the external constraints on the flow are varied, control parameters may cross the neutral surface reaching the critical values for which the initial configuration becomes unstable. Under these conditions perturbations behave as growing waves with different wavenumber and frequency, which are the theoretical counterparts of vortical or wave like structures observed experimentally.

Since in experiments the forcing of a system at a given frequency is often easier than a forcing at given wavenumber, an alternative solution of the dispersion relation in terms of real $\omega$ and complex $k$ has been conceived. The so-called *spatial modes* are sinusoidal disturbances with spatially evolving amplitude along $x$, say, of growth rate equal to the opposite of the imaginary part of $k$, $-k_i$.

Both temporal and spatial modes have meaning since the response of the system to a perturbation can be expressed in terms of their superposition. It is clear that the response is an effect, and, hence, it can't anticipate the producing cause. This is to say that *causality* has to be satisfied in the problem modeling. Temporal modes serve to the task because only $t > 0$ are considered and no ambiguity arises in defining growing modes. On the contrary, spatial modes analysis is not able to determine by itself if a $k_i < 0$ corresponds to an unstable wave propagating in the positive $x$ direction, or to a decaying wave propagating in the negative $x$ direction. Particular care is thus required in the use of this technique.

A correct formulation of a stability model requires, in fact, the solution of an initial value problem where the evolution of a perturbation is studied starting from the instant of time at which it is supposed to be introduced into the flow. The spatio-temporal dynamics of the response of the flow to the perturbation deserves great attention, since the way in which linear instabilities take perturbations into the nonlinear regime strongly depends on this aspect. Perturbations may propagate in both the directions at once while growing in time, or be advected by the flow. In the first situation the instability is said to be *absolute*, in the second *convective*. After an absolute instability has reached a certain $x$ position, growth in time persists, whereas a convective instability outstrips any spatial location and substantially leaves the medium in its unperturbed state. An unstable system for which convective instabilities are found, behaves as an amplifier of perturbations, whereas a system which is absolutely unstable behaves more as an oscillator. It has a sort of internal feedback mechanism so that any perturbation tends to contaminate the whole field without the need of boundary reflections.

The system description is obtained by associating to the dispersion relation in the spectral space $(k, \omega)$ the differential operator $D[-i\partial/\partial x, i\partial/\partial t]$ in the physical space $(x, t)$:

$$D[-i\partial/\partial x, i\partial/\partial t]\Psi(x, t) = 0. \tag{68}$$

If a prescribed forcing function $s(x, t)$ is turned on at $t = 0$, the system response $\Psi(x, t)$ is not freely evolving as in equation (68), but satisfies the forced differential equation

$$D[-i\partial/\partial x, i\partial/\partial t]\Psi(x, t) = s(x, t). \tag{69}$$

The causality principle states that for all $t < 0$

$$\Psi(x, t) = 0 \quad \text{if} \quad s(x, t) = 0. \tag{70}$$

If $s(x, t)$ is assumed to be the pulse $\delta(x)\delta(t)$, $\delta$ being the Dirac distribution, the system response $\Psi(x, t)$ is the system Green's function $G(x, t)$ [2, 4], and equation (69) becomes:

$$D[-i\partial/\partial x, i\partial/\partial t]G(x, t) = \delta(x)\delta(t). \tag{71}$$

Once $G(x, t)$ is known, any function $\Psi(x, t)$ representing the response to a different source $s$ can be obtained as convolution product between $G$ and $s$. Statement of stability properties does not require the determination of $G$ in detail, but only the evaluation of its asymptotic behavior. According to refs. [2, 4, 15] the system is *linearly stable* if

$$\lim_{t \to \infty} G(x, t) = 0 \tag{72}$$

on each ray $x/t = const$. It is *linearly unstable* if

$$\lim_{t \to \infty} G(x, t) = \infty \tag{73}$$

at least on one ray $x/t = const.$ Further distinction between absolute and convective instability comes from the behavior of $G$ along the ray $x/t = 0$. A linearly convectively unstable flow is an unstable one for which

$$\lim_{t\to\infty} G(x,\, t) = 0 \tag{74}$$

along the ray $x/t = 0$. It is linearly absolutely unstable if:

$$\lim_{t\to\infty} G(x,\, t) = \infty \tag{75}$$

along the ray $x/t = 0$.

In order to evaluate the relevant limits note that equation (71) in the spectral space takes the simple form:

$$D(k,\, \omega)G(k,\, \omega) = 1, \tag{76}$$

hence

$$G(k,\, \omega) = \frac{1}{D(k,\, \omega)} \tag{77}$$

and inverse Fourier and Laplace transforms give:

$$G(x,\, t) = \frac{1}{(2\pi)^2} \int_F \int_L \frac{e^{ikx - i\omega t}}{D(k,\, \omega)} d\omega dk. \tag{78}$$

$F$ and $L$ denote the integration contours in the complex $k$ and $\omega$ plane, respectively. These contours can not be chosen arbitrarily, since convergence of equation (78) and causality have to be satisfied. Since $G(x,\, t)$ has to be well behaved for $x \to \pm\infty$, the region of absolute convergence, where to chose the $F$ contour in the complex $k$ plane, is a strip including the real axis. $F$ can be initially taken to coincide with this last. Each solution of the dispersion relation for $k$ belonging to $F$, is thus a temporal mode. The assumption is made that each of the temporal modes has a bounded growth rate (if not causality cannot be respected see citeChom), so to assure $L$ to lie above the highest imaginary part of all the different $\omega_l(k_r)$ $(l = 1, 2, ...)$. This quantity from now on is denoted by $\omega_{i\,max}$. $L$ is in the region of absolute convergence of the $\omega$ plane.

Once the convergence requirements are fulfilled, Cauchy residue theorem can be applied for the evaluation of $G$. In the case $D(k,\, \omega)$ is analytic with respect to $k$ and $\omega$, the only singularities of the integrand are poles of $G(k,\, \omega)$, hence zeros of the dispersion relation. We refer for the moment to the simple case where only order one poles are found. Closure of $L$ at infinity, by an upper semi-circle for $t < 0$ and by a lower semi-circle for $t > 0$, and residue evaluation give a $G(x,\, t) = 0$ for $t < 0$ (no poles exist in the upper domain), and which, for $t > 0$ is:

$$G(k,\, t) = -i\sum_l \frac{e^{-i\omega_l(k)t}}{\frac{\partial D}{\partial \omega}(k, \omega_l(k))}. \tag{79}$$

Equation (78) becomes:

$$G(x,\,t) = -\frac{i}{2\pi} \sum_l \int_F \frac{e^{i[kx-\omega_l(k)t]}}{\frac{\partial D}{\partial \omega}[k,\,\omega_l(k)]}\,dk,\tag{80}$$

where the wave packet form of $G$ (superposition of waves of different wavenumbers) appears explicitly. Note that the sign of $\omega_{i\,max}$ determines the temporal evolution of $G$. If $\omega_{i\,max} > 0$ the system is unstable, if $\omega_{i\,max} < 0$ it is stable, neutral stability corresponding to the case $\omega_{i\,max} = 0$.

$F$ can be assumed to be different or to be deformed in the complex $k$ plane, and generalized temporal modes corresponding to the solution of $D(k,\,\omega) = 0$ for $k$ lying on a new contour can be found preserving the constraint of causality. In a similar way, once $L$ is chosen, generalized spatial modes can be traced. Remember $L$ is assumed to lie above $\omega_{i\,max}$. Two separate sets of generalized spatial branches will exist in the positive and negative $k$ half-plane. Since $F$ has to separate these two sets, its deformation is possible unless they collapse into each other, namely if a so-called *pinch point*, between spatial branches coming from opposite sides, forms. Assuming for simplicity that only one temporal branch exists, the pinch occurs at $k_0$, $\omega_0$ for which

$$D(k_0,\,\omega_0) = 0 \qquad \frac{\partial D}{\partial k}(k_0,\,\omega_0) = 0.\tag{81}$$

These two conditions are equivalent to search for the $k_0$ giving

$$\omega_0 = \omega(k_0) \qquad \frac{d\omega}{dk}(k_0) = 0,\tag{82}$$

the second equality stating $k_0$ to be a stationary point for the complex function $\omega(k)$. Rewrite equation (80) as

$$G(x,\,t) = -\frac{i}{2\pi} \int_{-\infty}^{+\infty} f(k) e^{-i\omega(k)t}\,dk,\tag{83}$$

where

$$f(k) = \frac{e^{ikx}}{\frac{\partial D}{\partial \omega}[k,\,\omega(k)]}.\tag{84}$$

Since $k_0$ is a stationary point for the complex function $\omega(k)$, method of steepest descendent, which is the complex counterpart of the method of stationary phase, states that the main contribution to $G(x,\,t)$ comes from points around $k_0$, where

$$\omega(k) = \omega_0 + \frac{1}{2}\frac{d^2\omega}{dk^2}(k - k_0)^2.\tag{85}$$

Substitution into equation (83) and choice of the steepest descendent path of the surface $\omega_i(k_r,\,k_i)$ passing through the point $k_0$ gives, after manipulation:

$$G(x,\,t) \simeq \frac{1}{\sqrt{2\pi}\,\frac{\partial D}{\partial \omega}(k_0,\,\omega_0)\sqrt{\frac{d^2\omega}{dk^2}(k_0,\,\omega_0)t}}\,e^{i(k_0 x - \omega_0 t) + \frac{i\pi}{4}}.\tag{86}$$

The wavenumber $k_0$ denotes the response $G$ to a pulse evaluated at a fixed $x$ value, hence along the ray $x/t = 0$. Remembering definitions (74) and (75), the sign of $\omega_{0i}$ determines the convective or absolute character of instabilities. Once $\omega_{i\,max}$ has been found to be greater than zero, hence instability of a system governed by some dispersion relation has been assessed, determination of $k_0$, $\omega_0$ allows identification of its nature. Note that the response on any ray $x/t = v$ is obtained by substituting $x' = x - vt$ and $t' = t$ into equation (78):

$$G(x', t') = \frac{1}{(2\pi)^2} \int_F \int_L \frac{e^{ikx' - i(\omega - kv)t'}}{D(k, \omega)} d\omega dk. \tag{87}$$

Upon introducing $\omega' = \omega - kv$ and $k' = k$, the above integral becomes:

$$G(x', t') = \frac{1}{(2\pi)^2} \int_F \int_L \frac{e^{ik'x' - i\omega't'}}{D'(k', \omega')} d\omega' dk', \tag{88}$$

where

$$D'(k', \omega') = D(k', \omega' + k'v). \tag{89}$$

The asymptotic response at a fixed station in the moving frame may then be deduced. Pinching now takes place at $(k_0', \omega_0')$ such that

$$\omega_0' = \omega'(k_0') \qquad \frac{d\omega'}{dk'}(k_0') = 0. \tag{90}$$

In terms of unprimed variables pinch occurs at $\tilde{k} = k_0'$ and $\tilde{\omega} = \omega_0' + k_0'v$ where

$$\tilde{\omega} = \omega(\tilde{k}) \qquad \frac{d\omega}{dk}(\tilde{k}) = v. \tag{91}$$

Similarly the asymptotic impulse response along the ray $x/t = v$ is:

$$G(x, t) \simeq \frac{1}{\sqrt{2\pi}} \frac{e^{i(\tilde{k}x - \tilde{\omega}t) + \frac{i\pi}{4}}}{\frac{\partial D}{\partial \omega}(\tilde{k}, \tilde{\omega})\sqrt{\frac{d^2\omega}{dk^2}(\tilde{k}, \tilde{\omega})t}}. \tag{92}$$

Instability is present if

$$\sigma_i = \tilde{\omega}_i - \tilde{k}_i \frac{x}{t} > 0, \tag{93}$$

$\sigma_i$ representing the temporal growth rate along the ray $x/t$. Note that equations (86) and (92) are derived within the hypothesis the dispersion relation has only first order zeros, hence behaving around $(\tilde{k}, \tilde{\omega})$ as:

$$\omega - \tilde{\omega} \sim (k - \tilde{k})^2. \tag{94}$$

In more complex situations residue evaluation must be carried out in opportune way, thus leading to different formulas. As a simple rule, if the dispersion relation has the form:

$$(\omega - \tilde{\omega})^n \sim (k - \tilde{k})^p, \tag{95}$$

the asymptotic behavior of $G(x,\,t)$ is given by:

$$\lim_{t\to\infty} G(x,\,t) \sim t^{\frac{(p-1)n}{p}-1} e^{i[\tilde{k}x-\omega(\tilde{k})t]}. \tag{96}$$

Depending on the value of $p$ and $n$, the algebraic modulation of amplitude may dominate at small time and in the case $\omega(\tilde{k})$ is real.

## 4   The liquid film

In searching for a dispersion relation between wavenumber and frequency of waves on liquid films one can resort, as a first step, to dimensional analysis. A liquid of density $\rho_1$ forming a thin sheet of uniform thickness $2b$ in a gas of density $\rho_2$ is considered. The ideal situation of the liquid at rest is taken under exam, as it corresponds to what is seen by an observer moving at the same velocity of the liquid. As previously pointed out, dispersion relations are simply modified in passing from a reference system to an other by the Doppler effect. The interface is assumed for the moment to be in a uniform tension state $\sigma$ so to consider the liquid as between two stretched inelastic membranes (elastic effect will be considered in chapter 5). Taking the frequency $\omega$ as the dependent variable in the mathematical function expressing the relationship between the physical variables of the phenomenon, independent variables are recognized in the wavenumber, the surface tension, the initial thickness and the liquid and gas density (gravity effects are not considered):

$$\omega = f(k,\,\sigma,\,b,\,\rho_1,\,\rho_2). \tag{97}$$

The object of dimensional analysis is to group several variables together to form a new variable which is nondimensional, hence independent of the measuring units, and to express it as a function of only nondimensional variables. In this way the number of independent variables reduces according to the Buckingamm's theorem (i.e. by rescaling of the length $L$, the mass $M$ and the time $T$ unite). The dimensional matrix related to the problem, formed by listing exponents of the primary dimensions of each variable is:

|   | $\omega$ | $k$ | $\sigma$ | $\rho_1$ | $\rho_2$ | $b$ |
|---|---|---|---|---|---|---|
| $T$ | -1 | 0 | -2 | 0 | 0 | 0 |
| $L$ | 0 | -1 | 0 | -3 | -3 | 1 |
| $M$ | 0 | 0 | 1 | 1 | 1 | 0 |

which has to be used in order to check for linear independence of the dimensions of the variable in terms of the chosen primary dimensions. This is done by finding the rank of the matrix, which is easily seen to be equal to 3. Subtracting it to the number of initial dimensional independent variables needed to represent the phenomenon, gives the number of nondimensional independent variables. Once a nondimensional frequency is defined it will be a function of two only nondimensional quantities, which are naturally chosen to be the wavenumber multiplied by the half thickness and the

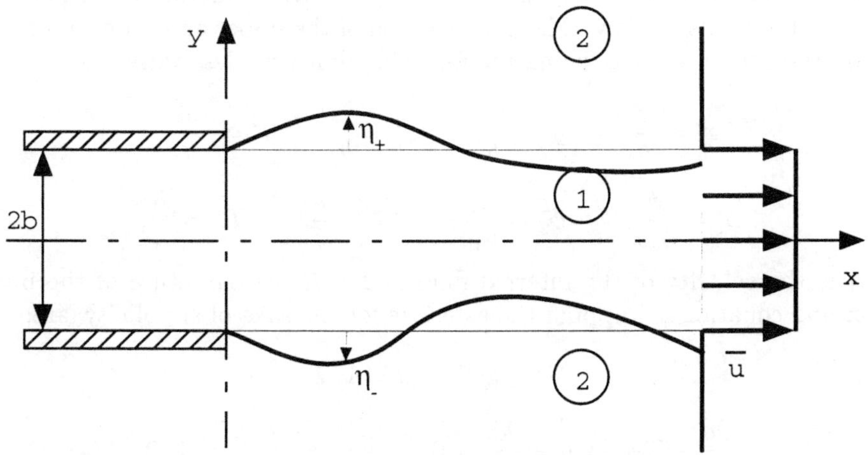



Figure 3: Representation of the liquid sheet and velocity profiles.

gas to liquid density ratio $\varsigma = \rho_2/\rho_1$. Other choises for the independent variable are possible especially if the zero $b$ limit should be considered. The searched dispersion relation will assume the form:

$$\omega \Big/ \sqrt{\frac{\sigma}{\rho_1 b^3}} = f(kb, \varsigma). \tag{98}$$

Study of wave propagation totally neglects the presence of the external gas, thus reducing further the number of independent variables. We let the reader play with this wonderful toy (see Fauve's contribution to this volume) and proceed performing the proper derivation of the dispersion relation in various asymptotic limits as discussed in the following for varicose and sinuous waves.

## 4.1   Dispersion relation

Let $(x, y, z)$ be a Cartesian coordinate system with the positive $x$ axis in the flow direction, and the $z$ axis as spanwise coordinate. A fluid of density $\rho_1$ (the water) comes out through a slit of ideal infinite length into a fluid of density $\rho_2$ (the air). Refer to figure 3 to have an immediate idea of the system under exam. The velocity profile for fluid 1 is uniform, the fluid 2 being at rest. Considering both the motions to not be affected by viscosity (a vortex sheet is assumed at the interface), be $\phi_1$ and $\phi_2$ the velocity potential of the internal and external fluid (defined as $\nabla\phi_i = -\mathbf{V}_i$, for $i = 1, 2$, being $\mathbf{V}_l = (u_l, v_l, w_l)$ the velocity vector). Closure to the potential equations is given by enforcement of the kinematics condition of no flow through the interface and the dynamic condition expressing the pressure jump as a function of surface tension according to the Laplace's law. The perturbation of the upper and lower interfaces are $\eta_+$ and $\eta_-$ respectively. Introducing the following dimensionless variables

$$x^* = \frac{x}{b}, \qquad y^* = \frac{y}{b}, \qquad z^* = \frac{z}{b}, \qquad t^* = \frac{t\bar{u}_1}{b},$$

$$\phi_1^* = \frac{\phi}{\bar{u}_1 b}, \qquad \phi_2^* = \frac{\phi_2}{\bar{u}_1 b}, \qquad \eta^* = \frac{\eta}{b}, \qquad R^* = \frac{R}{b},$$

where $\bar{u}_1$ is the velocity of the internal fluid and $1/R$ the curvature of the interface, the governing equations, dropping the asterisks for the sake of simplicity, are:

$$\nabla^2 \phi_l = 0 \qquad l = 1,\, 2 \tag{99}$$

$$\frac{\partial \eta}{\partial t} - \nabla \phi_l \cdot \nabla \eta + \frac{\partial \phi_l}{\partial y} = 0 \qquad y = \pm \eta \qquad l = 1,\, 2 \tag{100}$$

$$\frac{1}{2}(\nabla \phi_1)^2 - \frac{\varsigma}{2}(\nabla \phi_2)^2 - \frac{\partial \phi_1}{\partial t} + \varsigma \frac{\partial \phi_2}{\partial t} - \frac{1}{2} = \frac{1}{WeR} \qquad y = \eta_\pm. \tag{101}$$

$\varsigma$ is the density ratio

$$\varsigma = \rho_2/\rho_1 \quad \text{and} \quad We = \frac{\rho_1 \bar{u}_1^2 b}{\sigma} \tag{102}$$

is the Weber number ($\sigma$ the surface tension). The curvature is defined as:

$$\frac{1}{R} = \nabla \cdot \left( \frac{\nabla \eta}{[1 + (\nabla \eta)^2]^{1/2}} \right). \tag{103}$$

For details the reader is referred to the paper of de Luca and Costa [11].

As a first step, let us consider the case the external fluid is at rest. The perturbed flow variables are taken to have the form:

$$\phi_l = \bar{\phi}_l + \phi_l' \qquad l = 1,\, 2 \tag{104}$$

$$\eta_\pm = \bar{\eta}_\pm + \eta_\pm', \tag{105}$$

namely they are classically divided into steady mean flow quantities and unsteady disturbances of small amplitude. Symmetry properties of the mean flow allow the following definitions:

$$\phi_l' = \hat{\phi}_l(y) e^{i(\alpha x + \beta z - \omega t)} \qquad l = 1,\, 2 \tag{106}$$

$$\eta_\pm' = \hat{\eta}_\pm e^{i(\alpha z + \beta z - \omega t)}, \tag{107}$$

with $\omega$ as dimensionless frequency and $\mathbf{k} = (\alpha,\, \beta)$ as wavenumber vector in the $xz$ plane. The reference length for $\alpha$ and $\beta$ is the sheet half thickness $b$ and the reference time is $b/\bar{u}_1$. Substituting equations (104-105) into equations (99-101), neglecting perturbations quadratic terms and expanding both the mean and fluctuation quantities in a Taylor series about the unperturbed interface position so as to apply boundary conditions at $y = \pm\bar{\eta} = \pm 1$, one derives the disturbances equations. Further substitutions of equations (106-107) gives:

$$\frac{\partial^2 \hat{\phi}_l}{\partial y^2} - k^2 \hat{\phi}_l = 0 \qquad l = 1, 2 \tag{108}$$

$$-i\omega\hat{\eta} - i\alpha\hat{\eta}\frac{\partial \bar{\phi}_1}{\partial x} + \frac{\partial \hat{\phi}_1}{\partial y} = 0 \qquad y = \pm 1 \tag{109}$$

$$-i\omega\hat{\eta} + \frac{\partial \hat{\phi}_2}{\partial y} = 0 \qquad y = \pm 1 \tag{110}$$

$$i\omega\hat{\phi}_1 + i\alpha\hat{\phi}_1\frac{\partial \bar{\phi}_1}{\partial x} - i\varsigma\omega\hat{\phi}_2 + \frac{k^2}{We}\hat{\eta} = 0 \qquad y = \pm 1. \tag{111}$$

The general integral for $\hat{\phi}_1(y)$ is

$$\hat{\phi}_1(y) = A\cosh(ky) + B\sinh(ky), \tag{112}$$

while the solution for $\hat{\phi}_2(y)$, which has to vanish at large $y$ distances from the interface, assumes the form:

$$\hat{\phi}_2(y) = C[\cosh(ky) - \sinh(ky)]. \tag{113}$$

Linearity of the differential system allows the even and odd solutions for $\hat{\phi}_1$ to be considered separately. The odd solution corresponds to antisymmetric disturbances which displace each of the free surfaces in the same direction, giving rise to sinuous waves, the even solution corresponds to symmetric disturbances displacing the surfaces in opposite directions to form varicose waves. Accordingly the velocity potential for the sinuous modes is:

$$\hat{\phi}_{1s} = B\sinh(ky), \tag{114}$$

and that for varicose modes is:

$$\hat{\phi}_{1v} = A\cosh(ky). \tag{115}$$

Substitution of (114) and (115) into equations (109-111) leads to an algebraic equations system for which the existence of nontrivial solutions requires that:

$$\frac{(\omega - \alpha)^2}{\omega^2} = -f(k)\left(\varsigma - \frac{k^3}{\omega^2 We}\right), \tag{116}$$

where

$$f(k) = \frac{e^k + se^{-k}}{e^k - se^{-k}}, \tag{117}$$

being $s = 1$ for sinuous waves and $s = -1$ for varicose waves. The dimensionless velocity component along $x$ of the internal fluid equal to unity has been introduced.

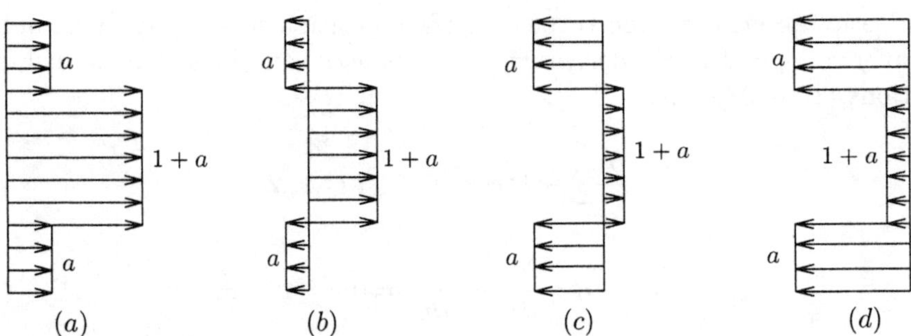

Figure 4: Evolution of the velocity profile with external flow parameter $a$. (a) $a \geq 0$: jet with co-flow. (b) $-0.5 \leq a \leq 0$: jet with counter-flow. (c) $-1 \leq a \leq -0.5$: wake with counter-flow. (d) $a \leq -1$: wake with co-flow, courtesy of [21].

Equation (116) is the searched dispersion relation. Generalization to keep into account possible motion of the external fluid is directly achieved by introducing the Doppler shift:

$$\omega \to \omega - \alpha a, \tag{118}$$

which gives the link between the wave frequency in the reference frame singled out by the slit exit and the wave frequency in a reference frame moving with the external fluid. Of course this transformation is not useful when only temporal instability is concerned but it is essential when spatial instability problem (i.e. spatially growing or decaying waves) is addressed. Equation (116) becomes:

$$\left[\frac{\omega - (1+a)\alpha}{\omega - \alpha a}\right]^2 = -f(k)\left[\varsigma - \frac{k^3}{(\omega - \alpha a)^2 We}\right]. \tag{119}$$

The external and internal streams are co-flowing when $a > 0$ or $a < -1$, and counter-flowing when $-1 < a < 0$. Jets correspond to $a > -0.5$ and wakes to $a < -0.5$. The particular values $a = 0$ and $a = -1$ are obviously associated with zero external flow jet and zero centerline velocity wake, respectively, as represented in figure 4.

The presence of an even uniform current on a wave emitted at any given frequency by a wave maker which single out the frame moving with it changes the wave vector (eventually complex). Possible current variations in a certain space direction enter into the phenomenon and the problem becomes one of wave propagation on non uniform media. If the variations happen over a length scale much greater than typical wavelengths, one may assume that at any given point in the streamwise direction waves have the same properties as a plane wave train on a local uniform current, within the *WKBJ* approximation. The nonparallel basic flow is thus treated as a locally parallel one. To the lowest order approximation in a small parameter characterizing the slow spatial evolution of the current, dispersion relation has the same form as on a uniform current, but variables have a local value.

## 4.2   Dispersive wave propagation on a thin film

We refer to situations where the density of the external fluid is negligible with respect to the density of the internal one, as for a liquid sheet moving in a surrounding gas. For $\varsigma = 0$ only real values of $\omega$ are found for real $k$, thus meaning that no wave growth can be expected, but only ordinary waves, according to the paper of Taylor [28]. Taking values of $a$ different from zero has the meaning of studying wave propagation into a moving reference frame. The case relative to a sheet ideally at rest, already considered in the dimensional analysis, is recovered with $a = -1$. Two different modes with evident dispersive character are found as solution of equation (119) in both the varicose and sinuous case. In the limit as $k \to \infty$, identical solutions are found, thus meaning that short sinuous and varicose waves move at the same speed.

### 4.2.1   Varicose waves

We refer initially to two-dimensional propagation ($\beta = 0$). The dispersion relation, written in terms of dimensionless frequency and wavenumber:

$$\omega^2 = \frac{\alpha^3 \tanh \alpha}{We} \tag{120}$$

is the same one obtains in studying capillary waves on a water depth equal to the sheet half thickness (Fauve's contribution to the present volume). Owing to the symmetry of varicose waves, the axis can be thought to as having the same role as a wall. In the short wave limit the classical dispersion relation for capillary waves on deep water, e.g. see Ligthill [20], is found:

$$\omega^2 = \frac{\alpha^3}{We}. \tag{121}$$

In ref. [20] it is obtained starting from the more general case of gravity waves on deep water in the presence of surface tension. It is shown to well describe surface waves when the wavelength $\lambda = 2\pi/\alpha$ is less than $\hat{\lambda}/4$, where $\hat{\lambda} = 2\pi\sqrt{\sigma/\rho_1 g}$, being $g$ the gravity acceleration. For water $\sigma = 0.074 \ Nm^{-1}$ and $\rho_1 = 1000 \ kgm^{-3}$, thus $\hat{\lambda} = 0.017 \ m$. Effect of gravity is indeed found to be negligible on waves shorter than about 4 $mm$, which is a characteristic capillary length.

The phase velocity of varicose waves is plotted in figure 5, together with the curves corresponding to the limits above considered. Group velocity is always greater than phase velocity, its ratio being plotted in figure 6. The curve relative to sinuous waves, discussed in the next subsection, is also represented. In the limit as $\alpha \to \infty$ the two curves tend to the same value. Varicose waves have an evident dispersive character even when $\alpha$ goes to zero since, even if they share the symmetry of the elastic mode of a membrane they differ by the nature of the restoring force which compensates the inertia. For the elastic membrane the restoring force is directly the change in surface tension associated with the stretching (thus linear in $\alpha$) whereas for the varicose mode of an anelastic thinfilm the restoring force is the pressure variation due to the curvature

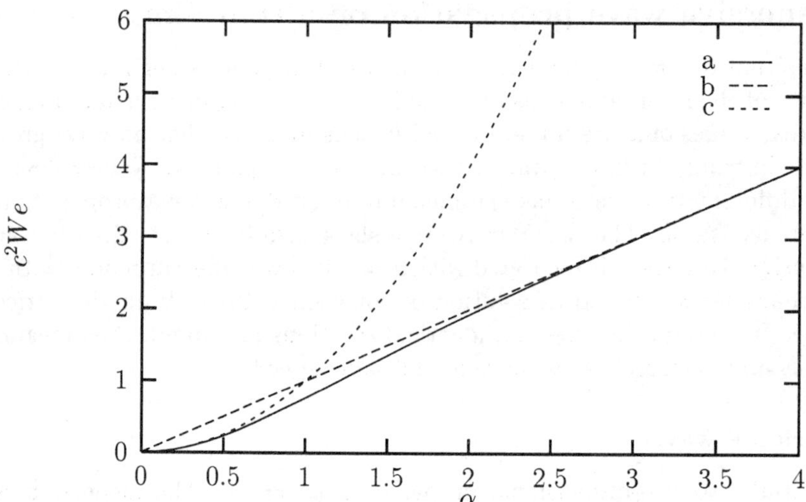



Figure 5: Phase velocity of varicose waves (a) and short (b) and long (c) wave limits.

which goes like $\alpha^2$ since the surface tension is constant. The kinetic energy continuously transforms into surface energy and vice versa through pressure effect due to curvature which is not efficient at large scale and explain the decrease to zero of the phase speed when the wavelength goes to infinity. Of course as soon as surfactants are added to the film (as in the next chapter) the elasticity will dominate the long wave dynamics of the varicose mode since pressure gradient (miltiplied by the film thickness) will be negligible compared to surface tension gradient.

It is of interest to evaluate the asymptotic time behavior of a wave packet of varicose waves in a reference frame moving with the external fluid. This is dominated by the wave $\alpha^*$ for which the group velocity:

$$\frac{d\omega}{d\alpha} = 1 + a \pm \frac{3\alpha\tanh\alpha + \alpha^2(1 - \tanh\alpha^2)}{2\sqrt{\alpha We}\tanh\alpha} \tag{122}$$

is equal to zero; since this is a growing function of $\alpha$ and

$$\lim_{k \to 0} \frac{d\omega}{dk} = 1 + a \pm \frac{1}{\sqrt{We}}, \tag{123}$$

solutions exist for the mode corresponding to the plus sign only if $a < -1 - (We)^{-1/2}$, and for the mode corresponding to the minus sign if $a > -1 - (We)^{-1/2}$. The relevant wavepacket amplitude, as obtained from equation (45) at $t = 1$, is plotted in figure 7,

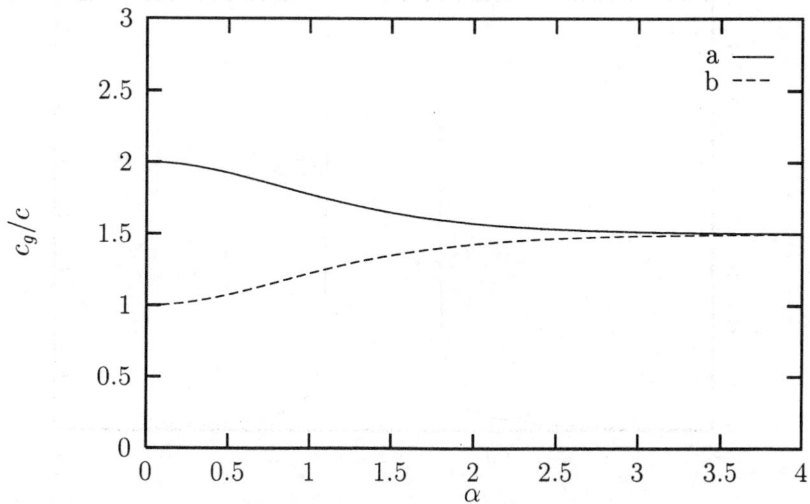

Figure 6: Group to phase velocity ratio of varicose (a) and sinuous (b) waves.

where it is evident that there exist a range of velocity of the external fluid for which no solution is found. The singular limits where the amplitude gets infinite, are due to the fact that, as $a$ approaches the values $-1 - (We)^{-1/2}$ and $-1 + (We)^{-1/2}$, the wave group velocity tends to a constant value, as in non dispersive systems. A perturbation in the form of a pulse should, in this singular case, preserve its shape. Higher order analyses would remove the singularity leading to different asymptotic behaviors.

When three dimensional propagation is considered, in the reference frame moving with the gas (in which the liquid has unitary positive velocity), the dispersion relation (equation (116)) rewrites as

$$(\omega - \mathbf{k} \cdot \mathbf{V}_1)^2 = \varpi^2(k). \qquad (124)$$

This can be solved for any given value of the frequency by considering the intersection of the plane $m = \omega - \mathbf{k} \cdot \mathbf{V}$ with the surface of revolution $m = \pm\varpi(k)$( [24]), and taking a diametrical section as shown in figure 8 for the case of stationary waves. If $\mathbf{k}$ is perpendicular to $\mathbf{V}_1$ the liquid motion does not affect the solution; for $\mathbf{k}$ in any other direction specified by the unit vector $\mathbf{e}$, solutions are represented by intersection points as $A$ and $B$. Choice of $\mathbf{e}$ gives waves corresponding to two points characterized by equal and opposite values of frequency and wavenumber. The relative phase velocity

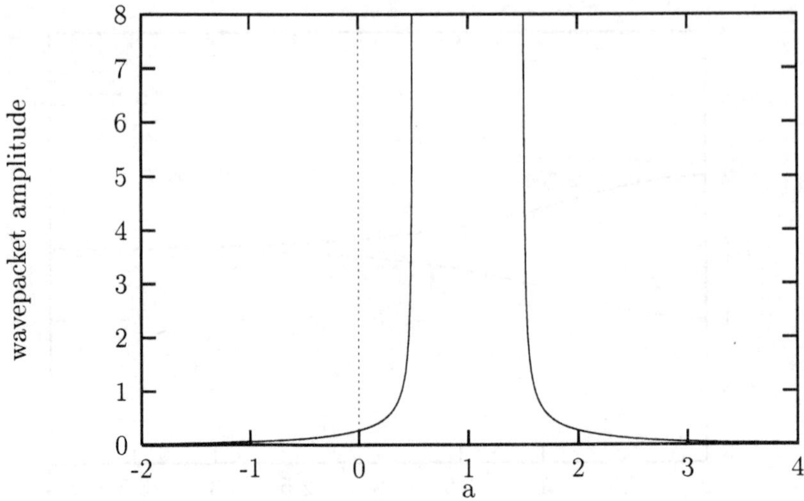

Figure 7: Wavepacket amplitude as a function of $a$. $We = 4$.

is the same and has a negative value. Indeed, the stationary character comes from the
fact that the phase speed in the direction of motion is equal and opposite to the liquid
speed. The symmetry of solution is lost for any other value of $\omega$. Two further waves
would eventually originate and solutions would exist also for $\mathbf{k}$ orthogonal to $\mathbf{V}_1$.

Wave patterns of varicose waves on a current were derived by Taylor [28] in the
limit as $\tanh k \simeq k$. Equations (61) and (67) write:

$$H(\alpha,\,\beta) = \alpha \pm \frac{k^2}{\sqrt{We}} = 0 \tag{125}$$

$$\frac{z}{x} = \frac{\pm 2\beta}{\sqrt{We} \pm 2\alpha} = \tan\xi, \tag{126}$$

where $\xi$ denotes the angle formed by the vector position $OP$ with the $x$ axis in a
$xz$ plane. Refer to figure 2 making $x$ and $z$ to coincide with $x_1$ and $x_2$ respectively.
Assuming $\gamma$ to be the angle between the wavenumber vector $\mathbf{k}$ in $P$ and the $x$ axis
(positive if measured clockwise), the following equality holds:

$$\mathbf{k} = (-k\cos\gamma,\, k\sin\gamma). \tag{127}$$

From equation (125)

$$\alpha = \mp\frac{k^2}{\sqrt{We}} = \mp k\cos\gamma, \tag{128}$$

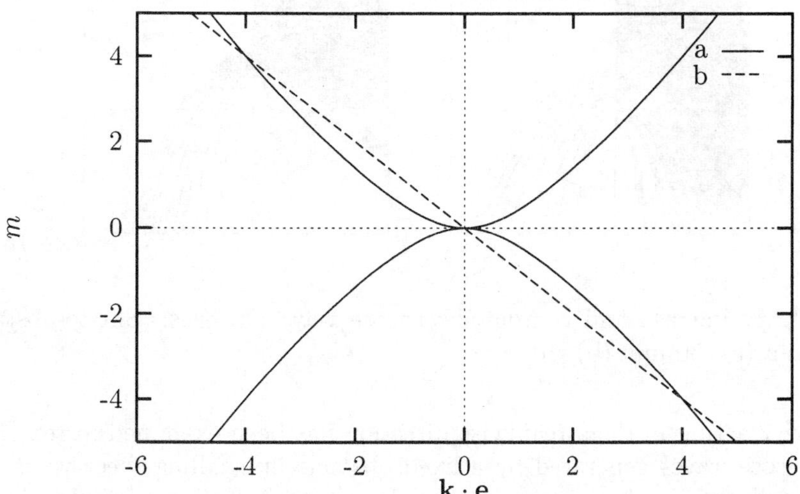


Figure 8: Three dimensional varicose waves for $We = 4$ as intersection of $m = \pm\varpi(k)$ (a) and $m = \omega - \mathbf{k} \cdot \mathbf{V}_1$ (b). Please not that the tangent in zero of the curve a is horizontal insuring the existence of an intersection point when the slope of b goes to zero.

hence

$$k = \sqrt{We}\cos\gamma. \tag{129}$$

Note that with the assumed positive value of $\gamma$ the only solution with physical meaning is the one corresponding to the plus sign in equation (125). Substitution of equation (129) into equation (126) gives

$$\tan\xi = \frac{2\sin\gamma\cos\gamma}{1 - 2\cos\gamma^2} = -\tan 2\gamma \tag{130}$$

and

$$\gamma = -\frac{\xi}{2}. \tag{131}$$

The phase can be evaluated as follows:

$$\theta = \mathbf{k}\cdot\mathbf{r} = r(-k\cos\gamma\cos\xi + k\sin\gamma sin\xi) = -r\sqrt{We}\cos\gamma\cos(\gamma + \xi) = -r\sqrt{We}\cos\frac{\xi^2}{2}. \tag{132}$$

Lines of constant $\theta$ are parabolas in the $x_1 x_2$ plane. Diminution of the sheet thickness

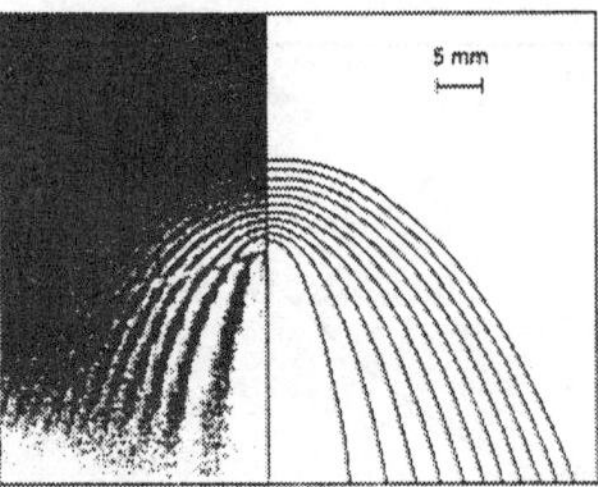
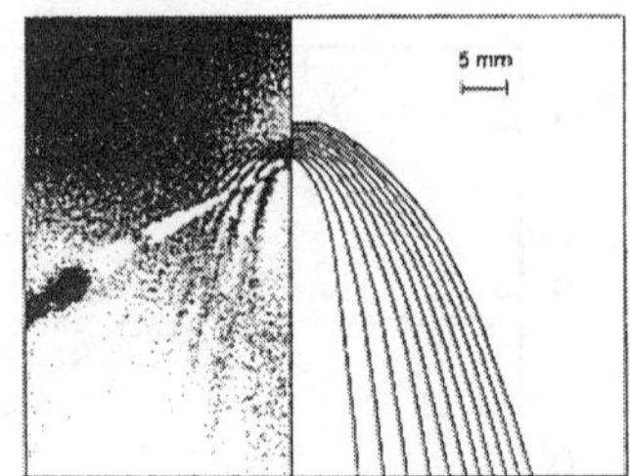



Figure 9: Experimental and computed varicose wave patterns. Slit-to-obstacle distance: 25 mm (a), 50 mm (b) [10].

due to gravity acting in the streamwise direction has been examined in ref. [10]. Stationary varicose waves generated by a fixed obstacle in a falling sheet were found to suffer a decrease of wavelength and amplitude in getting farther from the obstacle. A numerical determination of the wave patterns, starting from the dispersion relation in local form, was carried out. Results are shown in figure9 for two different locations of the obstacle with respect to the slit exit, together with the visualization taken from experiments on a water sheet.

### 4.2.2 Sinuous waves

Two-dimensional sinuous waves an a sheet ideally at rest obey the dispersion relation

$$\omega^2 = \frac{\alpha^3 \coth \alpha}{We} \tag{133}$$

from which is evident the non dispersive character of long waves. The analogy of sinuous long waves with the flapping mode found on a membrane is straightforward. The limiting value $We^{-1/2}$ is the dimensionless counterpart of the formula written in equation (14). Figure 10 shows the phase velocity and the short and long waves limits. The determination of stationary sinuous waves is achieved by examining figure 11. It is evident that intersection between the straight line and the curves is possible only if the tangent to the curve at $\mathbf{k} \cdot \mathbf{e} = 0$ is less inclined than the line itself. This condition in the limit as $k \to 0$ writes:

$$We \geq 1. \tag{134}$$

The equation for $H(\alpha, \beta)$ in the long wave limit is:

$$H(\alpha, \beta) = \alpha \pm \frac{k}{\sqrt{We}} = 0. \tag{135}$$

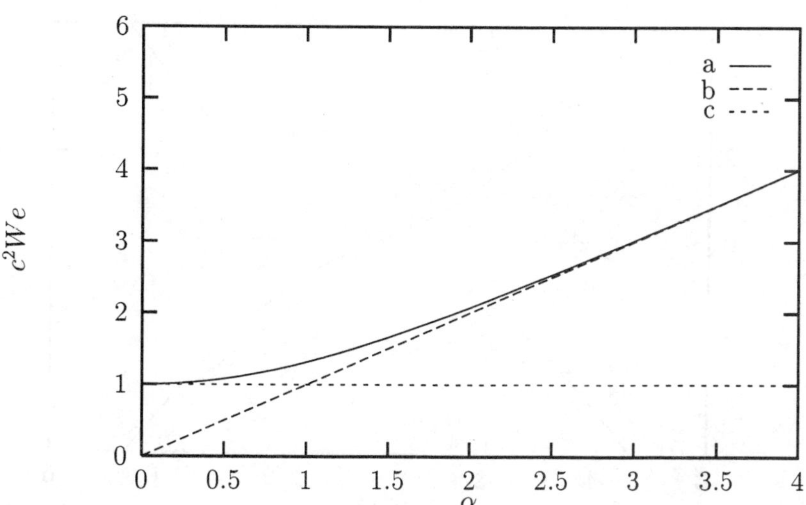

Figure 10: Phase velocity of sinuous waves (a) and short (b) and long (c) wave limits.

Since

$$\cos \gamma = \pm \frac{1}{\sqrt{We}}, \tag{136}$$

it is

$$\tan \xi = \frac{\pm \beta}{k\sqrt{We} \pm \alpha} = \frac{\pm \sin \gamma}{\sqrt{We} \mp \cos \gamma} = \pm \frac{\sqrt{1 - \frac{1}{We}}}{\sqrt{We}\left(1 - \frac{1}{We}\right)} = \pm \tan^{-1} \gamma. \tag{137}$$

It follows that

$$\xi = \pm \frac{\pi}{2} \pm \gamma. \tag{138}$$

The wavenumber vector is at each point orthogonal to the vector position. Stationary waves have straight fronts, inclined with respect to the $x_1$ axis of an angle

$$\xi = \sin^{-1} \frac{1}{\sqrt{We}}. \tag{139}$$

In analogy with shock waves, which can only exist in supersonic flow, stationary sinuous waves survive only in the region where $We > 1$. Remembering $We^{-1/2}$ to be the phase

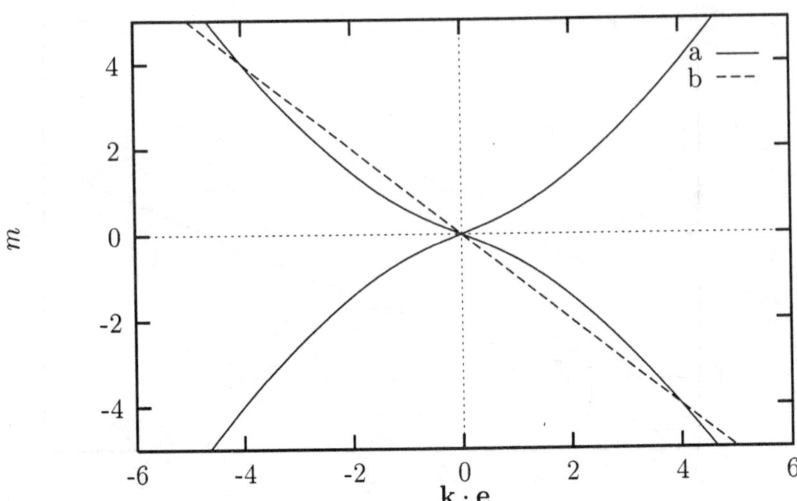


Figure 11: Three dimensional sinuous waves as intersection of $m = \pm\varpi(k)$ (a) and $m = \omega - \mathbf{k} \cdot \mathbf{V}_1$ (b).Please not that the slope in zero of the curve a is finite therefore the intersection with the curve b does not always exist

speed of long sinuous waves and the velocity of the liquid to be equal to the unity, the angle, as the Mach angle, is related to a ratio between the velocity of the flow and the velocity of propagation of waves. This considerations can be used as a tool to determine the maximum radius of a disk shaped sheet originating from two opposite impinging circular jets (figure 12 shows a large scale artiste view of the experiment realized at small scale by G.I. Taylor [28]). Since the liquid flow rate and radial momentum have to remain constant and equal respectively to $2\rho_1\pi a_0^2 u_0$ and $2\rho_1\pi a_0^2 u_0^2$ ($a_0$ is the radius and $u_0$ the velocity of the two circular jets), the velocity in the disk remains constant and the thickness decreases away from the center. The Weber number decreases from the center of the disk, until it reaches the unity. Here the flow undergoes a transition from 'supersonic' to 'subsonic', which can be thought to be responsible of the sheet to break. The value of the radius $r_{max}$ for which this occurs is obtained by equating the Weber number to the unity and expressing the thickness of the disk in terms of the known flow rate of the circular jets:

$$r_{max} = \frac{\rho_l a_0^2 u_0^2}{2\sigma}.$$

(140)

An additional mechanism has to be taken into account in the study of the sheet break up. This concerns the unstable nature of the response of the system to perturbations

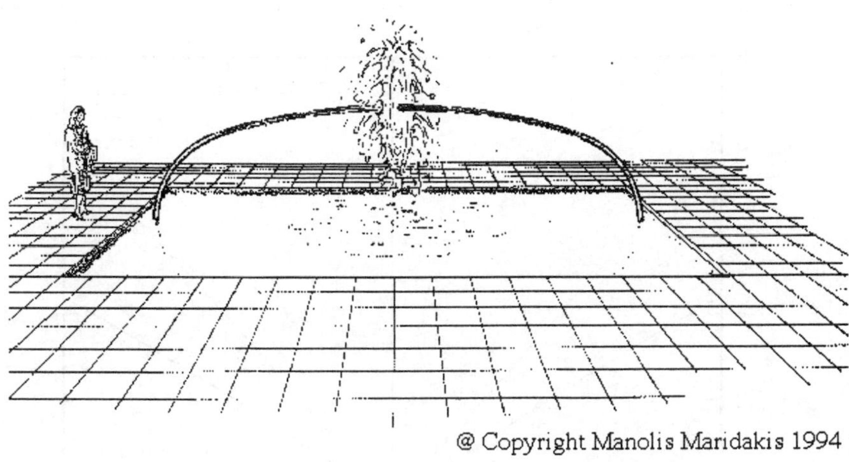


Figure 12: Monumental fountain realized by the artist Manolis Maridakis with the help of one of the present author (JMC) demonstrating a giant circular disk of water

due to the presence of the air, as explained in the next section. It involves either exponential amplification of initial perturbation in the consecutively unstable disk (this imposes the two impinging jet to be very laminar for giant disk to exist) or the break up of the jet due to the transition, at a particular distance from the center, from convective to absolute instability (this later idea being the natural extension of the transition from 'supersonic' to 'subsonic' mentioned above).

## 4.3  Instability of thin films

When the effect of the air is not neglected the neutral waves described above may be destabilized. The instability is, as in the case of the elastic membrane, due to the phase shift in the pressure term of the instationary potential flow induced in the air by the motion of the interface.

The neutral stability curve, i.e. the locus of $\omega_i = 0$ in the $kWe$ plane, is drawn in figures 13 and 14 for varicose and sinuous modes, respectively. Complex frequencies with positive imaginary parts, indicating the existence of temporal instabilities, are found as solution of the dispersion relation for $k$ and $We$ belonging to the region internal to the curve. The cut-off wavenumber for which instability is present gets smaller as $\varsigma$ decreases, becoming equal to zero for $\varsigma = 0$, in agreement with the previously made observation that neglecting of the external fluid density implies only ordinary waves to be found. For both varicose and sinuous modes the vertical axis in the diagram does not belong to the instability region, for $k = 0$ being $\omega = 0$ a double root, at each value

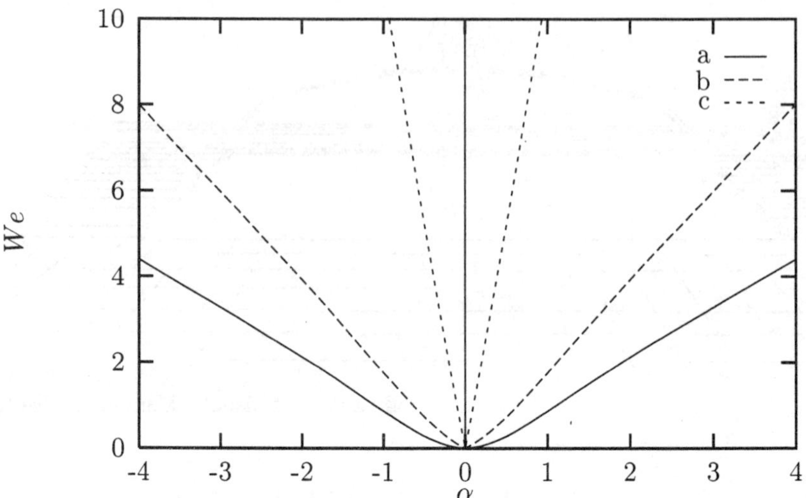

Figure 13: Neutral stability curve of varicose waves. $\varsigma = 10$ (a), $\varsigma = 1$ (b), $\varsigma = 0.1$ (c)

of the Weber number.

When sinuous modes are considered, there exists a limit value of the Weber below which no complex root of $\omega$ exists. This critical value is by no means affected by the gas-to-liquid density ratio. Moreover, the temporal stability analysis is not influenced by the velocity of the external fluid, since the imaginary part of $\omega$ is independent of $a$. This last enters as a further parameter in the definition of the instability nature.

The existence of exponentially growing sinuous waves only in the range of Weber greater than one can be related to the magnitude of the velocity sinuous waves exhibit with respect to the current. Contrary to varicose waves, sinuous waves have a minimum speed depending on the Weber. Since the current has been taken with a velocity equal to the unity, each sinuous wave tends to be stabilized when moving faster than the current. Regarding the sheet as an inelastic (the term due to $\sigma'$ is not considered) membrane of constant thickness in a reference frame on one of which side a current with negative unitary velocity flows, stability or instability can be related to the effect of the pressure gradient in the direction normal to the membrane. Within this reasoning the internal flow is in some sense frozen. We consider a small value of $\varsigma$, so to perform a perturbation analysis with respect to the case of $\varsigma = 0$ where only ordinary waves are found with a phase velocity equal to $We^{-1/2}$ in the limit as $\alpha$ tends to zero. In the same limit, which is consistent with observation of figure 14, the dispersion relation takes the form:

$$\omega^2 = \frac{\alpha^2}{We} + \varsigma\alpha\left(\frac{1}{We} - 1\right) \tag{141}$$

where the perturbation character of the term containing $\varsigma$ is evident. Going back to the differential counterpart of equation (141), so to have:

$$\frac{\partial^2\eta}{\partial t^2} = \frac{1}{We}\frac{\partial^2\eta}{\partial x^2} + i\varsigma\frac{\partial\eta}{\partial x}\left(\frac{1}{We} - 1\right), \tag{142}$$

shows the presence of a term

$$\Delta p = i\varsigma\frac{\partial\eta}{\partial x}\left(\frac{1}{We} - 1\right) \tag{143}$$

proportional to $(1/We - 1)$ which adds to the equation of flapping waves. The sign of $\Delta p$ depends on the relative magnitude of the current speed and the phase speed sinuous waves should have for $\varsigma = 0$, which can be indeed taken to not change due to the small value of $\varsigma$. Thus $\Delta p$ may act either in the same direction as surface tension preserving the unperturbed configuration, either in opposite direction so to make waves to grow progressively. It follows that in evaluating the sign of the pressure gradient due to the presence of the current, it is not sufficient to only refer to the gradient in the velocity of the fluid passing over the deformed membrane, as a stationary Bernoulli theorem

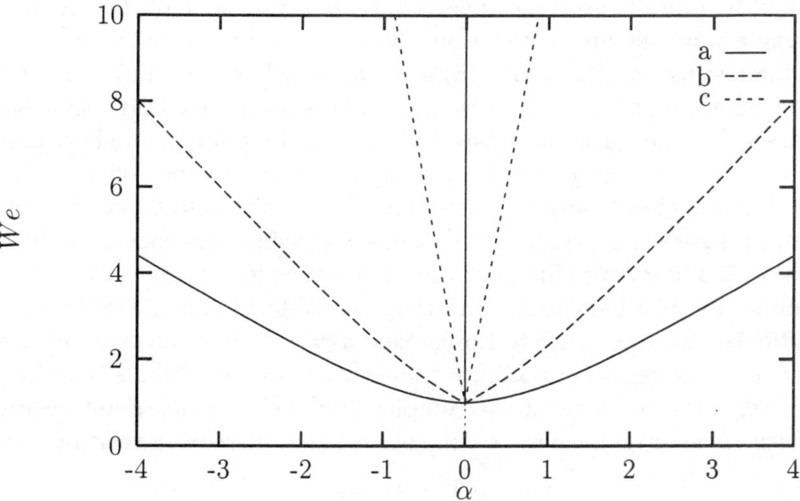


Figure 14: Neutral stability curve of sinuous waves. $\varsigma = 10$ (a), $\varsigma = 1$ (b), $\varsigma = 0.1$ (c)

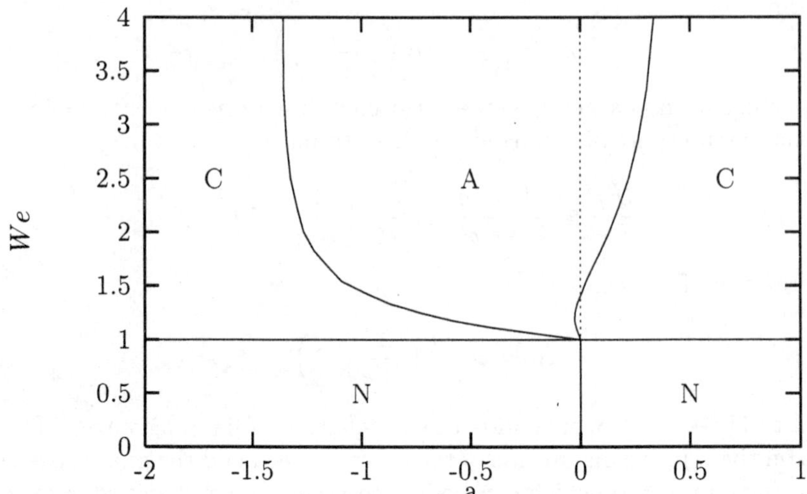


Figure 15: Regions of absolute and convective instability for sinuous waves.

would state, but it is necessary to also consider the role of the non-stationary term due to the membrane deformation itself. In analogous way, the absence of a minimum value of the phase speed of varicose waves, which always remain dispersive, can explain the existence of temporally growing waves whatever is the value of the Weber number, always existing a wave moving slower than the current being destabilized.

Further interesting results comes from an approach to the problem of stability based on the determination of the system response to a localized perturbation, as recently addressed by de Luca and Costa [11]. In searching for the highest pinch point singularity in the complex $\omega$ plane dominating the long time behavior of the system Green's function, a region of absolute instability is found, whose extension depends on $\varsigma$. As an example figure 15 represents the regions of absolute and convective instabilities in the parameter plane $(a, We)$ for $\varsigma = 1$ and sinuous modes. Regions were only neutral waves are found are also indicated. The range of Weber number less than the unity deserves particular care. We refer to the jet case $a = 0$. Even if no complex $\omega$ exists as solution for real $k$, the region of absolute convergence for the Green's function can not be extended below the real axis of the complex $\omega$ plane. A pinch point occurs just in correspondence of the origin, where, on the other hand, dispersion relation behaves as:

$$(\omega - \omega_0)^2 \sim (k - k_0)^3, \tag{144}$$

being $\omega_0 = k_0 = (0,0)$. It follows from equation (96) that an amplitude variation with time with a power equal to one third is indeed an algebraic growth for $G(x, t)$. The order of the singularity dominating the Green's function thus makes for the existence of an absolute instability region in the range of Weber number less than the unity.

When sheet of variable thickness are considered there may exist a point along the streamwise direction where the local Weber number cross the unity. The previously considered case of the circular sheet is one in which the Weber number decreases from greater to smaller than one as getting farther from the origin of the sheet. A transition from convective to absolute instability happens at a certain distance, which will give the limit size of the circular sheet. Rupture of the flow is indeed related to absolute instability.

If the falling sheet is considered, on the contrary, acceleration of the fluid in the vertical direction, and decrease of the thickness, make the Weber number eventually to grow from values smaller than the unity to values greater. A region of absolute instability may be present close to the slit exit section, thus explaining the rupture of the sheet observed in reducing the flow rate. Indeed the existence of a region of absolute instability is a necessary but not sufficient condition for the flow to break up. Being the greater the extension of this region the smaller the flow rate, rupture of the sheet may be believed to be due to the presence of a region of absolute instability of sufficiently great extent. Adding viscosity to the developed model for propagation of small amplitude disturbances would reduce the order of the singularity dominating the system response, thus removing the algebraic growth of perturbations. However, the time after which this effect occurs could be not sufficient to avoid nonlinear growth leading to the actual sheet break up, and short-time mechanisms of inviscid type dominate.

# 5  Soap film

Recent experiments first by Couder, then by Gharib and latter on by Goldburg's group among others use soap films as a two-dimensional (2D) fluid. The present chapter represents an attempt to define working conditions under which soap film flows obey classical 2D hydrodynamics. In particular we give an estimate of the smallest scale below which experimental data cannot be trusted because differential motions inside the film are no longer negligible. For larger scales, the necessary condition for obtaining generic behavior is that the typical velocity of the flow needs to be small compared to the Marangoni elastic wave velocity (elastic wave similar to the one described in the second chapter on the membrane). When this condition is fulfill the flow is governed by the 2D Navier-Stokes equations.

In Couder's experiment, a soap film is pulled out of a vessel and stretched onto a rectangular horizontal frame a few decimeters large. Using this membrane as a 2D towing tank, Couder studied the wake of a cylinder and 2D grid turbulence. Experimentally he observed that the thickness variations provided an excellent visualization of the film motion. When the film is lit by a white spot, rainbow iridescence appears

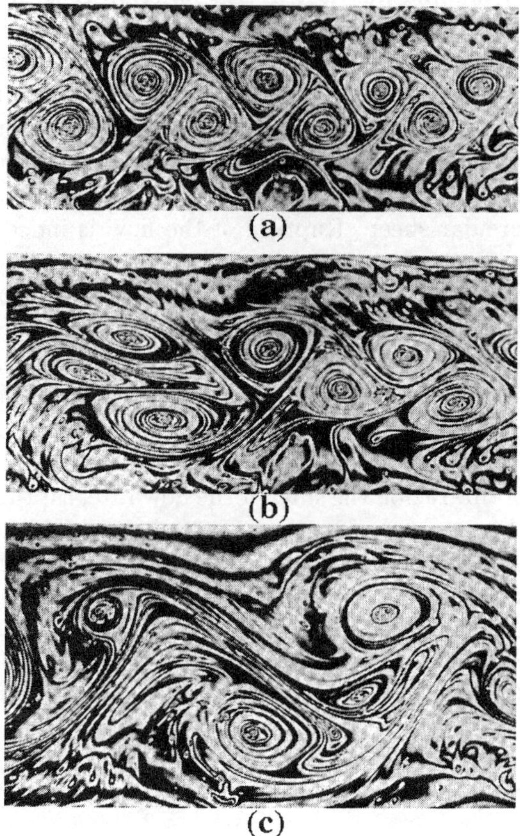

Figure 16: Time series of photographs of a turbulent wake in soap film from Couder &
Basdevant 1986. The visualization is obtained by the equal thickness fringe technique.

on the surface, the colors varying with the local thickness of the membrane. When a
monochromatic lamp (a yellow low pressure sodium lamp usually used for streets an
parking lots lighting) is used, bright fringes mark the film wherever its thickness is an
odd multiple of the quarter wavelength. Figure 16 presents photographs from Couder
and Basdevant (1986) on the cylinder wake instability with a pairing of same sign
vortices. Many details are revealed by visualization such as filaments between vortices
and fine scale structures inside vortex cores.

In Gharib's experiment the membrane is continually pulled out of a vessel onto a
frame by the positive surface tension gradient between the soap water vessel and a
plane water jet acting on the other end of the film. The experiment must use a jet

instead of a clear water vessel because the surface of the water must constantly be regenerated to maintain the surface tension difference. This apparatus has proven very flexible and many classical wind tunnel experiments involving 2D instability have been performed, such as shear flows, wakes and jets (see Gharib & Derango (1989)).

In the Goldburg's version of the soap tunnel, soap solution is injected onto two wires that form a 2D expansion zone. The film which moves down under the action of gravity is stretched and it accelerates till the air friction balances the gravity. This latest technique allows to realize soap tunnel 4 flours tall and 1 meter wide!

These Experiments on soap film hydrodynamics raise the important question of how far their interpretation in term of classical two-dimensional hydrodynamics may be trusted. First we establish the minimal model capturing the physics Soap film then using an asymptotic expansion technique, we show that classical two-dimensional hydrodynamics is recovered only in the small Mach number limit defined by $M^2 \ll 1$ where $M = U/v_L$ with $v_L$ the velocity of elastic Marangoni waves.

## 5.1 The model

The conditions under which soap films are used in Couder's experiment differ drastically from the conditions assumed in classical film theory. The surfaces of the membrane is very large (hundreds $cm^2$ and even several $m^2$ in Goldburg's set up). The film is stretched and objects dragged through it at speeds faster than 10 $cm/s$. Either the length (size 1 $cm$) or the time scale (1 $s$ to 0.1 $s$) is much larger than that described by recent studies of soap film (for example Joosten 1985 or Ivanov 1988). Therefore a new model must be developed following classical works, such as Plateau (1873), Boys (1890), Mysels et al (1959), Rusanov & Krotov (1979), Ivanov (1988). A description of the soap film properties used in the present article is contained in Couder, Chomaz and Rabaud (1989) and in Chomaz & Cathalau (1990).

### 5.1.1 Statement of the problem

A soap film may be viewed as a three phase flow (figure 17) composed of a three dimensional bulk phase of thickness $h$ and of two surface phases. Soap molecules, composed of a hydrophilic polar head associated with an hydrophobic carbon tail, tend to settle at the surface. Their surface concentration $\Gamma_1$ determines the surface tension $\sigma$ in the same way that the density of a gas determines its pressure (at a fixed temperature). The surface phases and the bulk phase are in equilibrium. The two surface phases are assumed to be in a fluid state.

For simplicity the soap film will be assumed flat and horizontal. For clarity the terms **across and crosswise** will refer to motion normal to the surface of the film and the term **planewise** will refer to motion in the plane of the film. Symmetric perturbations of the film will only be considered. We will assume that the soap film is sufficiently thick everywhere to neglect interaction between its two surfaces (thickness of order of 1 to 10 $\mu m$). Interactions with boundaries involving a meniscus and a contact angle,

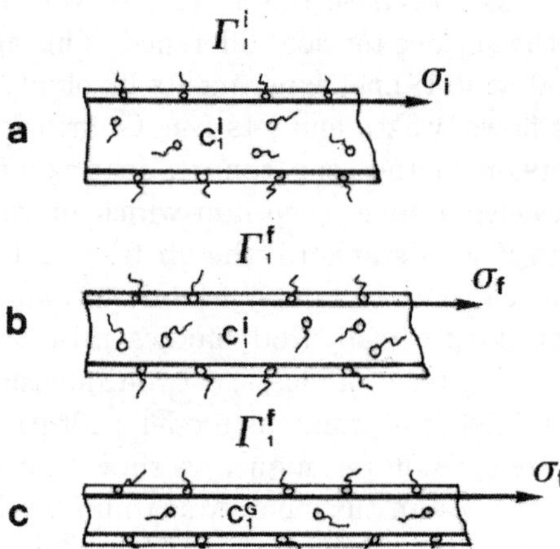

Figure 17: Evolution of a membrane under a stretch: (a) initial state, (b) instantaneous response (Marangoni elasticity), (c) long time response (Gibbs elasticity).

are not treated. Therefore the problem of vorticity generation in boundary layers with large variations of thickness is avoided. We emphasize that the internal pressure defect due to the meniscus at the contact with the boundary causes an abrupt variation of thickness and even a "black film" (a film of a few hundred Angstrom dominated by the interaction of the two surfaces). If this happens interactions between the two surfaces become essential and the film dynamics are outside the framework of this article. Therefore boundaries may be responsible for very peculiar behavior especially if black films are shedded from the boundary. In that case, the film will be formed of a mixture of thick ($> 1\mu m$) and very thin ($0.01\mu m$) film, which does not follow Navier Stokes equations. In my opinion this boundary effect constitutes the most dangerous potential source of experimental artifacts.

We describe the evolution of an initial velocity field in the plan of the film with a vorticity normal to the film. This reasonably corresponds to experiments where velocity shears are generated by a jet or by towing a thin disk. The motion in the plane of the film is assumed to take place in the spatial dynamical range $[L, l_K]$, with $L$ the largest scale of the flow and $l_K$ the smallest scale of the flow in the plane of the film. The characteristic scale $l_K$ is assumed large enough and the duration of the experiment small enough to validate a model where the variations of any quantities normal to the film are negligible.

### 5.1.2   Model for a uniform soap film at rest

*Chemical equilibrium*

We consider the case of a soap solution containing a single type of soap molecule and briefly recall the essential results, to make clear the following derivation. Quantitative data corresponding to solutions of sodium dodecyl sulfate (SDS) are presented and used to analyze the experimental results.

The properties of a soap solution are usually studied when a single surface phase is in contact with a nearly infinite bulk phase. In figure 18 is plotted the typical variation of the surface concentration at equilibrium, $\Gamma_1$, against the volume concentration of soap $c_1$ in the bulk phase. The surface tension is also presented versus $c_1$. The asymptote corresponds to the saturation of the surface in soap molecules and to the formation of micelles in the bulk.

For small concentrations, $\Gamma_1$ and $\sigma$ are proportional to $c_1$. Experiments give :

$$\Gamma_1 = kc_1 \ , \tag{145}$$

$k$ represents an equivalent thickness of the surface phase and equals $4\mu m$ for SDS The equation for $\sigma$ reads :

$$\sigma = \sigma_0 - RT\Gamma_1 \ , \tag{146}$$

with $T$ the temperature, $\sigma_0$ the water surface tension and $R$ the gas constant. In soap films, the bulk is no longer a reservoir and $c_1$ varies. The conserved quantity is the total amount of soap expressed as a total concentration $c_0$ :

$$c_0 = c_1 + 2\Gamma_1/h \ . \tag{147}$$

*Chemical kinetics*

The stability of soap films is defined by their responses to perturbations. When a membrane is stretched, it opposes the stretching by increasing its surface tension. The stability is controlled by the elasticity modulus $E$,

$$E = 2d\sigma/dlnA = -2hd\sigma/dh \ , \tag{148}$$

where $A$ is the surface area of the film. Two types of elasticity are considered :

i) the Marangoni elasticity $E_M$ specifies the response to disturbances acting on a time scale smaller than the chemical equilibrium time $\tau^*$ between the surfaces and the bulk. The surfaces behave like isolated phases and $A\Gamma_1$ is conserved. The Marangoni elasticity equals:

$$E_M = 2RT\Gamma_1 \ ; \tag{149}$$

ii) the Gibbs elasticity $E_M$ is associated with disturbances slower than the chemical relaxation. The two surface phases respond in equilibrium with the bulk phase. Taking

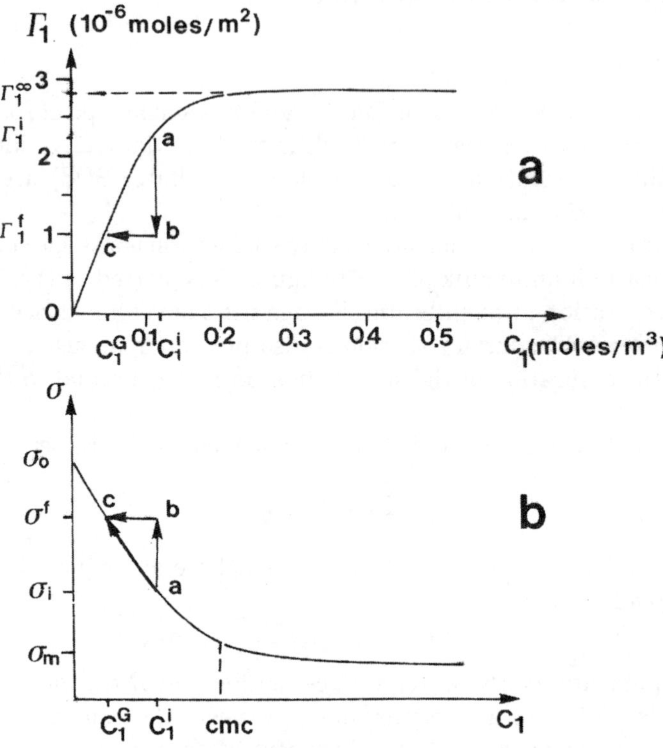

Figure 18: (a) The surface concentration $\Gamma_1$ as a function of the volume concentration $c_1$ (From Rusanov & Krotov). the segments ab, bc and ac represent respectively the Marangoni stretching, the return to equilibrium and the final Gibbs stretching they correspond to figure 17 (a), (b) and (c) states. (b) The isothermal surface tension $\sigma$ versus $c_1$.

into account (145 -148) we get :

$$E_G = 4RT\frac{c_0k^2h}{(h+2k)^2} \quad . \tag{150}$$

If the film is initially in equilibrium, $E_G$ and $E_M$ are linked by

$$E_G = E_M\frac{2k}{h+2k} = E_M\frac{2K}{1+2K} \quad , \tag{151}$$

where $K = k/h$. The parameter $K$ measures the quantity of soap molecules adsorbed on the surface compared to the quantity of soap molecules in solution in the bulk. For small $K$, the majority of soap molecules stays in the bulk, $c_1$ is nearly constant and $E_G$ is almost zero. Large $K$ corresponds to $E_G$ equal to $E_M$ and all the soap molecules adsorbed on the surfaces. The kinetic equation for $\Gamma_1$ is assumed linear in $\Gamma_1$ and $c_1$:

$$\frac{d\Gamma_1}{dt} = -(\Gamma_1 - kc_1)/\tau^* \quad , \tag{152}$$

which gives, applying (147):

$$\frac{d\Gamma_1}{dt} = -(\Gamma_1(\frac{h+2k}{h}) - kc_0)/\tau^* \quad . \tag{153}$$

### 5.1.3  Dynamics of extended non uniform soap film

*Limitation of the 2D approximation*

As stated, we consider motions in the plane of the soap film occurring in the scale range $[L, l_K]$, with $l_K \gg H$, $H$ being the mean thickness of the film. The dynamic at the smallest planewise scale $l_K$ is assumed to be dominated by viscosity. Therefore the characteristic advection time scale, $t_K$, associated with the scale $l_K$ is such that $t_K \sim l_K{}^2/\nu$. The hypothesis $l_K \gg H$ implies that $t_K \gg H^2/\nu = \tau_\nu$, the diffusion time of the velocity across the depth of the film. Because $t_K$ is much larger than $\tau_\nu$, the velocity may be assumed uniform at leading order across the membrane.

To derive the governing equations for the membrane we will take a Lagrangian point of view and consider a **film element** of planewise size smaller than $l_K$, the smallest scale of the flow in the plane of the film. This film element is made of two symmetric pieces of surface in the plane of the film and one curved surface normal to the film with an elementary volume of fluid in between. In general this materiel volume of fluid is deformed in a very complicated manner during its advection by the flow. Since the planewise size of the film element is smaller than $l_K$, the deformation of its two surfaces after a finite time $t$ will be negligible. But the deformation of the elementary volume depends on the velocity gradient normal to the film. Even if this velocity variations are small, their integrated effect over a finite time $t$ may deform noticeably the normal surface limiting the film element. An estimate of this normal deformation, $L_m$, after a certain time $t$ is computed below. When $l_K \gg L_m$, a film element of intermediate size between $l_K$ and $L_m$ stay undeformed all along its motion and the velocity is uniform inside this elementary volume. Straight forward masses and momentum balances may then be applied to this film element. In many real cases $L_m$ is smaller than $10H$ since the duration of an experiment (or the time for a film element to be advected away from the test section) does not exceed 100 turnover time $L/U$. Therefore the apparently drastic condition for the validity of the 2D approximation that $l_K \gg L_m$ is usually fulfill. More sophisticated models which would save the 2D approximation but take into account the differential transport by, for example, a modification of the dissipation operator, are not required and will be not treated here.

Let the film be horizontal and the gravity be neglected for simplicity. The coordinates are $z$ normal to the film and $(x, y)$ in the plane of the film. In order to estimate $L_m$, the horizontal deformation after a time $t$ of an initially vertical materiel line, we need to evaluate the vertical variation of the horizontal velocity inside the film.

In a soap film, surface tension acts on the two surfaces whereas, in the bulk fluid the only forces that play a role are the inertia, the pressure gradient and the viscosity.

First we show that the pressure gradient is negligible compared to the two other term.

In the bulk fluid the pressure is determined by the continuity equation at the surfaces located at $z = \pm h/2$ with $h$ the local thickness of the film:

$$p(h/2) - p_a = \sigma C \ , \tag{154}$$

where $C$ is the local curvature of surface and $p_a$, the atmospheric pressure. At leading order in $L/H$, $L$ being the typical horizontal scale considered, the Navier Stokes equation on $z$ gives :

$$\partial p / \partial z = 0 \ , \tag{155}$$

which mean that $p$ is uniform across the film and equal to $p_a + \sigma C$. The pressure gradient acts in the plane of the film and equals $\nabla(\sigma C)$, where $\nabla$ is the 2D horizontal gradient. Since $C \sim \nabla^2 h$ the pressure gradient is bounded from above by $\delta \sigma H / L^3$ with $\delta \sigma$ the mean variation of $\sigma$.

Following the horizontal momentum balance on a film element, the variation of $\sigma$ counterbalance the advection term integrated vertically across the film : $\nabla \sigma \sim \rho H \frac{\mathcal{D}\mathbf{u}}{\mathcal{D}t} \sim \rho H U^2 / L$, where $\mathbf{u}$ is the horizontal velocity, $\frac{\mathcal{D}}{\mathcal{D}t}$ the horizontal Lagrangian derivative and $\rho$ the density of the fluid. Therefore the pressure term is order $(H/L)^2$ smaller than the advection term and may be neglected in the Navier Stokes equation projected on the horizontal.

$$\frac{\mathcal{D}\mathbf{u}}{\mathcal{D}t} = \nu \frac{\partial^2 \mathbf{u}}{\partial z^2} \tag{156}$$

Continuity of the horizontal constrain at the surfaces of the film gives:

$$\nu \frac{\partial \mathbf{u}}{\partial z}(\pm h/2) = \rho \nabla \sigma \ . \tag{157}$$

The surface forces (the horizontal surface tension gradient) are transmitted to the bulk fluid by viscosity. Therefore when the film moves horizontally, there exists a 3D internal Poiseuille flow. This flow is driven by inertial forces of order $U^2/L$, due to the mean advection term ($U$ the mean velocity). The vertical variation of horizontal velocity due to this effect is of order $U^2 H^2 / \nu L$. After the time $t$ from the beginning of the experiment the differential displacement induced by this vertical variation of the horizontal velocity is of order $L_m$ such that $L_m / H = tU/L * UH/\nu$.

Whereas the viscous equilibrium is well established for the internal motion inside the soap film, it is not so for the soap concentration. The diffusivity $D$ of soap molecules is usually orders of magnitude smaller than the viscosity because the soap molecules possess long carbon chain and therefore are animated by a weaker Brownian motion than the molecules of water (for SDS $D/\nu = 4.10^{-4}$). The condition that the bulk soap concentration may be considered uniform across the film would impose $t_K \gg H^2/D$ equivalent to $(l_K/H)^2 D/\nu \gg 1$ which would be too restrictive (typically $l_K \gg 1mm$). As, at leading order, the velocity is horizontal and uniform on the vertical, the equation for the soap concentration, in the frame moving at the film element velocity, is dominated by the vertical diffusion. The vertical diffusion induces a relaxation similar to that introduced to model the equilibrium between the surfaces and the interstitial fluid. For simplicity the delay due to vertical diffusion will be included in the relaxation time $\tau^*$ and the horizontal diffusion neglected.

*Dynamical equations*

Let $\mathbf{x}$ be the 2D position on the soap surface. Following the discussion of the previous section, a soap film dynamics may be described by quantities with no crosswise dependency: $c_0(\mathbf{x})$, $c_1(\mathbf{x})$, $\Gamma_1(\mathbf{x})$, $\sigma(\mathbf{x})$, $h(\mathbf{x})$ and $\mathbf{u}(\mathbf{x})$, where $\mathbf{u}$ is the 2D velocity on the surface. Since we have assumed that the horizontal diffusion of molecules on the surface and in the bulk is negligible, equation (146) remains valid and (153) becomes:

$$\frac{\mathcal{D}\Gamma_1}{\mathcal{D}t} = -\Gamma_1 \nabla.\mathbf{u} - (\Gamma_1(\frac{h+2k}{h}) - kc_0)/\tau^* \ , \tag{158}$$

where $\frac{\mathcal{D}}{\mathcal{D}t}$ is the 2D Lagrangian derivative in the plane of the film. From incompressibility of the water we have:

$$\frac{\mathcal{D}h}{\mathcal{D}t} = -h\nabla.\mathbf{u} \ . \tag{159}$$

From conservation of soap inside an advected soap film element we deduce :

$$\frac{\mathcal{D}c_0}{\mathcal{D}t} = 0 \ , \tag{160}$$

In the experiment, the membrane is pulled out of a vessel too rapidly for lateral equilibrium to be established. Thus, the initial membrane is not uniform and the $c_0$ dependency on $\mathbf{x}$ comes from the way the membrane has been formed.

The equation for $\mathbf{u}$ derives from the momentum balance applied to a film element composed of the surfaces and of the bulk. The integrated pressure gradients due to curvature of the surfaces are, as already demonstrated, $\int_0^h (\sigma \nabla^2 h)dz$ of order $\delta\sigma H^2/L^3$ and therefore negligible compared to the gradient of $\sigma$ of order $\delta\sigma/L$, where $\delta\sigma$ is the typical variation of $\sigma$. Finally the equation for the motion of the film particle reads :

$$\rho h \frac{\mathcal{D}\mathbf{u}}{\mathcal{D}t} = 2\nabla\sigma + \mathbf{g}'\rho h + \mathbf{F_f} \ , \tag{161}$$

where $\mathbf{g}'$ is the gravity vector projected in the plane of the film and $\mathbf{F_f}$ represents the damping force, resulting from the bulk fluid viscosity $h\rho\nu\triangle\mathbf{u}$, from the surface viscosity

$\rho\nu_s\triangle\mathbf{u}$ and from the air friction. The air friction has been discussed by Couder et al. (1989) and is certainly a major potential source of spurious soap film behavior. For example, Couder et al. have shown that the air friction damps the energy of the large scales and so may change the turbulence energy cascade phenomenology. This effect will not be considered here and for clarity, we will carry out the analysis with $\mathbf{F_f} = h\rho\nu\triangle\mathbf{u}$ where $\nu$ represents now an effective viscosity.

## 5.2 Analysis of the dynamics

We have thus obtained a complete system of equations for the soap film motion defined by equations (146), (158) to (161), in the 2D approximation. We recall that this approximation is only valid down to the scale $L_m/H = (UH/\nu)\ (tU/L)$ which increases with time because of the integrated effect of internal Poiseuille flow. Eliminating $\sigma$ gives :

$$\begin{cases} \dfrac{\mathcal{D}\mathbf{u}}{\mathcal{D}t} & = & -\dfrac{E(\Gamma_1)}{\rho h \Gamma_1}\nabla\Gamma_1 + \mathbf{g'} + \nu\triangle\mathbf{u} \\[2ex] \dfrac{\mathcal{D}\Gamma_1}{\mathcal{D}t} & = & -\Gamma_1\nabla.\mathbf{u} - (\Gamma_1(\dfrac{h+2k}{h}) - kc_0)/\tau^* \\[2ex] \dfrac{\mathcal{D}h}{\mathcal{D}t} & = & -h\nabla.\mathbf{u} \\[2ex] \dfrac{\mathcal{D}c_0}{\mathcal{D}t} & = & 0 \end{cases} \qquad (162)$$

where $E(\Gamma_1)$ is the Marangoni elasticity :

$$E(\Gamma_1) = -2d\sigma/dln\Gamma_1\ , \qquad (163)$$

which reads in the perfect gas approximation (5) :

$$E(\Gamma_1) = 2RT\Gamma_1\ . \qquad (164)$$

A preliminary version of this system appears first in Chomaz & Cathalau (1990). We will now consider various regimes produced by this soap film model.

### 5.2.1 Classical limits

*Elastic waves*

Elastic waves, investigated by Lucassen et al. (1970), are the soap film equivalent of sound waves in gas. We obtain from (162) the Marangoni elastic waves making $(\tau^{*-1}, \nu, \mathbf{g'}) = (0, 0, 0)$ and performing a linear approximation for a uniform membrane

($c_0$ constant) :

$$\begin{cases} \dfrac{\partial \mathbf{u}}{\partial t} = -\dfrac{E(\Gamma_1)}{\rho H \Gamma_1}\nabla \Gamma_1 \\[2ex] \dfrac{\partial \Gamma_1}{\partial t} = -\Gamma_1 \nabla U \ . \end{cases} \tag{165}$$

These waves propagate at a speed $v_L = (E(\Gamma_1)/\rho H)^{1/2}$ which gives, in the perfect gas low approximation

$$v_L = (2RTc_0 k/\rho(H+2k))^{1/2} \ . \tag{166}$$

For SDS the waves travel at $4m/s$ in a $10\mu m$ thick film and at $13m/s$ in a $1\mu m$ thick film.

Elastic waves will obey the previous equation only if their period is shorter than the chemical relaxation time. A wave of wave vector $\alpha$, induces variation on a time scale $(\alpha v_L)^{-1}$ which must be smaller than $\tau^*$ to respect the Marangoni approximation. Therefore $(\tau^* v_L)^{-1}$ defines the lower limit for the modulus of the Marangoni elastic wave vector. This limits the wavelength of Marangoni elastic wave between 0 and $25cm$ for $H = 10\mu m$ and $\tau^* = 10^{-2}s$, and between 0 and $100m$ for $H = 1\mu m$ and $\tau^* = 1s$. Since the experiment deals with a typical size of $1cm$, the Gibbs elastic waves speed affecting longer wavelengths will not be important. This means that, to determine the possible effect of compressibility, the experimental velocity should be compared to the Marangoni elastic wave velocity and not to the Gibbs elastic wave velocity which is much smaller.

*Static equilibrium*

Setting $(\mathbf{u}, \frac{\mathcal{D}}{\mathcal{D}t}) = (0,0)$ we get :

$$\begin{cases} -\dfrac{E(\Gamma_1)}{\rho h \Gamma_1}\nabla \Gamma_1 + \mathbf{g}' = 0 \\[2ex] \Gamma_1\left(\dfrac{h+2k}{h}\right) - kc_0 = 0 \ , \end{cases} \tag{167}$$

thus with $E(\Gamma_1)/\Gamma_1 = 2RT$

$$\frac{2RT}{\rho h}\nabla\left(\frac{khc_0}{h+2k}\right) = \mathbf{g}' \ . \tag{168}$$

For $\mathbf{g}'$ oriented in the negative $y$ direction, $h$ decreases with $y$ to insure stability. Thus $c_0$ must increase with $y$. The variation of $h$ has two origins:

    - a compressibility effect equivalent to the isothermal atmospheric equilibrium

    - a soap concentration induced stratification for inhomogeneous soap film equivalent to a temperature induced stratification in the atmosphere.

As we will see, the analogy between the field $c_0$ and a temperature field in a Boussinesq fluid is exact in the linear approximation. Couder et al. (1989) have computed the thickness profiles for constant $c_0$ using (168).

### 5.2.2  The incompressible approximation for uniform membrane

We consider the dynamics of a uniform membrane neglecting any gravity effect. In the same way as for classical flows, the dominant balance principle applied to the system (162) with no gravity shows that the relative variations of $\Gamma_1$ and $h$ are at maximum of order $(U/v_L)^2$. $U/v_L$ defines a Mach number $M$ for the soap. Variations of $\Gamma_1$ and $h$ are equivalent to compressibility effects and are small for small soap Mach number.

Scaling $(\mathbf{u}, \mathbf{x}, t, h, \Gamma_1)$ by $(U, L, L/U, H, C_0 kH/(H+2k))$, where $C_0$ is the value of $c_0$ assumed constant for simplicity, we obtain :

$$\begin{cases} \dfrac{\mathcal{D}\mathbf{u}}{\mathcal{D}t} &= -M^{-2}h^{-1}\nabla\Gamma_1 + Re^{-1}\triangle\mathbf{u} \\[2ex] \dfrac{\mathcal{D}\Gamma_1}{\mathcal{D}t} &= -\Gamma_1\nabla.\mathbf{u} - \left( (\dfrac{h+2K}{h})\Gamma_1 - (1+2K)c_0) \right)/\tau \\[2ex] \dfrac{\mathcal{D}h}{\mathcal{D}t} &= -h\nabla.\mathbf{u} \ , \end{cases} \qquad (169)$$

with $K = k/H$, $Re = UL/\nu$, $\tau = \tau^* U/L$. The same symbol has been kept for the field and their scaled form to avoid proliferation of notation.

We assume $M^2$ to be a small quantity and we use classical asymptotic expansion technique. The variables are expressed as series of $M^2 = \epsilon$, $a = \sum_j \epsilon^j a_j$.

The various order in the expansion give :

order zero :

$$\Gamma_{10} = h_0 = 1 \ , \qquad (170)$$

and

$$\begin{cases} \dfrac{\partial\mathbf{u}_0}{\partial t} + \mathbf{u}_0.\nabla\mathbf{u}_0 &= -\nabla\Gamma_{11} + Re^{-1}\triangle\mathbf{u}_0 \\[2ex] \nabla.\mathbf{u}_0 &= 0 \ . \end{cases} \qquad (171)$$

The system (171) corresponds to the incompressible 2D Navier Stokes equation where $\Gamma_{11}$ is a first order term that plays the role of pressure and obeys a Poisson equation:

$$\triangle\Gamma_{11} = -\nabla(\mathbf{u}_0.\nabla\mathbf{u}_0) \ . \qquad (172)$$

The boundary equations depend on the particular assumptions that has been made (effect of the meniscus, geometry, air or jet entrainment...). <u>First order</u> We seek equations for $h_1$, $\Gamma_{11}$. From (167) we get :

$$\Gamma_1\dfrac{\mathcal{D}h}{\mathcal{D}t} - h\dfrac{\mathcal{D}\Gamma_1}{\mathcal{D}t} = \left( \Gamma_1\dfrac{h+2K}{h} \right)/\tau \ . \qquad (173)$$

At first order, we obtain :

$$(\dfrac{\partial}{\partial t} + \mathbf{u}_0\nabla)(h_1 - \Gamma_{11}) = (-2h_1 K + (1+2K)\Gamma_{11})/\tau \ . \qquad (174)$$

We can identify two contributions to the thickness variation by introducing the formal field $e_1$ which measure the departure from the instantaneous variation of thickness due to the variation of "pressure" (term $\Gamma_{11}$, defined by (172)):

$$h_1 = +\Gamma_{11} + e_1/2K \ , \tag{175}$$

with

$$(\frac{\partial}{\partial t} + \mathbf{u}_0\nabla)e_1 = (\Gamma_{11} - e_1)/\tau' \tag{176}$$

The introduced field $e_1$ has no equivalent in gas. The chemical kinetic forces the thickness follows a "pressure" perturbation on a time scale $\tau'$. For small $\tau'$, $e_1$ equals the pressure and $\Gamma_{11} + e_1/2K$ represents the Gibbs response of $h_1$ to a variation of $\Gamma_{11}$:

$$h_1 = \frac{1 + 2K}{2K}\Gamma_{11} \ . \tag{177}$$

For large $\tau'$, the right hand side of (VIIIa) is close to zero and so is $e_1$. Then $h_1$ follows a Marangoni type variation

$$h_1 = \Gamma_{11} \ . \tag{178}$$

For $\tau'$ of order 1, $e_1$ is not obvious as illustrated in Chomaz & Cathalau and more numerical simulations are required to understand its behavior. Moreover the weight of $e_1$ is $1/2K$ which means that the thicker the film is, the more important the $e_1$ contribution to $h_1$ becomes.

### 5.2.3  Simulation of the evolution of the membrane thickness on a prototype flow

We have computed the thickness variation given by the above analysis on a prototype flow: the temporally evolving wake. The correspondence between the experiment on wakes ([8]) and numeric is only qualitative because the experiment deals with a spatially developing instability of the wake whereas the numeric treats the temporal evolution of an initially homogeneous wake profile. This subtlety is not supposed to modified the behavior of the various contributions to the thickness variation. Therefore we select one single prototype evolution of the Bickley wake and compute the evolution of each contribution to the thickness as a function of the single parameter $\tau'$ the chemical relaxation time.

Numerical integration on the 2D Navier Stokes equation is similar to the one presented by Basdevant $et$ $al$ (1981) (see also Basdevant and Sadourny (1983) ; Couder and Basdevant (1986) ; Chomaz $et$ $al$ (1988)). It integrates in time the 2D Navier Stokes equations in their vorticity $\omega$ formulation ($\omega\mathbf{e}_z = \nabla \times \mathbf{u}_0$).

$$\frac{\partial\omega}{\partial t} + \mathbf{u}_0.\nabla\ \omega = S(\omega) + F(\omega) \ , \tag{179}$$

where $S$ and $F$ respectively represent the eventual external forcing and the dissipation terms. Spatial derivatives are evaluated using a spectral decomposition. For simplicity,

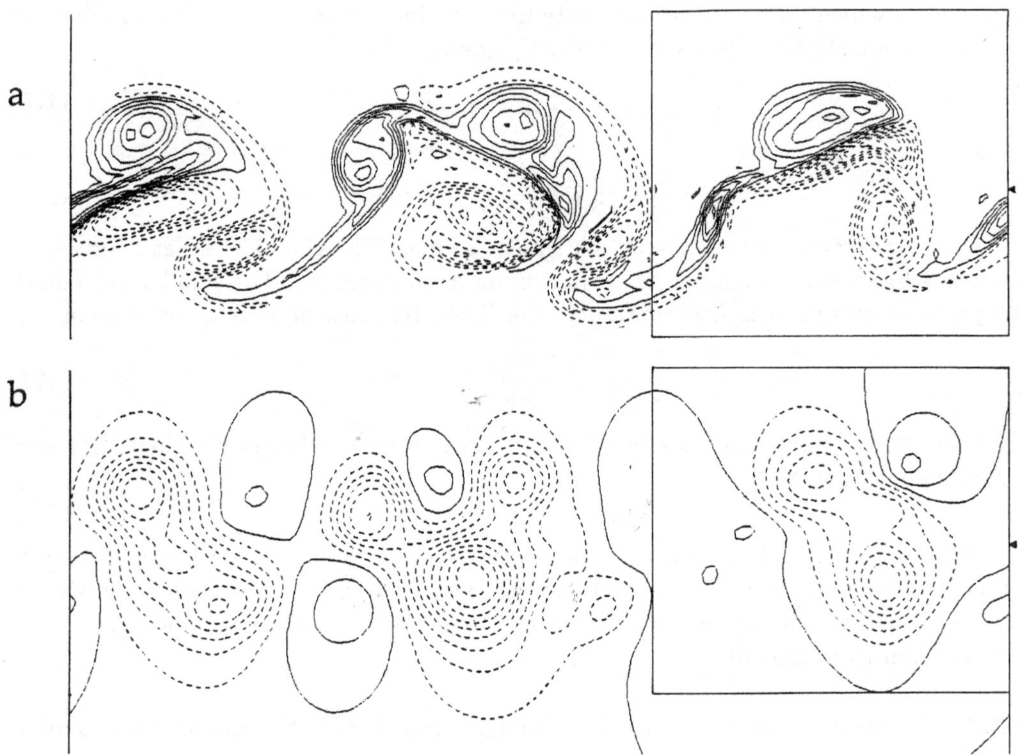

Figure 19: a) vorticity and b) pressure at time $t = 16$ for the prototype experiment with $L = 0.17$. , the vorticity vary between -5.3 and 5.3 and the pressure between -0.29 and 0.081. The close up on the right will serve to show details of the field $e_1$ since it contains both a typical pairing and stirring events.

the physical domain is chosen to be a periodic square $2\pi \mathrm{x} 2\pi$ discretized on a 128x128 (or more if required by precision) regular collocation grid. Non linear terms are evaluated at collocation points and their Fourier series are truncated at wave number $N_m = 59.9$ in a circular manner (higher for higher resolution). With respect to time discretization, the time scheme is second order centered (leap-frog) for nonlinear terms and exact integration for linear dissipative terms. An Euler integration replaces the leap-frog one every 80 time steps to insure stability. Equation (179) is in a dimensionless form therefore the typical value of $\mathbf{u}$ is set to 1 in the simulation. For the $128^2$ resolution the time step, $\Delta t$, used is either 0.02 or 0.01. To serve as a reference, figure 149 presents the detailed isovorticity and isopressure contours at time $t = 16$ of the prototype experiment.

For the prototype experiment we have computed each contribution to the thickness $\Gamma_{11}$ and $e_1$. The parameter $K$ modifies the respective weight of the contribution but not the dynamics. It compares the amount of soap molecules adsorbed on the surface to the amount of molecules in the internal fluid. Small $K$ corresponds to a film where the internal fluid acts as a large reservoir of soap molecules and therefore chemical kinetics and equilibrium dominate the variation of thickness through the field $e_1$. Large $K$ corresponds to a film dominated by the two surface phases where chemical kinetics is negligible. In that case the $e_1$ contribution is negligible. On the contrary the chemical kinetics time scale $\tau'$ affect both the dynamics and the amplitude of $e_1$ as may be observed on equation (176).

*Dynamic pressure field : $\gamma_1$*

The field $\gamma_1$ is the most trivial contribution to the thickness variation of a soap film under 2D motions. Figure 149 b shows the pressure field associated to the vorticity field (figure 149 a. We recognize the two properties of the pressure : it retains the large scale variations of the vorticity and it gets a tendency to isotropy. Therefore, the field $\gamma_1$ visualizes large scale vortices more or less as circular depression with an intensity depending on the strength and the shape of the vortex. The amplitude of this contribution (order -0.3) serves as a reference to compare with $e_1$.

*The relaxing field $e_1$*

The field $e_1$ directly corresponds to the relaxation toward the chemical equilibrium. It has been shortly discussed in Chomaz and Cathalau (1989). For $\tau' \ll 1$ the soap film is locally in equilibrium all the time and $e_1 = \gamma_1$, the total contribution $\gamma_1 + e_1/2K$ represents the variation of thickness due to Gibbs elasticity. For $\tau' \gg 1$ the soap film exhibits no relaxation as its surfaces are isolated from the bulk fluid, and the $e_1$ amplitude is nearly zero.

Figure 150 represents the field $e_1$ obtained on the prototype experiment for different values of $\tau'$. Each field is represented using 10 isolines between the minimum and the maximum of the field which are written on the figure. As predicted small $\tau'$ corresponds to an $e_1$ field nearly equal to the pressure field, $\gamma_1$, in shape and amplitude whereas large $\tau'$ gives $e_1$ field nearly nil.

The numerical simulation shows that for $\tau'$ of order 1 the amplitude of the field is large and seems visually correlated to the vorticity $\omega$. To confirm this intuition we compute the correlation of $e_1$ with various functions of $\omega$. We obtain the best results using $\omega^2$ and even a better correlation if we only consider zone of vorticity square higher than 5 % of the vorticity square maximum. The following table summarizes the information collected on the prototype experiment.

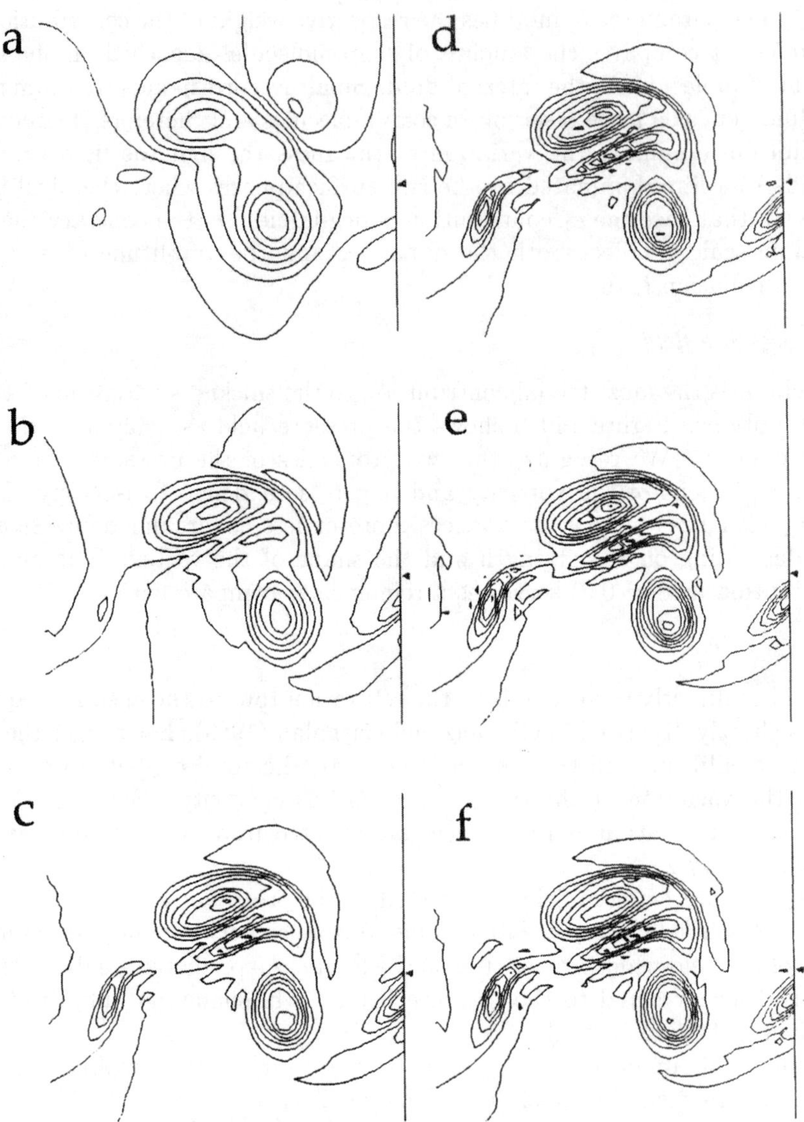

Figure 20: Close-up of the $e_1$ field obtained at $t = 16$ on the prototype experiment for a) $\tau' = 0.2$, b) $\tau' = 2$, c) $\tau' = 10$, d) $\tau' = 20$, e) $\tau' = 100$, f) $\tau' = 1000$ (see the table for the amplitude)

| $\tau'$ | 0.2 | 2 | 10 | 20 | 100 | 1000 |
|---|---|---|---|---|---|---|
| Correlation with $\omega^2$ | 0.65 | 0.90 | 0.95 | 0.94 | 0.94 | 0.93 |
| Correlation with $\omega^2$ with a 5% threshold | 0.70 | 0.93 | 0.98 | 0.98 | 0.98 | 0.98 |
| Maximum amplitude of $e_1$ | 0.29 | 0.26 | 0.14 | 0.083 | 0.020 | 0.002 |

The amplitude has to be compared to -0.29 the amplitude of the pressure contribution $\gamma_1$. We see that the correlation between $e_1$ and $\omega^2$ increases with $\tau'$ and is already equal to 93 % at $\tau' = 2$. At the same time value the $e_1$ amplitude equals -0.26 and starts to decrease with $\tau'$.

The value of $\tau'$ is therefore fundamental to evaluate variation of thickness due to the chemical relaxation. This effect may explain the correlation observed between the soap film thickness (visualization) and the vorticity. It needs two characteristics to be dominant: $\tau'$ of order 1 to have a non zero contribution correlated to $\omega^2$ and $K$ small, $K$ small means a strong reservoir effect of the interstitial fluid (thick film) to get a large contribution.

But other contributions may have competitive effects. Pictures of soap film (figure 16) indicates that the thickness of the film was not uniform even before the start of the experiment. The fringes, visible far from the wake region, correspond to variations of thickness linked to the gravity inside the film as the film is bent to sustain its own weight. They have two origins : an isoconcentration variation of thickness (equivalent to the compressible variation of density of an isothermal atmosphere) and a variation of thickness due to variation in the total soap concentration (equivalent to the thermal stratification of the atmosphere). Those contributions deserve further studies.

# References

[1] C. M. Bender and S. A. Orszag, *Advanced mathematical methods for scientists and engineers*, McGraw-Hill, Inc., New york, 1978.

[2] A. Bers, Linear waves and instabilities in *Physique des Plasmas*, edited by C. DeWitt and J. Peyraud, Gordon and Breach, New York, 117, 1975.

[3] C.V. Boys, *Soap bubbles and the forces which mould them*, Society for promoting Christian knowledge, London and Anchor books, New York, 1890.

[4] R. J. Briggs, *Electron-stream interaction with plasmas*, Addison-Wesley Publishing Company, London, 1969.

[5] J.-M. Chomaz and B. Cathalau, *Soap films as two-dimensional classical fluids*, Physics Review A, **41** (4), 2243, 1990.

[6] J.-M. Chomaz, P. Huerre and L. G. Redekopp, *A frequency selection criterion in spatially developing flows*, Studies in Applied Mathematics, **84**, 119, 1991.

[7] Y. Couder, *The observation of a shear flow instability in a rotating system with a soap membrane*, Le Journal de Physique - Lettres, **42** (19), 429, 1981.

[8] Y. Couder and C. Basdevant, *Experimental and numerical study of vortex couples in two-dimensional flows*, Journal of Fluid Mechanics **173**, 225, 1986.

[9] Y. Couder, J.M. Chomaz and M. Rabaud, *On the hydrodynamics of soap films*, Physica D, **37**, 384, 1989.

[10] L. de Luca and M. Costa, *Stationary waves on plane liquid sheets falling vertically*, European Journal of Mechanics, B/Fluids, **16**, 75, 1997.

[11] L. de Luca and M. Costa, *Instability of a spatially developing liquid sheet*, Journal of Fluid Mechanics, **331**, 127, 1997.

[12] P.G. Drazin and W.H. Reid, *Hydrodynamic Stability*. Cambridge university Press, Cambridge, 1981.

[13] M. Gharib and P. Derango, *A liquid film tunnel to study 2D flows*, Physica D, **37**, 406, 1989.

[14] J.W. Gibbs, *The collected works*, Longmans Green, New York, 1931.

[15] P. Huerre and P.A. Monkewitz, *Local and global instabilities in spatially developing flows*, Annual Review of Fluid Mechanics, **22**, 473, 1990.

[16] I.B. Ivanov, *Thin liquid films*. Marcel Dekker, Inc., New York & Basel, 1988.

[17] J.G.H. Joosten, *Solitary waves and solitons in liquid films*, Journal of Chemical Physics, **82**, 2427, 1985.

[18] H. Kellay, X. Wu and W. Golburg, *Experiments with turbulent soap films*, Physics Review Letters, **74**, 3875, 1995.

[19] H. Lamb, *Hydrodynamics*, 4th ed, Cambridge University Press, Cambridge, 1916.

[20] J. Lighthill, *Waves in Fluids*, Cambridge University Presss, Cambridge, 1978.

[21] T. Loiseleux, J.-M. Chomaz and P. Huerre, *The effect of swirl on jets and wakes: linear instability of the Rankine vortex with axial flow*, Physics of Fluids, **10** (5), 1120, 1998

[22] J. Lucassen, M. Van Den Tempel, A. Vrij and F. Hesselink, Proc. K. Ned. Akad. Wetensch. B, **73**, 109, 1970.

[23] K.J. Mysels, K. Shinoda and S. Frankel, *Soap films, studies of their thinning*, Pergamon Press, New-York, 1959.

[24] D. H. Peregrine, *Interaction of water waves and currents*, Advances in Applied Mechanics, **16**, 9, 1976.

[25] J. Plateau, *Statique expérimentale et théorique des liquides soumis aux seules forces moléculaires*, Gauthier Villars, Paris, 1870.

[26] A.L. Rusanov and V.V. Krotov, *Gibbs elasticity of liquid films*, Progress in Surface and Membrane Science, **13**, 415, 1979.

[27] H. B. Squire, *Investigation of the stability of a moving liquid film*, British Journal of Applied Physics, **4**, 167, 1953.

[28] G. I. Taylor, *The dynamics of thin sheets of fluid, II - waves on fluid sheets*, Proceedings of the Royal Society of London - A, **253**, 296, 1959.

[29] G. B. Whitam, *Linear and nonlinear waves*, Wiley interscience, New york, 1973.

# PATTERN SELECTION IN SURFACE TENSION DRIVEN FLOWS

**H.A. Dijkstra**
Utrecht University, Utrecht, The Netherlands

## Abstract

When a motionless liquid layer is heated from below, spontaneous convection appears
when the vertical temperature gradient exceeds a critical value. Under slightly supercritical conditions, the liquid organizes into steady regular polygonal patterns, for
example rolls or hexagons. If the liquid has an upper free surface open to ambient air,
both buoyancy gradients and surface tension gradients may be responsible for these
flows. The latter effect is dominating in thin layers and in a micro-gravity environment
and in that case usually hexagonal patterns are observed. In this chapter, an introduction is provided into the physics of these flows by giving an (incomplete) overview
of theoretical, experimental and numerical results which have been obtained over the
last decades. Focus is on the existence of the critical temperature gradient and the
selection of steady patterns near critical conditions.

# 1 Cellular convection

## 1.1 Setting of the problem

The study of the physical problems in the area of cellular convection are motivated by results from a conceptually simple experiment (Fig. 1a). A container which may have rectangular or circular cylindrical shape is filled with a relatively viscous liquid, such as silicone oil. Above the upper surface of the liquid is an ambient gas, for example air and the temperature far the gas-liquid interface is nearly constant. When the initially motionless liquid is heated from below, the liquid remains motionless below a critical value of the vertical temperature gradient. The heat transfer through the layer is only by heat conduction. When the temperature gradient slightly exceeds the critical value, the liquid is set into motion and after a while the flow organizes itself into cellular patterns. The motion of the liquid can also be detected by measuring the horizontally averaged vertical heat flux. A measure for the increase of heat transport due to convection is the Nusselt number $Nu$ which is unity in case of conduction only. In Fig. 1b, $Nu$ is plotted as a function of a measure of the vertical temperature gradient. The onset of convection in the liquid is shown by the increase of $Nu$ above unity.

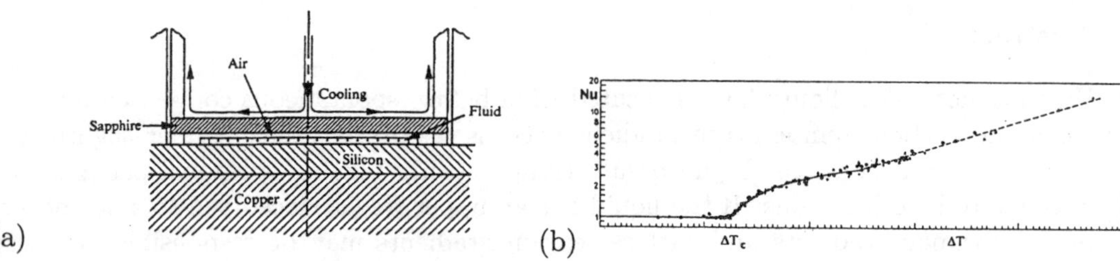


Figure 1: *a) Sketch of the experimental set-up (from [1]); the liquid is situated on the (heated) silicon block and separated from the (cooled) sapphire block by a small air gap. b) Plot of the Nusselt number (see text) as a function of the vertical temperature gradient $\Delta T$ (from [2]); $Nu = 1$ if the heat transport is by conduction only and $Nu$ increases if there is convection in the liquid; $\Delta T_c$ is the critical temperature gradient.*

The classical example of patterns in cellular convection is the hexagonal pattern as observed at the beginning of this century by Bénard [3]. As shown by the picture of Fig. 2 of a modern experiment [4] the hexagons are fairly regular. The liquid is rising at each center of a hexagon and descends along each side. The regular pattern adjusts to the boundaries of the container through larger boundary cells which have a less uniform size. The hexagonal pattern is always found in relatively thin liquid layers which have the upper free surface in contact with an ambient gas.

The experimental results induce several questions. First one wants to understand the physical origin of the critical vertical temperature gradient. Why is the liquid motionless up to a certain gradient and then suddenly set in motion ? It will turn out that the flow arises through an instability, i.e. small perturbations become amplified when critical conditions are exceeded. In section 1.2, the basic physics of the instabilities will be described. A second point of interest is why the liquid motion organizes into regular patterns and why the hexagonal pattern seems to be preferred. Through the following chapters, these problems will be formulated more precise and theory is presented which provides answers to different details of each problem.

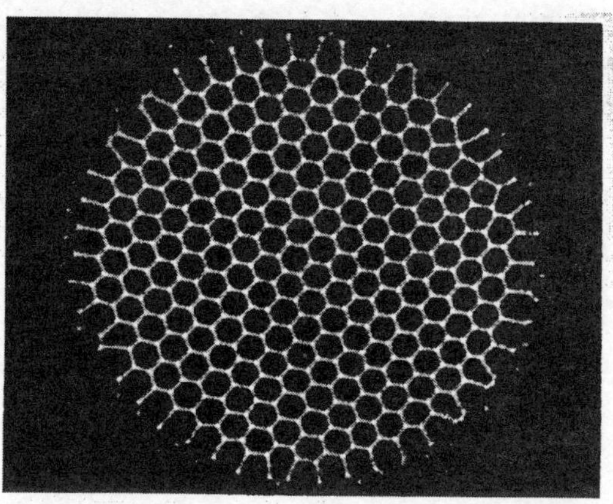

Figure 2: *Top view of a hexagonal pattern obtained in a liquid layer (1.9 mm deep) heated from below (from [4]). The image is a schadowgraph picture and bright lines indicate cold liquid.*

## 1.2   Rayleigh-Bénard-Marangoni convection

The density of the silicone oils as used in the experiments above depends on the temperature. In most cases, a linear equation of state is adequate over the temperature range considered, i.e.

$$\rho = \rho_0(1 - \alpha_T(T - T_0)) \tag{1.1}$$

where $\rho_0$ and $T_0$ are reference values and $\alpha_T$ is the coefficient of thermal expansion. When gravity is present, temperature differences will cause density differences and consequently a buoyancy driven flow is generated. A well-known example is the flow near a plate which is put vertically into a liquid of homogeneous temperature $T_0$. When the plate is heated with respect to the liquid up to a temperature $T_1 > T_0$, the liquid near the plate will be lighter than its environment and rises.

To explain the existence of the critical temperature gradient in cellular convection often the following mechanism is proposed [5]. Consider a motionless liquid layer of depth $d$ heated from below having a constant vertical temperature gradient $\beta = \Delta T/d$. A spherical liquid volume of radius $R$ within this motionless solution is moved upwards (for example from point 2 to a point 1 in Fig. 3) by some infinitesimally small perturbation. If the velocity of the volume is indicated by $V$, then the viscous drag $F_d$ of the volume can (for small velocities) be approximated as $F_d = C_d R V \mu$, where $\mu$ is the dynamic viscosity of the liquid and $C_d$ is a proportionality constant. As the volume moves upwards, it becomes warmer than its environment. The time scale over which thermal anomalies adjust due to diffusion is $\tau_\kappa = R^2/\kappa$, where $\kappa$ is the thermal diffusivity of the liquid. Hence, the temperature within the volume at a time $t$ is equal to that of its surroundings at time $t - \tau_\kappa$. The temperature difference $\delta T$ at time $t$ between the volume and its surroundings is $\delta T = \tau_\kappa V \beta$ and this gives a buoyancy force $F_b = C_b \rho_0 \alpha_T \, \delta T \, R^3 g$, where $C_b$ is a proportionality constant and $g$ is the gravitational acceleration.

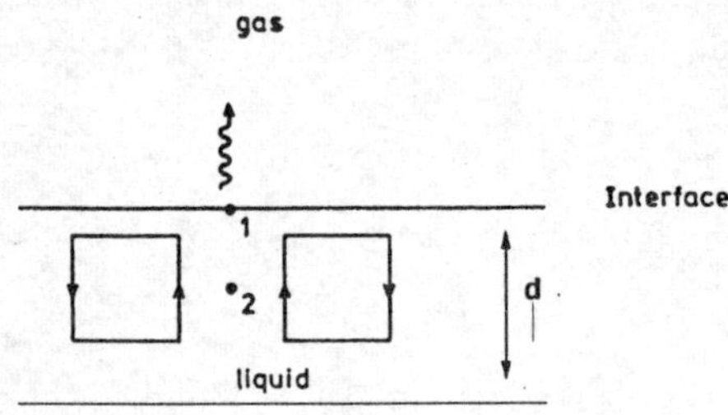


Figure 3: *Sketch to explain the buoyancy driven instability in a liquid layer of depth d heated from below.*

The ratio of $F_b$ and $F_d$ increases with the radius of the volume, such that largest values occur for a volume with a radius equal to the depth of the layer, i.e. $R = d$. In this case, the ratio is proportional to the Rayleigh number $Ra$ defined by

$$Ra = \frac{g \alpha_T \beta d^4}{\nu \kappa} \qquad (1.2a)$$

where $\nu = \mu/\rho_0$ is the kinematic viscosity of the liquid. The Rayleigh number can also be expressed as a ratio of time scales, i.e.

$$Ra = \frac{\tau_\nu \tau_\kappa}{\tau_b^2} \qquad (1.2b)$$

with $\tau_\nu = d^2/\nu$, $\tau_\kappa = d^2/\kappa$ being the diffusive time scales associated with transport of heat and momentum and $\tau_b$ is the advective time scale $d/V_b$ based on the buoyancy induced velocity scale $V_b = d\sqrt{g\alpha_T\beta}$. Since the process with the smallest time scale is dominant, a large value of $Ra$ indicates the domination of the buoyancy force over viscous drag giving rise to motion in the liquid. If motion occurs on a scale $d$ in Fig. 3, this movement is amplified because by continuity, warmer liquid is drawn from below to point 2, making it warmer than it already was. The existence of a critical temperature gradient is equivalent to the existence of a critical value of $Ra$.

Lord Rayleigh [6] approached this problem mathematically and indeed demonstrated that such a critical value exists. However, there were several discrepancies between the results predicted by the analysis of Lord Rayleigh and the experiments of Bénard [3]. The values of the critical temperature gradient and the wavelength of the expected pattern did not match. Furthermore, the correlation between the deflection of the free air-liquid surface and the vertical velocity just below the interface did not correspond. In the experiments there was upward flow below depressions of the interface whereas Lord Rayleigh's theory predicted downward flow below depressions [7].

It took several decades before Block [8] and Pearson [9] realized that another mechanism can be responsible for the existence of a critical temperature gradient. In most experiments of Bénard [3], a free surface was present separating the liquid from the ambient gas above. The surface tension of the silicone oils depends on the temperature at the gas-liquid interface and in most cases a linear relation is adequate, i.e.

$$\sigma = \sigma_0(1 - \gamma_T(T - T_0)) \tag{1.3}$$

Any surface tension gradient induces a shear stress on the interface directed from points of low to locations of high surface tension. Since the liquid is viscous, this shear stress induces bulk motion. The generation of interfacial stresses by surface tension gradients and the resulting bulk flow is called the Marangoni-effect after the Italian Marangoni [10].

Surface tension does not only depend on temperature, but also on the concentration of a component which may be dissolved in the liquid, for example acetone in water. A similar relation (1.3) holds for most components, with a coefficient $\gamma_C$ instead of $\gamma_T$. A nice example of liquid motion induced by surface tension gradients are tears in a glass of strong wine [11]. Here, because the alcohol is more readily depleted near the glass boundary, its concentration is lower with respect to that at the center of the glass. Hence, a surface tension gradient is set-up with higher surface tension near the glass wall and liquid is pulled upward along the wall. Eventually, the film becomes unstable and tears develop which fall down along the wall.

To explain how the Marangoni-effect can also lead to flow in a liquid heated from below, in Fig. 4 also a volume of liquid is considered which, due to an infinitesimal perturbation, moves upward towards the interface at a point 2. Since its temperature is higher, the surface tension at the interface is slightly decreased. Hence, a flow develops from points 2 towards a neighbouring point 1. By continuity, this flow has to be compensated by liquid from below and hence the original perturbation is amplified when the surface tension induced stress exceeds the drag force. This can again be formulated as a critical value of a dimensionless number, the Marangoni number Ma, which is written, similar to (1.2b), as

$$Ma = \frac{\tau_\nu \tau_\kappa}{\tau_s^2} \tag{1.4a}$$

where the diffusive time scales are as above, but now $\tau_s$ is the advective time scale $d/V_s$ associated with the surface tension induced velocity $V_s = \sqrt{\frac{\sigma_0 \gamma_T \beta}{\rho_0}}$. Hence, when the time scale of motion due to the surface tension gradients exceeds those of viscous and thermal dissipation, motion will result. When the time scales are explicitly substituted in (1.4a), the Marangoni number $Ma$ becomes

$$Ma = \frac{\sigma_0 \gamma_T \beta d^2}{\rho_0 \nu \kappa} \qquad (1.4b)$$

In most experiments, the value of the critical temperature gradient depends on both on the buoyancy and surface tension driving mechanisms [12]. This holds also for the resulting convection, which in general is termed Rayleigh-Bénard-Marangoni convection. Since, for constant $\beta$, $Ra$ is proportional to $d^4$ and $Ma$ to $d^2$, the buoyancy (surface tension) mechanism will dominate in thick (thin) layers. There is one special situation where the surface tension mechanism always dominates. In micro-gravity conditions, density differences introduce hardly any flow since the acceleration due to gravity is reduced to less than 1% of its value on earth. Hence, nearly pure surface tension driven convection, termed Bénard-Marangoni convection, is found. In liquid layers where the upper surface is a rigid lid, no surface tension gradients can develop and the resulting convection is pure buoyancy driven and termed Rayleigh-Bénard convection.

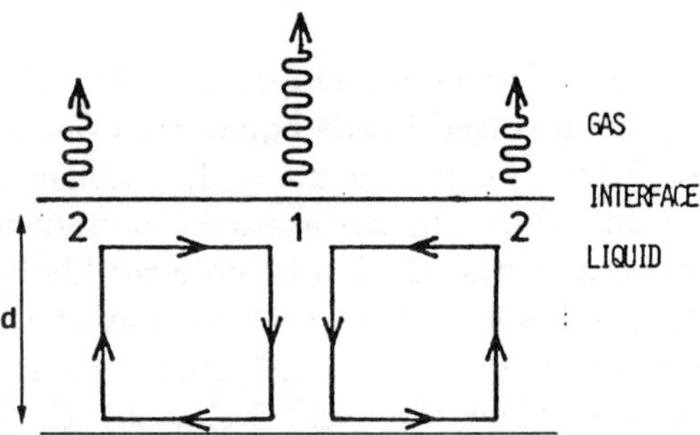


Figure 4: *Sketch to explain the surface tension gradient driven instability in a liquid layer of depth d heated from below.*

## 1.3   Phenomenology of the patterns

Once the critical temperature gradient is exceeded, certain perturbations will grow and finally the whole liquid is set into motion. In a typical experiment, a step-wise change in heating is applied and one waits for a long time to look at the response. The characteristic time scale of adjustments of horizontal gradients in temperature is the horizontal thermal diffusion time scale $L^2/\kappa$, where $L$ is the horizontal length scale of the liquid layer. Typical experiments [1] are performed with highly viscous silicone oils with $\nu = 10^{-4}\ [m^2/s]$ and thermal diffusivity $\kappa = 10^{-7}\ [m^2/s]$. The depths of the layers are typically a few millimeter and the horizontal extension of layer is about 50 times the layer depth. This implies thermal diffusion time scales in the order of a few days.

The flow in the liquid is visualized, for example by looking at the movement of tracer particles (such as aluminum flakes) or by looking at a shadowgraph pattern [13]. In most cases, the vertical heat flux over the layer is also measured by a series of thermistors. The onset of convection is not an unambigious spontaneous event and cannot be measured very accurately [14]. It can be seen as a slow but significant increase of the Nusselt number above its conduction value ($Nu = 1$).

For slightly supercritical conditions steady flow patterns appear after a very long time. In rectangular boxes, patterns can be of several type, for example two-dimensional rolls and three-dimensional rolls. However, the most famous pattern found is the hexagonal pattern (Fig. 2) found originally by Bénard [3]. There is an interesting problem concerning the wavelength dependence of this pattern with increasing value of $Ma$. The number of cells is proportional to the dimensional wavenumber $k$ of the patterns, which for hexagonal cells is given by $k = 4\pi/3L_H$ where $L_H$ is the side length of a hexagon. Hence, from counting the number of hexagons, the wavenumber of the pattern can be determined.

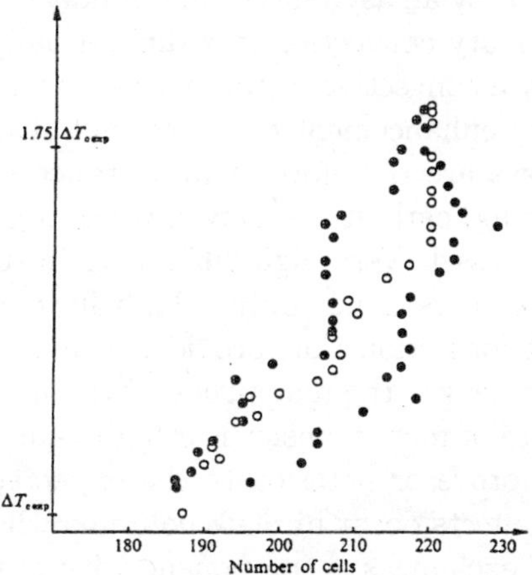


Figure 5: *The number of hexagonal cells as a function of the vertical temperature differ-ence $\Delta T$ over a liquid layer of depth 1.9 [mm] (from [1]). Different marking corresponds to different experiments where the bottom temperature is either increased or decreased.*

When the vertical temperature difference is increased, the number of hexagons increases (Fig. 5) implying that the wavelength of the surface tension driven hexagonal pattern decreases. This is one of the intriguing questions of pattern selection in convective flows. The increase of wavenumber found for Bénard-Marangoni convection is contrary to the decrease in wavenumber found for Rayleigh-Bénard flows [4]. In the latter, the wavelength of the cells tends to increase with increasing vertical temperature gradient. To explain why the hexagonal pattern is preferred in shallow layers and why the wavelength of the pattern decreases with $Ma$ is one of the classical problems in cellular convection [4].

## 1.4  Relevance of surface tension driven convection

Apart from the intriguing problem of organization of macroscopic flows, surface tension driven flows have relevance to many problems in science and engineering [15]. Since the surface tension depends on temperature and the composition of the liquids at the interface, practically everywhere where gas-liquid interfaces are in the picture, effects of surface tension gradients become visible. Strong wine creeps up the wall of a glass, fast movements occur near the wick of a burning candle and small surface-crawling insects can be seen moving rapidly to safety to waterplants on the back of a film drawn by surface forces once the water is suddenly polluted by surface active components.

With respect to applications, surface tension driven flows are classified in macro-scale and micro-scale flows. This classification does not refer to the ultimate form of the convection but to its origin. Macro-scale convection is brought about by macroscopic scale asymmetry, e.g. by geometric asymmetry of the bulk phases about the interface (causing the 'wine-tears' [11]) or by an asymmetry in boundary conditions. An example of the latter is the thermocapillary convection in a differentially heated cavity. Micro-scale convection is triggered by a convective instability as we have considered above and can be responsible for a strong enhancement of the overall heat and/or mass transfer.

Surface tension driven flows are therefore of importance in many sections of the process industry. In steel making, carbon is removed from iron by blowing oxygin into the molten iron in a Bessemer-vessel. Very high differences in surface tension can occur due to oxidation. In this case there is not only a high flow rate along the interface, but even spontaneous droplet formation that provides a large mass transferring area. In two-phase mass transfer processes, the magnitude of the interface between different phases and the consequent rate of mass transfer is a high valued design criterion. Surface tension gradients can improve or deteriorate the expected rate of mass transfer dramatically [16]. Important effects occur in plate columns where a liquid flows downward over rigid particles and exchanges a component with a gas which flows upward (for example, the water/acetone/air system). Within the stagnant phase between the particles, Bénard-Marangoni convection occurs. In addition, small films of liquid are drawn up the packing material, a similar effect as in the wine glass, increasing the area of mass transfer enormously. Further areas of application are combustion, welding and the containerless processing of crystals.

# 2   The critical temperature gradient

In this section, the onset of convection due to the Rayleigh-Bénard-Marangoni instability is studied for a liquid layer which is horizontally unbounded. In section 2.1 the mathematical description of the system is presented which has a simple motionless solution (section 2.2). The critical conditions for this motionless solution and the most unstable flow patterns are determined for a flat gas-liquid interface in section 2.3. The influence of interface deformation is discussed in section 2.4.

## 2.1   Formulation of the model

Consider a horizontally unbounded liquid layer (Fig. 6) of an incompressible Newtonian liquid of mean depth $d$ which is heated from below. Above the interface there is an ambient gas and heat is transferred from the liquid to the gas.

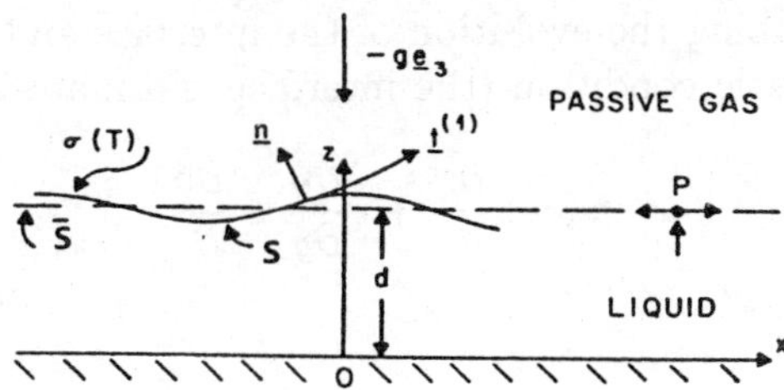


Figure 6: *Sketch of the liquid layer of mean depth d and with an interface S which is heated from below (from [17]). In the picture, $\bar{S}$ is the equilibrium flat interface, $\mathbf{t}^{(1)}$, $\mathbf{t}^{(2)}$ and $\mathbf{n}$ are tangent vectors and normal to the interface and $\mathbf{e}_3$ is the unit vector in the vertical.*

The equations describing the evolution of this system are the continuity equation

$$\nabla \cdot \mathbf{v} = 0 \qquad (2.1a)$$

the momentum balances,

$$\rho_0 \left[ \frac{\partial \mathbf{v}}{\partial t} + \mathbf{v}.\nabla \mathbf{v} \right] = -\nabla p + \mu \nabla^2 \mathbf{v} - \rho g \mathbf{e}_3 \qquad (2.1b)$$

and the thermal energy balance

$$\rho_0 C_p \left[ \frac{\partial T}{\partial t} + \mathbf{v}.\nabla T \right] = \lambda_T \nabla^2 T \qquad (2.1c)$$

In these equations, $(x, y, z)$ are the Cartesian coordinates of a point in the liquid layer, $t$ denotes time, $\mathbf{v} = (u, v, w)$ is the velocity vector, $p$ denotes pressure and $T$ is the temperature, respectively. Finally, $\rho_0$, $g$, $C_p$, $\mu$ and $\lambda_T$ are the reference density, the acceleration due to gravity, the specific heat, the dynamic viscosity and the thermal conductivity, respectively. The thermal diffusivity $\kappa$ and kinematic viscosity $\nu$ are given by $\nu = \frac{\mu}{\rho_0}$ and $\kappa = \frac{\lambda_T}{\rho_0 C_p}$. All these quantities will be assumed constant. In the equations above, the Boussinesq approximation is applied which is adequate here since the density variations are small with respect to $\rho_0$. In this approximation, density variations are only considered in the body force term of (2.1b) and apart from this, the liquid is considered incompressible.

Let the gas-liquid interface $\mathbf{S}$ be parametrized by $z = d + \eta(x, y, t)$. The tangents to the interface $\mathbf{t}_1$ and $\mathbf{t}_2$, and the unit normal $\mathbf{n}$ are given by

$$\mathbf{t}_1 = (1, 0, \frac{\partial \eta}{\partial x}) \; ; \; \mathbf{t}_2 = (0, 1, \frac{\partial \eta}{\partial y});$$

$$\mathbf{n} = (1 + (\frac{\partial \eta}{\partial x})^2 + (\frac{\partial \eta}{\partial y})^2)^{-\frac{1}{2}} (-\frac{\partial \eta}{\partial x}, -\frac{\partial \eta}{\partial y}, 1) \tag{2.2}$$

The equations describing the evolution of the interface and its interaction with the bulk are [18] the kinematic condition (the interface is a material surface)

$$w = u \frac{\partial \eta}{\partial x} + v \frac{\partial \eta}{\partial y} + \frac{\partial \eta}{\partial t} \tag{2.3a}$$

the normal stress balance

$$\mathbf{n} \cdot \mathbf{T} \cdot \mathbf{n} = 2H\sigma \tag{2.3b}$$

the tangential stress balance (for i = 1,2)

$$\mathbf{t}_i \cdot \mathbf{T} \cdot \mathbf{n} = \mathbf{t}_i \cdot \nabla \sigma \tag{2.3c}$$

where $\mathbf{T}$ is the stress tensor for a Newtonian liquid which is given in components by

$$\mathbf{T}_{ij} = -p\delta_{ij} + 2\mu D_{ij} \; ; \; D_{ij} = \frac{1}{2}(\frac{\partial v_i}{\partial x_j} + \frac{\partial v_j}{\partial x_i}) \tag{2.3d}$$

and $2H$ is the mean curvature of the interface

$$2H = \frac{\frac{\partial^2 \eta}{\partial x^2}(1 + (\frac{\partial \eta}{\partial y})^2) - 2\frac{\partial \eta}{\partial x}\frac{\partial \eta}{\partial y}\frac{\partial^2 \eta}{\partial x \partial y} + \frac{\partial^2 \eta}{\partial y^2}(1 + (\frac{\partial \eta}{\partial x})^2)}{(1 + (\frac{\partial \eta}{\partial x})^2 + (\frac{\partial \eta}{\partial y})^2)^{\frac{3}{2}}} \tag{2.3e}$$

The lower boundary is a very good conducting boundary and therefore the temperature is constant. Moreover, no-slip conditions apply and hence,

$$z = 0 : T = T_B \; ; \; \mathbf{v} = 0 \tag{2.4a}$$

At the interface, heat is transferring from the liquid to the gas. This is usually modelled by the Newtonian cooling law

$$z = d + \eta \; : \quad -\lambda_T \mathbf{n} \cdot \nabla T = h(T - T_A) \tag{2.4b}$$

where $h$ is an interfacial heat transfer coefficient and $T_A$ is the temperature of the gas far from the interface. The equations are closed by prescribing the equation of state and the temperature dependence of surface tension, i.e.

$$\rho = \rho_0(1 - \alpha_T(T - T_0)) \; ; \; \sigma = \sigma_0(1 - \gamma_T(T - T_0)) \tag{2.5}$$

Given initial conditions, the evolution of the system is determined by the equations (2.1) to (2.5).

## 2.2  Motionless solution

For $\bar{\mathbf{v}} = 0$ and $\bar{\eta} = 0$ (flat interface) there is a steady state given by

$$\bar{T}(z) = T_B - \beta z \; ; \; \beta = \frac{h(T_B - T_A)}{\lambda_T + hd} \tag{2.6a}$$

The quantity $\beta$ is the vertical temperature gradient over the layer which was already used in the definition of the Rayleigh and Marangoni numbers. The pressure is readily determined from (2.1b) and one obtains

$$\bar{p}(z) = p_0 + \rho_0 g([\alpha_T(T_B - T_0) - 1]z + \alpha_T \beta \frac{z^2}{2}) \tag{2.6b}$$

This motionless solution is characterized by only conductive heat transfer and is easily realized in laboratory experiments.

## 2.3  Linear Stability Analysis

In a linear stability analysis, sufficient conditions for stability of the basic state (2.6) are determined. Perturbation velocities $\tilde{\mathbf{v}}$, temperature $\tilde{T}$ and interface $\tilde{\eta}$ are superposed on the basic state and thereafter the equations are linearized in the perturbation quantities. The linearized equations (2.1) become (tildes are omitted for clarity)

$$\nabla \cdot \mathbf{v} = 0 \tag{2.7a}$$

$$\rho_0 \frac{\partial \mathbf{v}}{\partial t} = -\nabla p + \mu \nabla^2 \mathbf{v} + \alpha_T g \rho_0 T \tag{2.7b}$$

$$\frac{\partial T}{\partial t} - w\beta = \kappa \nabla^2 T \tag{2.7c}$$

The pressure can be elimated from the equations (2.7b) which gives (with (2.7a))

$$\rho_0 \frac{\partial \nabla^2 w}{\partial t} = \mu \nabla^4 w + \alpha_T g \rho_0 \nabla_H^2 T \tag{2.7d}$$

with $\nabla^2_H = \frac{\partial^2}{\partial x^2} + \frac{\partial^2}{\partial y^2}$ being the horizontal Laplace operator. First, the easiest case of a non-deformable interface will be considered.

### 2.3.1　Flat interface

For the case $\eta = 0$, the linearized equations at the interface $z = d$ become

$$w = 0 \tag{2.8a}$$

$$\mu\frac{\partial u}{\partial z} = \frac{\partial \sigma}{\partial x} \;\; ; \;\; \mu\frac{\partial v}{\partial z} = \frac{\partial \sigma}{\partial y} \tag{2.8b}$$

$$-\lambda_T\frac{\partial T}{\partial z} = hT \tag{2.8c}$$

The equations (2.7c,d) and (2.8) are non-dimensionalized using scales $\frac{\kappa}{d}$ for velocity, $\frac{d^2}{\kappa}$ for time, $\beta d$ for temperature and $d$ for length. This leads to the non-dimensional problem

$$\frac{\partial T}{\partial t} - w = \nabla^2 T \tag{2.9a}$$

$$Pr^{-1}\frac{\partial \nabla^2 w}{\partial t} = \nabla^4 w + Ra\nabla^2_H T \tag{2.9b}$$

$$z = 0 : T = w = 0 \tag{2.9c}$$

$$z = 1 \; : \; w = 0 \; ; \; \frac{\partial^2 w}{\partial z^2} = Ma\,\nabla^2_H T \; ; \; \frac{\partial T}{\partial z} = -Bi\,T \tag{2.9d}$$

where $\nabla^2_H$ is the horizontal Laplace operator. In the equations, the dimensionless numbers $Pr$ (Prandtl), $Ra$ (Rayleigh), $Ma$ (Marangoni) and $Bi$ (Biot) appear which are defined as

$$Ra = \frac{\alpha_T g\beta d^4}{\nu\kappa}; \; Ma = \frac{\gamma_T\beta\sigma_0 d^2}{\rho_0\nu\kappa}; \; Pr = \frac{\nu}{\kappa}; \; Bi = \frac{hd}{\lambda_T} \tag{2.10}$$

The meaning of the stability parameters $Ra$ and $Ma$ was already explained in the previous section.

The problem (2.9) is separable in time and space and solutions of the form

$$w(x, y, z, t) = W(z)\phi(x, y)e^{\lambda t}$$

$$T(x, y, z, t) = \Theta(z)\phi(x, y)e^{\lambda t} \tag{2.11}$$

exist, where $\lambda = \lambda^R + i\lambda^I$ is the complex growth factor and the function $\phi$ satisfies the equation

$$\nabla^2_H\phi + k^2\phi = 0 \tag{2.12}$$

With $\phi(x,y) = e^{i(k_x x + k_y y)}$ and $k = \sqrt{(k_x^2 + k_y^2)}$ being the wavenumber of a particular disturbance, the problem for $W$ and $\Theta$ becomes

$$Pr^{-1}\lambda(D^2 - k^2)W = (D^2 - k^2)^2 W + Ra\, k^2\, \Theta \qquad (2.13a)$$

$$\lambda\Theta - W = (D^2 - k^2)\Theta \qquad (2.13b)$$

$$W(0) = 0 \; ; \; DW(0) = 0 \; ; \; \Theta(0) = 0 \qquad (2.13c)$$

$$W(1) = 0 \; ; \; D^2W(1) + Ma\, k^2\Theta(1) = 0 \; ; \; D\Theta(1) + Bi\, \Theta(1) = 0 \qquad (2.13d)$$

where $D = \frac{d}{dz}$.

The set of equations (2.13) defines an eigenvalue problem, $\lambda$ being the eigenvalue and $(W, \Theta)$ being the eigensolution. It will be assumed that there exists an infinite denumerable sequence of eigenvalues $\lambda_n$ which may be ordered in such a way that $\lambda_{n+1}^R > \lambda_n^R$, $n = 0, 1, 2, \ldots$. For pure Rayleigh-Bénard convection, the problem is self-adjoint and the assumption has been proven, the eigenvalues all being real. If for some choice of parameters, all eigenvalues $\lambda$ are in the left complex halfplane the motionless state is linearly stable and the perturbation amplitudes decay to zero. However, if at least one eigenvalue $\lambda$ has a positive real part, then the perturbation grows and the motionless state is unstable. Hence the transition from stable to unstable state occurs at values of the parameters such that $\lambda$ crosses the imaginary axis. The transition state $\lambda^R = 0$ is called the neutral state.

The principle of exchange of stability is said to be valid if at neutral stability also $\lambda^I = 0$ [19]. The neutral state is then stationary rather than oscillatory. Assuming this principle to be valid, the parameter values defining the neutral state (called the critical values) can be calculated by putting $\lambda = 0$ in the equations (2.13); the $Pr$ number drops out of the equations. It has been proven (e.g. [20]) that the principle of exchange of stabilities is valid for the case of pure Rayleigh-Bénard convection, free sidewalls and arbitrary conditions on the bottom wall. For pure Bénard-Marangoni convection ($Ra = 0$), the principle has been proven by Vidal and Acrivos [21]. For a more general case, the principle has not been proven although their are some partial results [22]. Possible oscillatory neutral states are excluded here.

The equations (2.13) at neutral stability were solved analytically for $Ra = 0$ by Pearson [9] and the general case was solved by Nield [23] using a Fourier series method. The result of Pearson [9] is

$$Ma = \frac{8k(k\, cosh\, k + Bi\, sinh\, k)(k - sinh\, k\, cosh\, k)}{(k^3\, cosh\, k - sinh^3\, k)} \qquad (2.14)$$

In Fig. 7, the neutral curves defined by (2.14) are plotted for different values of $Bi$. As can be seen, each curve has a minimum for $Ma$ as a function of $k$, called the critical

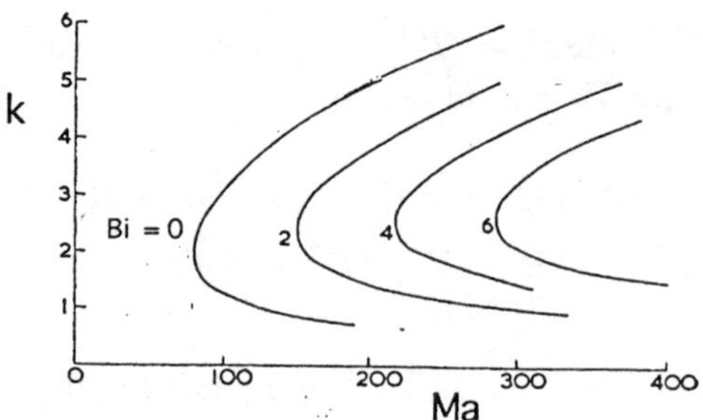


Figure 7: *Neutral curves of Ma versus wavenumber k for several values of Bi (from [9]) according to (2.14).*

value for that particular wavenumber. The smallest critical value, $Ma_c = 79.61$, is found for $Bi = 0$ with a critical wavenumber $k_c = 1.993$. This indicates a dimensionless wavelength of the cell patterns of about $3d$, such that the horizontal dimensions of the resulting cells are larger than the liquid depth. A zero value of $Bi$ implies that no heat is lost by the perturbations, maintaining an optimal value of the surface tension gradients which cause the flow. For $Bi \to \infty$, the interface becomes isothermal and no instability is possible; hence $Ma_c \to \infty$.

When buoyancy is taken into account and $Ra > 0$, both driving mechanisms cooperate in causing flow which is clear from the instability mechanisms described above. Consequently, $Ma_c$ decreases when $Ra$ increases and there exists a value of $Ra$ for which $Ma_c = 0$. If $Bi = 0$, this value becomes $Ra = 669$, which is slightly larger than the famous critical value $Ra_c = \frac{27\pi^4}{4}$, calculated by Lord Rayleigh [6] for buoyancy driven convection between two perfectly conducting, stress free boundaries. Hence, for each value of $Bi$, there exists a curve in the $(Ra, Ma)$-plane of critical values $(Ra_c, Ma_c)$. For $Bi = 0$ and $Bi = \infty$ these curves are shown in Fig. 8. Critical wavenumbers are in the range between 2 and 3, the latter giving a wavelength of about $2d$.

### 2.3.2  The effect of interface deformation, Ra = 0

Up to here, we did not consider the interface to be deformable. There are several papers dealing with this surface deformation, e.g. Scriven and Sternling [24], Smith [25] and Takashima [26]. When the interface deformation is considered, its perturbation has the form

$$\eta(x, y, t) = Z\phi(x, y)e^{\lambda t} \qquad (2.15a)$$

In linearizing the equations (2.3) at the interface, one makes use of

$$(\bar{T}+\tilde{T})(x,y,d+\tilde{\eta},t) \approx \bar{T}(x,y,d,t)+\tilde{T}(x,y,d,t)+\frac{\partial \bar{T}}{\partial z}(x,y,d,t)\,\tilde{\eta}(x,y,t)+\ldots \quad (2.15b)$$

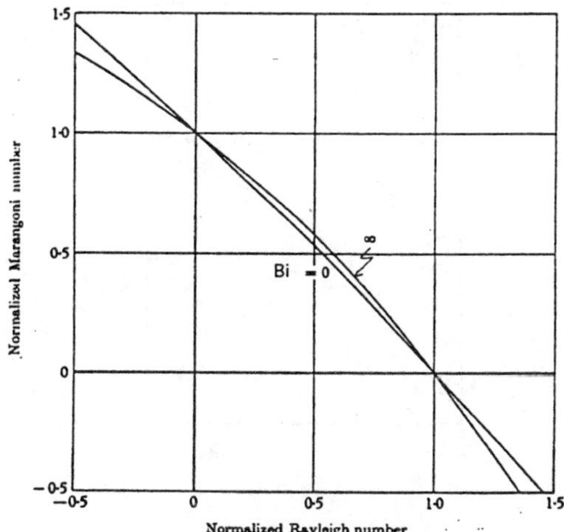


Figure 8: *Plot of normalized Rayleigh $(\frac{Ra}{Ra_c})$ and Marangoni $\frac{Ma}{Ma_c}$ numbers indicating the critical values $Ma_c$ and $Ra_c$ for two extreme values of $Bi$ (from [23]).*

In the normal stress balance, the pressure now introduces a gravity contribution as a restoring force on the interface. This pressure can be eliminated and using the continuity equation, all horizontal velocity gradients are expressed into those of the vertical velocity. For $Ra = 0$, the equations (2.13c-d) now become [26]

$$W(1) - \lambda Z = 0$$

$$Cr \left[ Pr^{-1}\lambda - (D^2 - 3k^2) \right] DW(1) + (Bo + k^2)\, k^2 Z = 0$$

$$D^2 W(1) + Ma\, k^2\, (\Theta(1) - Z) = 0$$

$$D\Theta(1) + Bi\, \Theta(1) - Bi\, Z = 0 \tag{2.16}$$

The value of the Crispation number $Cr = \frac{\mu\kappa}{\sigma_0 d}$ (also called Capillary number) gives an indication of the extent of interface deformation from the equilibrium interface. When the equilibrium interface is flat, the limit $Cr \to 0$ is the large mean surface tension limit and therefore corresponds to zero deformation. In this limit, $Z = 0$ and the equations (2.16) reduce to the equations (2.13c-d). The other new parameter is the Bond number $Bo = \frac{\rho_0 g d^2}{\sigma_0}$, measuring the ratio of restoring forces due to gravity and surface tension on the interface.

The linear stability problem with (2.16) was solved in Takashima [26] under the assumption of the principle of exchange of stability and the result is

$$Ma = \frac{8k(k\, cosh\, k + Bi\, sinh\, k)(sinh\, k\, cosh\, k - k)(Bo + k^2)}{8Cr\, k^5\, cosh\, k + (Bo + k^2)(sinh^3\, k - k^3 cosh\, k)} \tag{2.17}$$

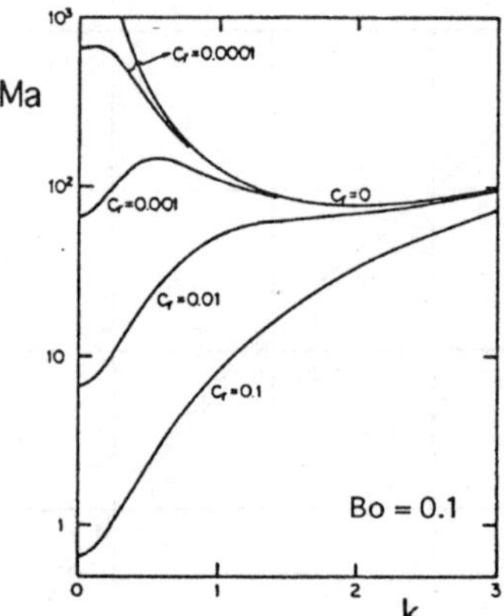



Figure 9: *Neutral stability curves for various values of Cr when $Bi = 0$ and $Bo = 0.1$. The region below each curve represents a stable conduction state (from [26]).*

For the values $Bo = 0.1$ and $Bi = 0$, neutral curves for different values of $Cr$ are plotted in Fig. 9. When $Cr < 8.3 \ 10^{-4}$, the same qualitative results as in Fig. 7 are obtained, i.e. those for $Cr = 0$. However, for $Cr > 8.3 \ 10^{-4}$, the critical wavenumber decreases abruptly from about 2.0 to 0.0 and $Ma_c$ rapidly decreases as $Cr$ increases. Hence, the free interface deformation becomes important for large enough $Cr$, the precise transition value depending on the value of $Bo$.

## 2.4   Discussion

In typical experiments on earth, the vertical temperature difference is increased in small steps. The linear stability analysis above provides sufficient conditions for instability of the motionless solution. In pure Rayleigh-Bénard convection (Ma = 0) with a non-deformable interface, the critical vertical temperature gradient is determined by the value of $Ra_c$. This stability boundary simultaneously provides sufficient conditions for stability, i.e. for $Ra < Ra_c$, the motionless is not only stable to very small perturbations but also to perturbations of arbitrary amplitude [20]. For pure B'enard-Marangoni convection (Ra = 0) with a non-deformable interface, the critical conditions are determined by the value of $Ma_c$. However, in this case this boundary does not provide sufficient conditions for stability. The latter conditions can be estimated by determining nonlinear stability boundaries – which guarantee stability to arbitrary disturbances – i.e. the energy stability boundary ($Ma_e$) [27]. For a flat interface, it is found that $Ma_e < Ma_c$ and hence an interval exists ($Ma_e < Ma < Ma_c$) where motion can be induced if perturbations have sufficiently large amplitude. The effect of interface deformation on the energy stability boundaries has been investigated in Davis and Homsy [28] and Castillo and Velarde [29].

In combined Rayleigh-Bénard-Marangoni convection, critical conditions depend on both parameters $Ra$ and $Ma$ with critical curves such as in Fig. 8. Since both $Ra$ and $Ma$ depend linearly on the temperature gradient, their ratio remains constant. In the $(Ra, Ma)$ plane, a critical line can be drawn, which is characterized by an angle $\psi$, with $tan\,\psi = \frac{Ma}{Ra}$ and scales with $d^{-2}$. For large layer thickness, i.e. small $\psi$, the instability is buoyancy controlled and for small layer thicknesses (or under micro-gravity conditions), the surface tension mechanism will dominate. If the critical curve is crossed, motion will initially set in with convection patterns characterized by a wavenumber $k_c$. All patterns which are solutions of the Helmholtz equation (2.12) are equally unstable. In this way, linear theory is degenerate and does not provide information about the shape of the pattern which is most likely to be observed.

The influence of the free surface becomes important when $Cr$ exceeds a critical value depending on the Bond number $Bo$. If this value is exceeded, long waves are destabilized (the critical wavenumber $k_c$ approaches zero (Fig. 9)). The critical value of $Cr$ decreases with decreasing value of $Bo$, and hence the layer thickness can always be made small enough such that the long wave instability becomes relevant. In the limit $Bo \rightarrow 0$, the motionless solution is always unstable to long waves if $Ma > 0$ and $Cr > 0$, which was a result originally obtained by Sternling and Scriven [24].

The relation between free surface deflections and the vertical movement just below the interface can be used to distinguish between buoyancy driven and surface tension driven mechanisms. For a stationary instability ($\lambda = 0$), the vertical velocity at the interface $W(1) = 0$. However, if $DW(1) < 0$ then the liquid immediately below the interface is moving upwards. Hence, if the ratio $Z/DW(1) > 0$, then there is upflow below interface depressions. For pure surface tension driven convection (with $Ra = 0$), this ratio is given by

$$\frac{Z}{DW(1)} = \frac{2Cr}{sinh^2 k - k^2} \tag{2.18}$$

which is always positive. Hence, there is upflow beneath depressions and downflow beneath elevations. This is just opposite to what is known from pure buoyancy driven convection [7] and very strongly reinforces that the hexagonal patterns observed by Bénard [3] were mainly surface tension driven.


# 3    Pattern selection in the infinite layer

We have seen in the previous section, that motion in a liquid heated from below occurs when a critical vertical temperature gradient is exceeded. Although the critical temperature gradient could be determined exactly, the linear stability problem was degenerated with respect to the shape of the pattern of the flow arising from the instability. Linear theory is not able to predict either the flow pattern nor its final amplitude.

Nonlinear theory is needed to explain how exponentially growing disturbances equilibrate to finite amplitude. In this chapter the particular roll and hexagon patterns are described in more detail (section 3.1). Then the nonlinear equilibration of these patterns is considered within weakly nonlinear theory using perturbation methods (section 3.2) and numerical simulations (section 3.3).

## 3.1   Roll and hexagon patterns

All possible patterns of wavenumber $k$ which were amplified at slightly supercritical conditions satisfied the Helmholtz equation (2.12), i.e.

$$\nabla_H^2 \phi + k^2 \phi = 0 \tag{3.1}$$

The solutions of this equation are given by

$$\phi(x, y) = \sum_{n=-N, n\neq 0}^{N} c_n \phi_n(x, y) \tag{3.2a}$$

with

$$\phi_n(x, y) = e^{i\mathbf{k}_n \cdot \mathbf{r}} \tag{3.2b}$$

where for each $n$, the wavevector $\mathbf{k} = (k_x, k_y)$, $k^2 = |\mathbf{k}_n|^2$ and $\mathbf{r} = (x, y)$. Furthermore, to obtain real solutions $\phi$ there is an additional constraint on the coefficients $c_n$, i.e. (with * indicating complex conjugate)

$$c_{-n} = c_n^* \; ; \quad \sum_{n=-N, n\neq 0}^{N} |c_n|^2 = 1 \tag{3.2c}$$

For $N = 1$, we obtain two-dimensional cellular patterns, which are called parallel roll-cells or simply rolls. For cells whose axis is aligned with the $y$-axis, $\mathbf{k}_{-1} = k(-1, 0)$ and $\mathbf{k}_1 = k(1, 0)$ and $c_{-1} = \frac{1}{\sqrt{2}}$ and $c_1 = \frac{1}{\sqrt{2}}$ such that

$$\phi = c_{-1}\phi_{-1} + c_1\phi_1 \equiv \phi_R(x, y) = \sqrt{2}\, \cos kx \tag{3.3}$$

Hence, the wavenumber of the patterns is $k$, the wavelength is given by $\frac{2\pi}{k}$ and a sketch of the roll pattern is given in Fig. 10a. Obviously a similar pattern exists with rolls whose axis is aligned with the $x$-axis

For the case $N = 3$ the hexagonal pattern is obtained. Here, the wavevectors and the coefficients $c_i$ are given by

$$\mathbf{k}_1 = k(0, 1) \; ; \; \mathbf{k}_2 = \frac{k}{2}(\sqrt{3}, -1) \; ; \; \mathbf{k}_3 = \frac{k}{2}(-\sqrt{3}, -1)$$

$$c_1 = c_2 = c_3 = \frac{1}{\sqrt{6}} \tag{3.4a}$$

with $\mathbf{k}_{-i} = -\mathbf{k}_i$ and $c_{-i} = c_i$. The wavevectors make an angle of $120°$ with eachother and the planform function $\phi_H(x, y)$ becomes

$$\phi \equiv \phi_H(x, y) = \sqrt{\frac{2}{3}} \, \left(2 \cos \frac{\sqrt{3}kx}{2} \cos \frac{ky}{2} + \cos ky\right) \qquad (3.4b)$$

A sketch of this hexagonal pattern is provided in Fig. 10b. Liquid ascends in the center of the hexagon and descends along the hexagonal boundaries.

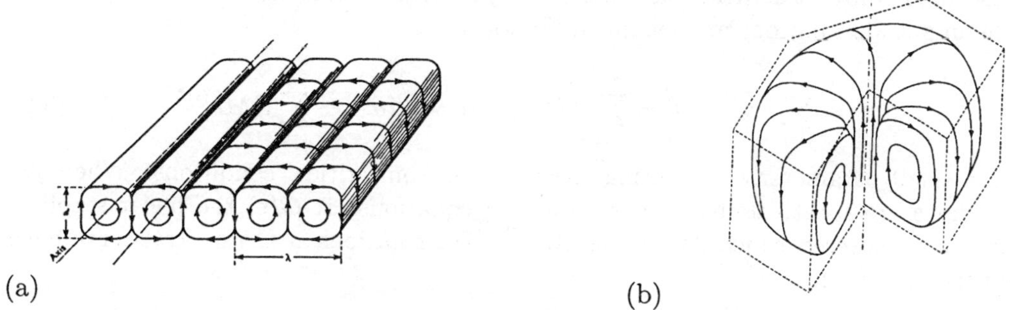

(a)
(b)

Figure 10: *Sketches (from [4]) of a roll pattern (a) with the axis aligned with the y-as and the flow in a single hexagon (b).*

## 3.2 Weakly nonlinear analysis

In experiments it is found that for slightly supercritical conditions, steady flow pattern appear. This motivates to search for steady finite amplitude solutions for parameter values close to the critical conditions by perturbation methods. We first show results with the classical Gorkov-Malkus-Veronis method as in Cloot and Lebon [30] and then continue with a more modern approach using amplitude equations [31]. Both approaches are applied to a model neglecting any interface deformation.

### 3.2.1 Gorkov-Malkus-Veronis method

When the governing equations for deviations with respect to the conduction solution as presented in chapter 2 are non-dimensionalized in the same way as in section 2.2 and the pressure is eliminated, these become

$$Pr^{-1}\left[\frac{\partial}{\partial t}\nabla^2 w - \frac{\partial^2}{\partial x \partial z}(\mathbf{v} \cdot \nabla u) - \frac{\partial^2}{\partial y \partial z}(\mathbf{v} \cdot \nabla v) + \nabla_H^2(\mathbf{v} \cdot \nabla w)\right] = Ra\nabla_H^2 T + \nabla^4 w$$

$$(3.5a)$$

$$\frac{\partial T}{\partial t} + \mathbf{v} \cdot \nabla T = w + \nabla^2 T \qquad (3.5b)$$

There are two additional equations for $u$ and $v$, which can be solved once $w$ is known. The boundary conditions are

$$z = 0 : w = \frac{\partial w}{\partial z} = T = 0 \qquad (3.5c)$$

$$z = 1 : w = 0 \; ; \; \frac{\partial^2 w}{\partial z^2} - Ma\nabla^2_H T = 0 \; ; \; \frac{\partial T}{\partial z} + Bi\, T = 0 \qquad (3.5d)$$

In the analysis, the Rayleigh number is fixed and the Marangoni number is increased to the value of $Ma$ at neutral conditions, say $Ma^{(0)}(k)$. Near neutral conditions, steady finite amplitude solutions are sought of the form

$$\mathbf{v} = \sum_{i=1}^{K} \epsilon^i \mathbf{v}^{(i)} \; ; \; T = \sum_{i=1}^{K} \epsilon^i T^{(i)} \; ; \; Ma = Ma^{(0)} + \sum_{i=1}^{K} \epsilon^i Ma^{(i)} \qquad (3.6)$$

with a small parameter $\epsilon$ measuring the distance from critical conditions. When the expansions are substituted into the governing equations, at $\mathcal{O}(\epsilon)$ the linear stability problem is recovered which determines $Ma^{(0)}$. The solutions at the neutral curve have the particular form

$$(w^{(1)}, T^{(1)}) = (W^{(1)}(z), \Theta^{(1)}(z))\, \phi(x, y) \qquad (3.7)$$

with $\phi$ satisfying (3.1) and $W^{(1)}$ and $\Theta^{(1)}$ satisfying the equations (2.13). The minimum of the neutral curve is again the critical value, i.e. $Ma^{(0)}(k_c) = Ma_c$.

At $\mathcal{O}(\epsilon^2)$, the following set of equations is obtained

$$\nabla^4 w^{(2)} + Ra\nabla^2_H T^{(2)} =$$

$$Pr^{-1}\left[\nabla^2_H(\mathbf{v}^{(1)} \cdot \nabla w^{(1)}) - \frac{\partial^2}{\partial x \partial z}(\mathbf{v}^{(1)} \cdot \nabla u^{(1)}) - \frac{\partial^2}{\partial y \partial z}(\mathbf{v}^{(1)} \cdot \nabla v^{(1)})\right] \qquad (3.8a)$$

$$w^{(2)} + \nabla^2 T^{(2)} = \mathbf{v}^{(1)} \cdot \nabla T^{(1)} \qquad (3.8b)$$

and the tangential stress balance becomes (at $z = 1$)

$$\frac{\partial^2 w^{(2)}}{\partial z^2} - Ma^{(0)}\nabla^2_H T^{(2)} = -Ma^{(1)}\nabla^2_H T^{(1)} \qquad (3.8c)$$

The other boundary conditions in (3.5c,d) remain the same at each order of approximation.

The linear operator in the left hand side of (3.8) is singular, because from the linear stability problem we know that it has non-trivial solutions. For existence of a solution to (3.8), the right hand side has to be orthogonal to the eigensolutions of the adjoint linear operator. This part is quite technical, but it can be shown that this orthogonality condition (Fredholm alternative) leads to a condition on $Ma^{(1)}$, which is different for each pattern characterized by the number $N$ in (3.2a). More specific, $Ma^{(1)} = 0$ for 2D-rolls ($N = 1$) (and also for 3D-rolls ($N = 2$)) but nonzero for hexagonal patterns

($N = 3$). Under the condition of values of $Ma^{(1)}$ as obtained above, the second order solutions $(w^{(2)}, T^{(2)})$ can be obtained for each $N$. Hence, in the weakly nonlinear domain, stationary solutions of both rolls and hexagons exist.

The next step is to determine the stability of each of these steady solutions to arbitrary perturbations. Let $(\tilde{\mathbf{v}}, \tilde{T})$ denote these perturbations, then their evolution equations are obtained by linearizing the governing equations around the finite amplitude steady states $(\mathbf{v}, T)$ from (3.6). These linear equations admit solutions proportional to $e^{\tilde{\lambda} t}$. Because the steady state is expanded in terms of $\epsilon$, also the perturbations can be expanded in the same parameter, i.e.

$$\tilde{\mathbf{v}} = \sum_{i=0}^{K} \epsilon^i \tilde{\mathbf{v}}^{(i)} \; ; \; \tilde{T} = \sum_{i=0}^{K} \epsilon^i \tilde{T}^{(i)} \; ; \; \tilde{\lambda} = \sum_{i=0}^{K} \epsilon^i \tilde{\lambda}^{(i)} \tag{3.9}$$

The eigenvalues $\tilde{\lambda}^{(i)}$ are determined by a similar analysis as that to determine $Ma^{(0)}$ and are evaluated for three different cases by Cloot and Lebon [30]. First, disturbances with the same wavenumber as the neutral state are considered, but having different wave vectors. In this way, one determines the stability of hexagons to the same hexagonal pattern which may be slightly shifted over the domain. Second, the stability of hexagons to rolls was investigated, each pattern still having the same wavenumber. Finally, the stability of the hexagonal pattern to perturbation patterns (either rolls or/and hexagons) with different wavenumbers are considered. The intersection of each stability domain gives the final stability domain of a certain pattern.

The results for $Bi = 0$ and $Pr = 7$ are summarized in Fig. 11 over the range of $Ra$ considered. For $Ra = 669$ ($Ma_c = 0$), an area with wavenumbers larger than the critical one exists for which hexagons are stable. It is well-known that stable rolls also exist with wavenumbers larger than the critical one [32]. When $Ra$ is decreased, the critical value $Ma_c$ increases and at $Ra = 500$, the band of stable hexagons has moved towards smaller wavenumber than the critical one. Finally, in pure surface tension driven convection ($Ra = 0$), it is found that the band of stable wavenumbers is always larger than $k_c$ and that also under subcritical conditions, finite amplitude hexagonal patterns are stable. This is in agreement with results found in a simpler model by Scanlon and Segel [33].

It is observed that regions of stable hexagons having wavenumbers larger as well as smaller than the critical wavenumber appear, depending on the value of $Ra$. For typical experiments [1], the Rayleigh number is relatively small (about 100) and the wavenumbers found do not agree with the predicted wavenumbers using weakly nonlinear theory – which are too large – but the theory correctly predicts smaller cells. The stability results on the hexagons do not qualitatively change when $Bi$ or $Pr$ are increased. Only the values of the $Ma$ numbers shift, but the location of the stable band of wavenumbers with respect to the critical wavenumber remains the same.

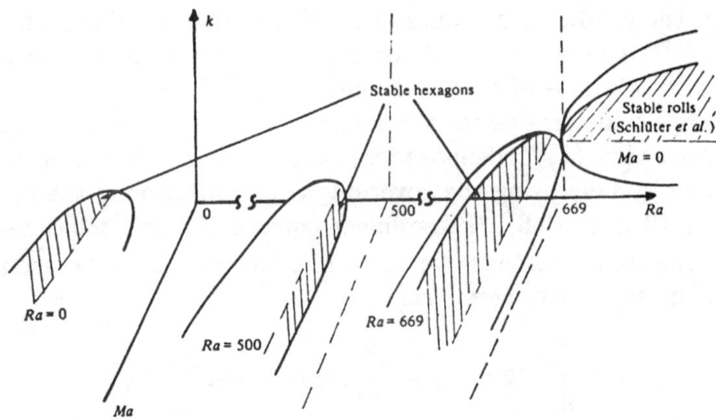


Figure 11: *Composite plot of the wavenumber range of stable hexagons in the $(Ma, Ra)$-plane (from [30]).*

## 3.2.2   The generalized Ginzburg-Landau equations

A more general approach has been followed by Bestehorn [31]. For the case of $Pr \to \infty$ and $Bi = 0$, the weakly nonlinear structure of solutions is determined near a point on the critical line $(Ma_c, Ra_c)$ as in Fig. 8. The quantities are expanded into series of eigenfunctions of the linear stability problem and Galerkin projection is used to obtain an infinite set of ordinary differential equations governing the amplitudes of the particular eigenfunction components. This set of equations is reduced by the principle of adiabatic elimination, i.e. putting all time derivatives to zero for all modes which are sufficiently strongly damped. The amplitudes of the latter 'slaved' modes are expressed into those of the dynamically active ones, resulting (near criticality) in simple amplitude equations for specifically chosen patterns. The stability of these patterns to a general class of disturbances is then considered.

The main results of the analysis are shown in Fig. 12 showing plots of a measure of supercriticality ($\epsilon$) versus the wavenumber $k$. All drawn curves in the figures indicate stability boundaries for different type of instabilities and the dashed curve indicates the wavenumber which has a maximal growth rate according to linear theory. The lower drawn curve is the neutral stability boundary. In Fig. 12a, $\psi = 0$ (defined by $tan\ \psi = Ma/Ra$) and there is only Rayleigh-Bénard convection. There are no stable hexagons, but only rolls are stable within the wavenumber range defined by the downward hatched area. As the angle $\psi$ is increased and surface tension gradients contribute to the development of the flow, areas of stable hexagonal patterns appear (upward hatched areas in Fig. 12b and 12c). For even larger $\psi$, the region of stable rolls disappears and only stable hexagons appear (Fig. 12d). The black dots in this figure represent the data from the experiments in Koschmieder and Switzer [1].

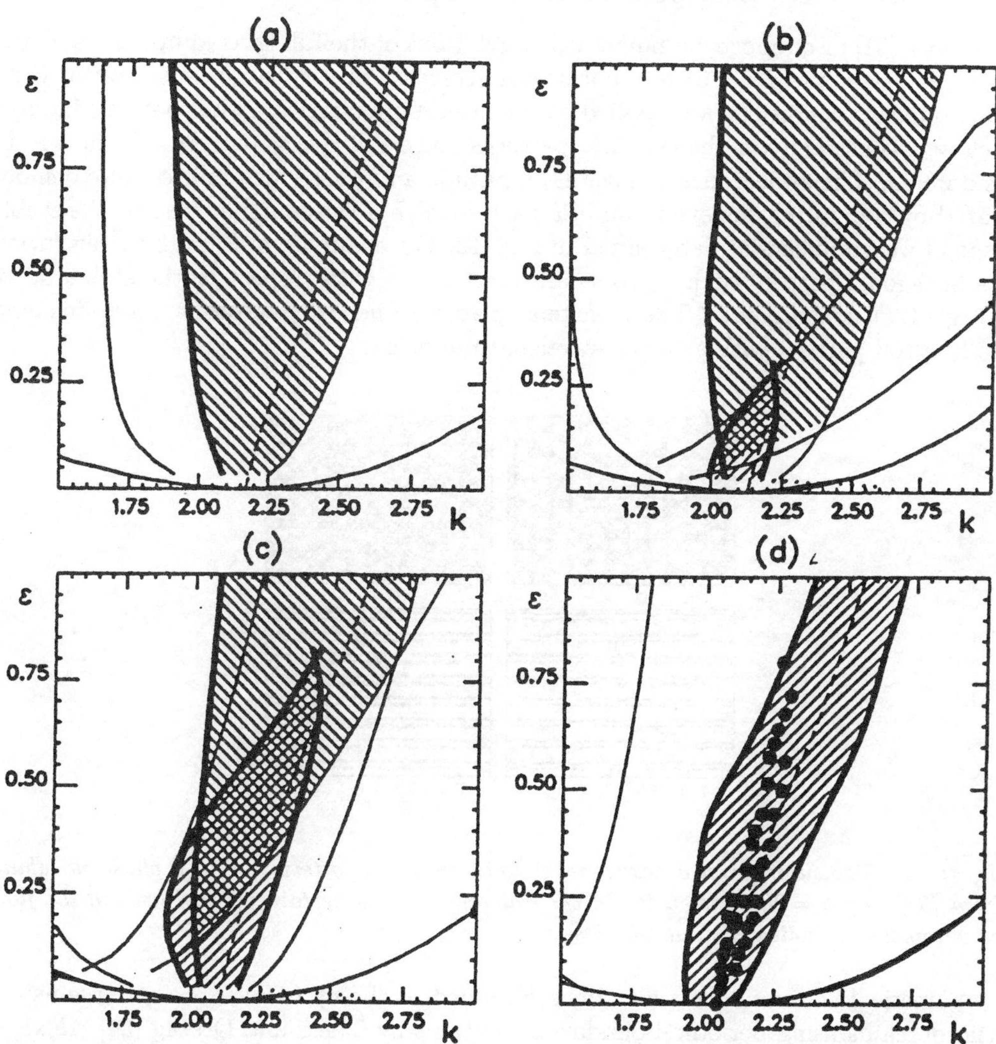

Figure 12: *Domain of stable rolls (hatched towards southeast) and hexagons (hatched towards southwest) as determined from the analysis of Bestehorn [31] for different angles (in degrees) of $\psi = arctan\ (Ma/Ra)$. a) $\psi = 0$  ;  b) $\psi = 10$  ;  c) $\psi = 20$  ;  d) $\psi = 70$ In panel d), the dots are experimental results from [1].*

## 3.3   Numerical computations of supercritical flows

Bestehorn [31] also reports on numerical simulations of the full three-dimensional equations ($Pr \to \infty$ and $Bi = 0$) in a horizontal geometry of very large aspect ratio. Periodic boundary conditions are applied at the lateral boundaries. Note that this induces already a constraint on the possible patterns since not every pattern will fit within the domain. The appearance of several amplitude and phase instabilities were demonstrated by computing the evolution of a pattern with a wavenumber outside the stable range of wavenumbers as determined in Fig. 12. For example, in Fig. 13 the evolution of a hexagonal pattern with a too small wavelength (with respect to the stable range in Fig. 12c) is computed. The hexagonal pattern undergoes a phase instability and finally a roll-pattern with a larger wavelength appears.

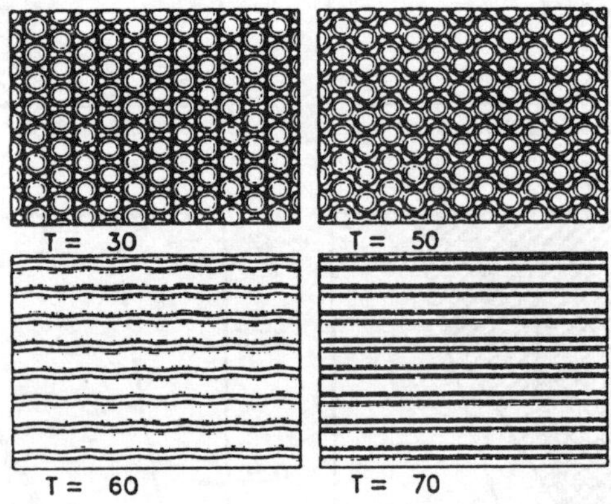


Figure 13: *Transition from a hexagonal pattern to a roll pattern due to a phase instability (from [31]) for $\psi = 20$. Shown is the vertical velocity at a certain vertical level of the flow, the parameter $T$ indicates dimensionless time.*

Numerical solutions have also been obtained in the case $Pr \to \infty$ in large aspect ratio domains using periodic boundary conditions by Thess and Orszag [34]. Hexagonal patterns are stable already below onset conditions, just as determined in Cloot and Lebon [30]. Moreover, although roll patterns are found, Thess and Orszag [34] claim that the hexagonal pattern is preferred because it is found for random initial conditions. Computations done in a lateral strip support an increase of wavenumber of the hexagonal pattern with increasing $Ma$, just as in experiments [1]. For even larger $Ma$, patterns appear with imperfections and finally become time-dependent. This strongly nonlinear regime has also been investigated by Thess and Orszag [34], but will not be further described here.

## 3.4   Discussion

Weakly non-linear theory for a horizontally unbounded layer of liquid (e.g. [31]) gives results which are in good agreement with the experimental results (e.g. [1]). The theory shows (Fig. 12d) that for Bénard-Marangoni convection (small $Ra$), hexagonal patterns are stable steady states over quite a range of $Ma$. Hexagonal convection is already stable below critical conditions as determined by linear stability theory, showing that just below the critical temperature gradient, finite amplitude perturbations can induce motion in the liquid. This is in agreement with the difference in linear and energy stability bounds, as discussed in section 2.4.

The theory clearly indicates a preference for hexagonal convection in pure Bénard-Marangoni convection and preference for rolls in pure Rayleigh-Bénard convection. In each case, a band of wavenumbers of the patterns seems to be allowed, which is bounded by (secondary) instabilities of the patterns. In Fig. 12d, it is clear that the center of this band shifts to larger wavenumber (i.e. to smaller wavelength), if $\epsilon$ is increased. This is in agreement with experimental results [1] where smaller cells are found for increasing temperature gradient.

Despite this success, there are two issues which remain a bit unsatisfactory. First issue is that there is no clear physical mechanistic view why the flow system displays this behavior. For example, why is the hexagonal pattern preferred in experiments and theory for pure Bénard-Marangoni convection ? The answer from the theory that this pattern is more stable than other patterns asks for a clarification of the physical mechanisms of the secondary instabilities of the patterns. For example, one would like to have a physical mechanism of the transition seen in Fig. 13, where a hexagonal pattern evolves into a roll pattern due to a secondary instability. Although much work on these secondary instabilities has been done for simpler patterns (e.g. rolls) similar analyses for the hexagonal pattern seem to be lacking.

A second issue is that the band of allowed wavenumbers in weakly nonlinear theory seems to be larger than that found by experiments (e.g. Fig. 12d). Since the wavelength is (within experimental error) independent of how the temperature gradient is increased (Fig. 5), it is viewed as unique [4] by experimentalists. One reason of the restriction of the band of allowed wavenumbers may be the presence of lateral walls in experiments. The lateral walls, although far away, may have a strong influence on localizing the band of stable wavenumbers. The effect of these lateral walls is therefore considered next.

# 4   The influence of lateral walls

From pictures such as Fig. 2, one observes that the cell pattern at the sidewalls is different from that in the interior of the pattern. Boundary cells are present in which the velocity approaches zero at the rigid wall. The question is therefore how these boundary cells influence the band of stable hexagons as known from the theory for the infinite layer. The influence of the presence of rigid sidewalls is considered in more detail in this chapter. In section 4.1 experimental results are reported on Bénard-Marangoni convection in relatively small containers for which it is certain that the lateral walls strongly determine the possible patterns. In section 4.2, linear and weakly nonlinear theory is presented for the case where the sidewall boundary conditions are idealized as being 'slippery'. In that case, each sidewall is a streamline of the flow but the tangential velocity is nonzero. The real 'rigid' sidewall case, where the tangential velocity also vanishes on the lateral walls is presented in section 4.3.

## 4.1  Experimental results

When lateral walls are considered, another parameter enters the problem, i.e. the aspect ratio $A$ which is defined as the ratio of liquid depth and a typical horizontal length scale of the liquid layer. In three dimensions and rectangular boxes, actually two aspect ratios ($A_x$ and $A_y$) appear. It has proven interesting to approach the subject of pattern formation from the small aspect ratio regime. Motivation for a number of studies have been the beautiful patterns obtained experimentally by Koschmieder and Prahl [35]. These were performed in square containers and cylindrical containers in the aspect ratio range covering $A = 4 - 10$ (compare this to the value $A \approx 50$ in the experiments of Koschmieder and Switzer [1]). The critical $Ma$-numbers found experimentally approach $Ma_c = 79.8$ for $A \to \infty$. For small $A$ the value of $Ma_c$ increases rapidly because the box becomes too small to allow wavenumbers of perturbations for which an optimal amplification, according to both mechanisms in section 1, can occur.

For the horizontally square boxes and small $Bi$ a number of steady patterns found are shown in Fig. 14. In Fig. 14a ($A = 1.82$), a 1-cell solution is found with liquid coming up at the center and flowing down near the walls. This pattern is found over quite an aspect ratio range up to $A = 4$. At $A = 5.68$, a 2-cell pattern is obtained with liquid flowing downward along a diagonal of the box (Fig. 14b). The pattern at $A = 6.48$ is really remarkable since it consists (Fig. 14c) of one cell filling a quarter of the box together with two wedge-shaped symmetrically oriented cells filling up the rest of the box. At larger aspect ratio, a four cell pattern appears (Fig. 14d) for $A = 6.48$, a five cell for $A = 8.4$ (Fig. 14e), a strange six cell pattern at $A = 8.1$ (Fig. 14f) and finally an eight cell pattern at $A = 8.75$ (Fig. 14g).

## 4.2  Slippery Sidewalls

Several studies have tackled the problem of pattern selection in small aspect ratio containers. In Rosenblat et al. [36], the sidewalls are assumed to be stress free or 'slippery', implying that apart from kinematic conditions, the tangential stress vanishes at each sidewall (instead of the tangential velocity). Furthermore, there is no heat flux through the sidewalls. Hence, if the walls are located at $x = 0, L_x$ and $y = 0, L_y$ and length is non-dimensionalized with the liquid depth $d$, these boundary conditons become (with $A_x = \frac{L_x}{d}$ and $A_y = \frac{L_y}{d}$)

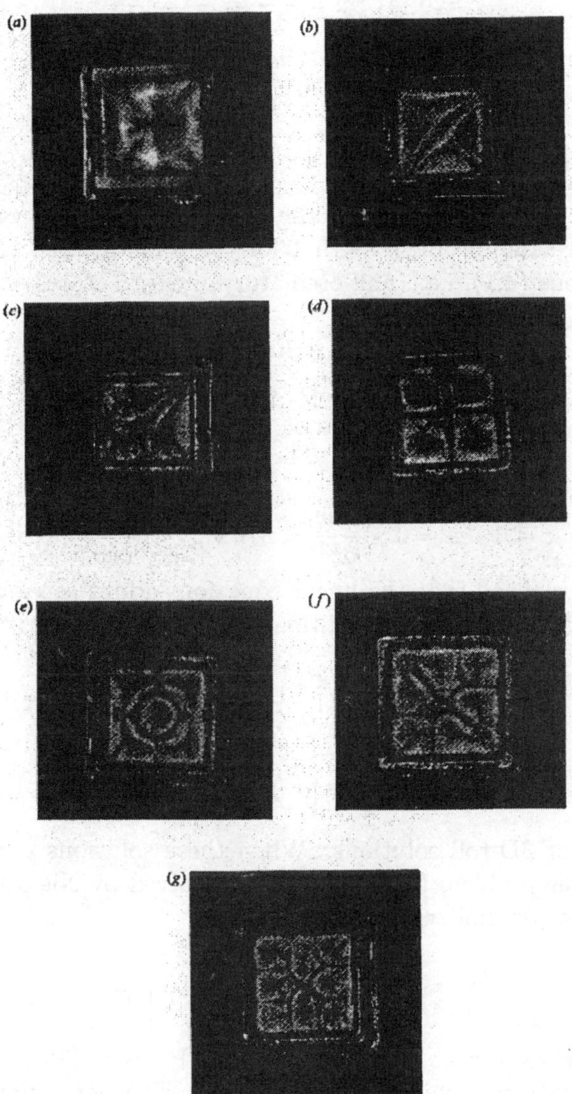

Figure 14: *Steady flow patterns as obtained by Koschmieder and Prahl [35] in horizontally square containers. a) Ma = 380, Ra = 228, A = 1.82.   b) Ma = 54, Ra = 33, A = 5.68.   c) Ma = 80, Ra = 42, A = 6.18.   d) Ma = 78, Ra = 38, A = 6.36.   e) Ma = 67, Ra = 19, A = 8.40.   f) Ma = 72, Ra = 22, A = 8.08.   g) Ma = 63, Ra = 16, A = 8.75.*

$$x = 0, A_x : u = \frac{\partial w}{\partial x} = \frac{\partial v}{\partial x} = \frac{\partial T}{\partial x} = 0$$

$$y = 0, A_y : v = \frac{\partial w}{\partial y} = \frac{\partial u}{\partial y} = \frac{\partial T}{\partial y} = 0 \tag{4.1}$$

### 4.2.1 Linear Stability

When the linear stability problem of the conduction state is considered (only for a flat interface), the same problem as that for the infinite layer appears, but now with boundary conditions (4.1), i.e. (non-oscillatory) neutral states are determined by

$$\nabla^4 w + Ra\nabla_H^2 T = 0 \tag{4.2a}$$

$$\nabla^2 T + w = 0 \tag{4.2b}$$

$$z = 0 : T = w = \frac{\partial w}{\partial z} = 0 \tag{4.2c}$$

$$z = 1 : w = 0 \;;\; \frac{\partial^2 w}{\partial z^2} = Ma\nabla_H^2 T \;;\; \frac{\partial T}{\partial z} = -Bi\, T \tag{4.2d}$$

It is immediately realized that the linear equations admit separable solutions – which in addition satisfy the boundary conditions (4.1) – of the form

$$w(x,y,z) = \cos\left[\frac{m_1\pi x}{A_x}\right] \cos\left[\frac{m_2\pi y}{A_y}\right] W(z)$$

$$T(x,y,z) = \cos\left[\frac{m_1\pi x}{A_x}\right] \cos\left[\frac{m_2\pi y}{A_y}\right] \Theta(z) \tag{4.3}$$

representing 2D- or 3D-roll solutions. When these solutions are substituted into the equations, a similar problem for $(W, \Theta)$ as that solved by Nield [23] appears, but now with an effective wavenumber $k$ given by

$$k^2 = \left[(\frac{m_1}{A_x})^2 + (\frac{m_2}{A_y})^2\right] \pi^2 \tag{4.4}$$

Consequently, the results as in Nield [23] and Pearson [9] carry over but only for wavenumbers of patterns which fit into the box. The particular roll-patterns can be identified by the two integers $(m_1, m_2)$. For the case $Ra = Bi = 0$, the neutral stability curves for the most unstable modes are shown in Fig. 15 as a function of $A_x$ for a fixed value of $A_y = 1.0$. In this case, the rolls aligned with the $y$-axis are most unstable (only the modes $(m_1, 0)$, with increasing $m_1$ for larger $A_x$). Hence, for $1.0 < A_x < 2.2$, the $(1, 0)$ mode is most unstable followed by an interval $(2.2 < A_x < 3.8)$ for which the $(2, 0)$ mode is most unstable. For larger values of $A_y$, also 3D-roll patterns may become involved and a summary of these results is given in Fig. 16, which shows the areas of the particular critical modes in aspect ratio space $(A_x, A_y)$. As can be seen,

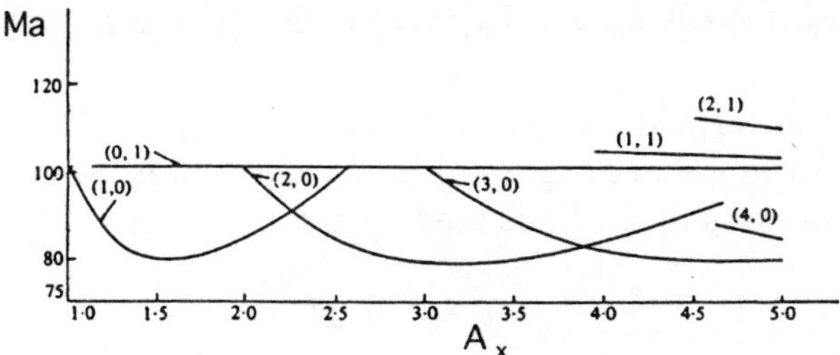

Figure 15: *Neutral stability curves for roll patterns* $(m_1, m_2)$ *in rectangular containers of aspect ratio* $A_x$, $A_y = 1.0$ *for* $Bi = 0$ *(from [36]).*

there exist values of the aspect ratios for which two modes are simultaneously critical, for example the $(1, 0)$ and $(2, 0)$ modes at $A_x = 2.2$, $A_y = 1.0$ (Fig 15).

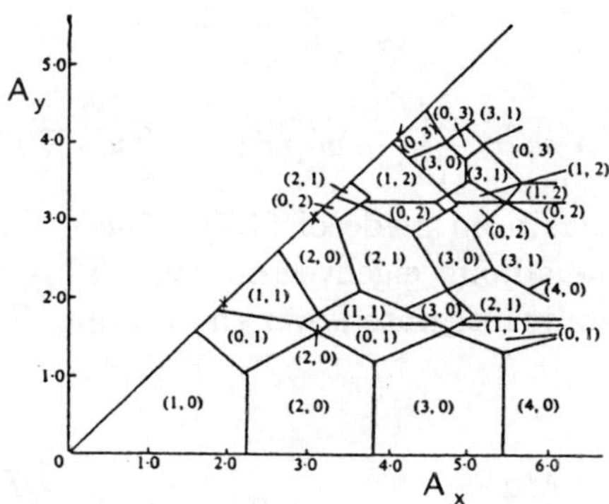

Figure 16: *Composite plot showing the critical modes* $(m_1, m_2)$ *at onset in the aspect ratio space* $(A_x, A_y)$ *(from [36]).*

### 4.2.2   Weakly nonlinear theory

To determine finite amplitude convection patterns, Rosenblat et al. [36] follow an eigenfunction expansion procedure, using the eigenfunctions of the linear stability problem. It is applied to the case $A_y = 1.0$ such that only $x$-rolls are relevant (Fig. 16) at onset. As eigenfunctions, they use those corresponding to the critical values of Ra (with notation $Ra_{mj}$) for fixed $Ma$, because these eigenfunctions constitute a complete set (Rosenblat et al. [37]). The eigenfunctions are indicated by

$$w_{mj}(x, z) = W_{mj}(z)\ cos\ k_m\ x\ ;\ T_{mj}(x, z) = \Theta_{mj}(z)\ cos\ k_m\ x\ ;\ k_m = \frac{m\pi}{A_x} \tag{4.5}$$

where $j$ indicates the vertical wavenumber and $m = m_1$.

All dependent variables are expanded into a horizontal mean (indicated by a bar) and a deviation from this mean (indicated by a prime), e.g.

$$T = \overline{T} + T' \tag{4.6}$$

The evolution equations for the averaged and primed quantities are then derived for the case $Ra = Bi = 0$. Since the mean velocity field introduced by convection is zero ($\overline{\mathbf{v}} = 0$) the equations for the mean field reduce to

$$\frac{\overline{\partial T}}{\partial z} = \overline{w'T'} \tag{4.7}$$

and those for the fluctuations to

$$\nabla^2\mathbf{v}' - \nabla p' = Pr^{-1}(\frac{\partial \mathbf{v}'}{\partial t} + [(\mathbf{v}' \cdot \nabla \mathbf{v}')]') \tag{4.8a}$$

$$\nabla \cdot \mathbf{v}' = 0 \tag{4.8b}$$

$$\nabla^2 T' + w' = \frac{\partial T'}{\partial t} + w'\overline{(w'T')} + [(\mathbf{v}' \cdot \nabla T')]' \tag{4.8c}$$

where to obtain (4.8) use has been made of (4.7). The scalar product $<,>$ is taken of (4.8a) and (4.8c) with the adjoint eigenvectors $(\mathbf{v}^*_{mj}, T^*_{mj})$ at $Ma = Ma_c$ and $Ra = Ra_{mj}$ and thereafter the sum is integrated over the volume. This gives for each $m$ and $j$,

$$\frac{(Ma - Ma_c)}{Ma} < T^*_{mj}w' > -\frac{Ma}{Ma_c} Ra_{mj} < w^*_{mj}T' > = < T^*_{mj}\frac{\partial T'}{\partial t} + Pr^{-1}\mathbf{v}^*_{mj} \cdot \frac{\partial \mathbf{v}'}{\partial t} > +$$

$$< T^*_{mj}\left[w'\ \overline{(w'T')} + [(\mathbf{v}' \cdot \nabla T')]'\right] + Pr^{-1}\mathbf{v}^*_{mj} \cdot [(\mathbf{v}' \cdot \nabla \mathbf{v}')]' > \tag{4.9}$$

To determine finite amplitude solutions near criticality for a simple eigenvalue $Ma_c$, where one mode goes unstable, quadratic interactions of this mode determine the class of modes for which the amplitude is considered through (4.9). For example, for $A_x = 1.0$, the mode $(1, 0)$ is critical at $Ma_c = 79.8$. At this point $R_{11} = 0$ by construction and the quadratic interaction of this eigenmode generates a response in its first harmonic represented by the eigenmode associated with $R_{21} \neq 0$. The class of modes considered is therefore $\{11, 21\}$ and the dependent variables are expanded into

$$(\mathbf{v}', T') = A_1(\mathbf{v}_{11}, T_{11}) + A_2(\mathbf{v}_{21}, T_{21}) \tag{4.10a}$$

Equation (4.9) is used to obtain equations for the amplitudes $A_1$ and $A_2$. In a final step it is realized that the second mode decays rapidly, such that it is slaved to the first mode. Its time derivative is put equal to zero and an amplitude equation of the form

$$a_1 \dot{A}_1 = (Ma - Ma_c)A_1 - a_2 A_1^3 \qquad (4.10b)$$

results, with coefficients $a_i$ depending on the parameters of the problem [36].

### 4.2.3   Bifurcation theory

The amplitude equation (4.10b) is a typical example of problems studied in bifurcation theory. Many introductory textbooks to bifurcation theory and its applications are available nowadays (Kuznetsov [38]; Nayfeh and Balachandran [39]). A bifurcation diagram is a graph in which the variation of the solutions of a particular problem are displayed in the state-control space.

The simplest type of bifurcations are those which involve only one parameter in the system under study and are therefore called codimension-one bifurcations. An example of such a simple bifurcation occurs in the one-dimensional autonomous dynamical system

$$x' = f(x, \mu) = \mu x - x^3 \qquad (4.11)$$

where $x$ is the state variable and $\mu$ the control parameter and the prime indicates the time-derivative. For $\mu < 0$, there is only one stationary solution (or fixed point) $\bar{x} = 0$, but for $\mu > 0$, three fixed points exist, i.e. $\bar{x} = 0$, $\bar{x} = \mu^{\frac{1}{2}}$ and $\bar{x} = -\mu^{\frac{1}{2}}$. Hence, the number of fixed points changes as $\mu$ crosses zero. To determine the stability of the fixed points, the sign of the derivative $\frac{\partial f}{\partial x} = \mu - 3x^2$ must be considered at each of the fixed points. For $\bar{x} = 0$, it follows that $\frac{\partial f}{\partial x} = \mu$ indicating that $\bar{x} = 0$ is stable for $\mu < 0$ but unstable for $\mu > 0$. At both additional fixed points existing for $\mu > 0$, it follows that $\frac{\partial f}{\partial x} = -2\mu$, showing that these are both stable.
The bifurcation diagram of the equation (4.11) is shown in Fig. 17a as a graph of $x$ against $\mu$ where stable fixed points are drawn, while unstable states are dashed. At $\mu = 0$, the system undergoes a qualitative change, since the number of fixed points changes from one to three. A bifurcation occurs which is called a pitchfork bifurcation. Two other bifurcation diagrams, those for the transcritical bifurcation and the saddle node bifurcation are shown in Fig. 17b and Fig. 17c. In the caption the simplest one dimensional systems (as in (4.11)) which exhibit such a bifurcation are shown.

Whereas in the previous bifurcations, the number of fixed points changed as a parameter was varied, it is also possible that the character of the solution changes from stationary to oscillatory as a single parameter is changed. An example of a simple dynamical system undergoing such a transition is the two-dimensional autonomous system given by

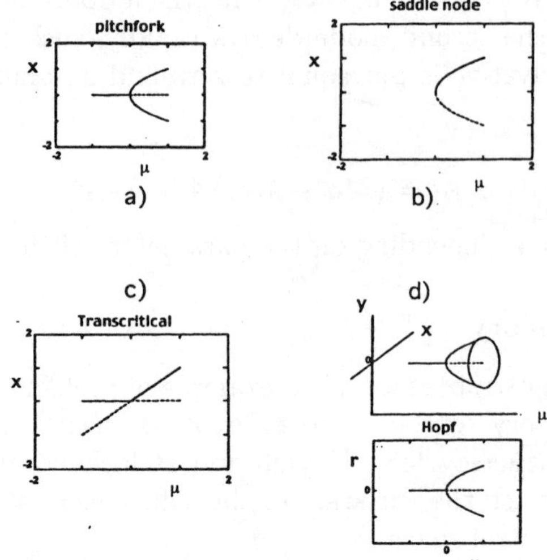

Figure 17: *Overview of the four possible codimension-one bifurcations. a) pitchfork bifurcation, $x' = f(x,\mu) = \mu x - x^3$. b) limit point or saddle node, $x' = f(x,\mu) = \mu - x^2$. c) transcritical bifurcation, $x' = f(x,\mu) = \mu x - x^2$. d) Hopf bifurcation.*

$$x' = \mu x - \omega y - x(x^2 + y^2) \tag{4.12a}$$

$$y' = \mu y + \omega x - y(x^2 + y^2) \tag{4.12b}$$

By the transformation $x = r\,cos\,\theta, y = r\,sin\,\theta$ it is transformed into

$$r' = \mu r - r^3 \tag{4.12c}$$

$$\theta' = \omega \tag{4.12d}$$

Comparing the first equation in (4.12c) with (4.11), it is observed that a pitchfork bifurcation occurs at $\mu = 0$ in the $(r, \mu)$ plane. For $\mu < 0$, only one stable fixed point exists, which corresponds to a stationary solution of the original equations. However, for $\mu > 0$ the stable nontrivial fixed points now correspond to a periodic solution of the original equations (4.12a,b) with a frequency $\omega$, according to (4.12d). The bifurcation diagram for this case is shown in Fig. 17d. At $\mu = 0$, the system undergoes a qualitative change and the bifurcation is called a Hopf bifurcation. More complicated bifurcations may arise as more than one parameter in the system is changed. For example, codimension-two bifurcations may arise through intersection of two codimension-one bifurcations as a second parameter is varied.

### 4.2.4   Competition of rolls

For the amplitude equation (4.10b) there is only one solution (the conduction solution) for $Ma < Ma_c$, which looses stability at $Ma = Ma_c$. For $Ma > Ma_c$, two symmetry related solutions exist, each having one convection cell (corresponding to the $(1,0)$ mode), but which rotate each in opposite direction: a pitchfork bifurcation occurs. In Rosenblat et al. [36], also two cases are analysed where linear stability theory is ambiguous, i.e. at a value of a certain aspect ratio, two modes become unstable at the same conditions. For example at $A_x^d = 2.2$, the modes $(1,0)$ and $(2,0)$ both become unstable at $Ma_c = 90.2$. Because the $(n,0)$ mode is again represented by the eigenfunction corresponding to $R_{n1}$, the nonlinear interactions of both modes give rise to contributions from eigenfunctions corresponding to $R_{21}$ and $R_{41}$ and moreover their mutual interaction gives rise to the one for $R_{31}$. Hence the class of modes is $S = \{11, 21, 31, 41\}$ and the solutions are sought through

$$(\mathbf{v}', T') = \sum_{i=1}^{4} B_i(\mathbf{v}_{i1}, T_{i1}) \tag{4.13a}$$

After adiabatic elimination of the modes 31 and 41, the resulting amplitude equations become

$$\dot{B}_1 = b_1 \left[ \Gamma - (\theta + 1)\Delta \right] B_1 - b_2 B_1 B_2 - b_3 B_1^3 - b_4 B_1 B_2^2$$

$$\dot{B}_2 = c_1 \left[ \Gamma + (1 - \theta)\Delta \right] B_2 - c_2 B_1^2 - c_3 B_2^3 - c_4 B_1^2 B_2 \tag{4.13b}$$

where again coefficients $b_i, c_i$ depend on parameters. Here, $\Delta = \frac{1}{2}(Ma_1 - Ma_2)$, with $Ma_i$ being the value of $Ma$ for which the $(i, 0)$ mode becomes unstable, $\Gamma = Ma - Ma_c$ and $\theta = sign(A_x - A_x^d)$.

For $A_x$ slightly larger than 2.2, i.e. $A_x = 2.4$, Fig. 15 shows that the $(2,0)$ mode becomes unstable followed by the $(1,0)$ mode at slightly larger $Ma$. The bifurcation diagram is shown in Fig. 18 (for $Pr = 10$ and $Bi = 0$) with the amplitude $B_1$ in Fig. 18a and $B_2$ in Fig. 18b. Both instabilities are related to the primary bifurcation points $P_{01}$ and $P_{02}$. At $P_{01}$ the two-cell patterns destabilize the conduction solution and both patterns with upflow and downflow in the center (both signs of $B_2$ in Fig. 4.5b) are stable steady states. However, as the pattern with center upflow remains stable, the one with downflow is destabilized at a secondary pitchfork bifurcation $S_{01}$. Along the branch connecting $S_{01}$ and $P_{02}$ the pattern changes from a two-cell downflow to a one cell pattern (the $(1,0)$ mode), which indeed bifurcates from $P_{02}$. Along this connecting branch, a Hopf bifurcation $H_0$ occurs giving rise to oscillatory behavior. The bifurcation diagrams as sketched above were recomputed by Dijkstra [40], using the full nonlinear equations and it was found that the bifurcation diagram in Fig. 18 has a large domain of validity.

An extension of the results of Rosenblat et al. [36] has been given by Dauby et al. [41]. They realized that the aspect ratios can be chosen such that two roll solutions can become unstable simultaneously and their linear combination is able to represent a

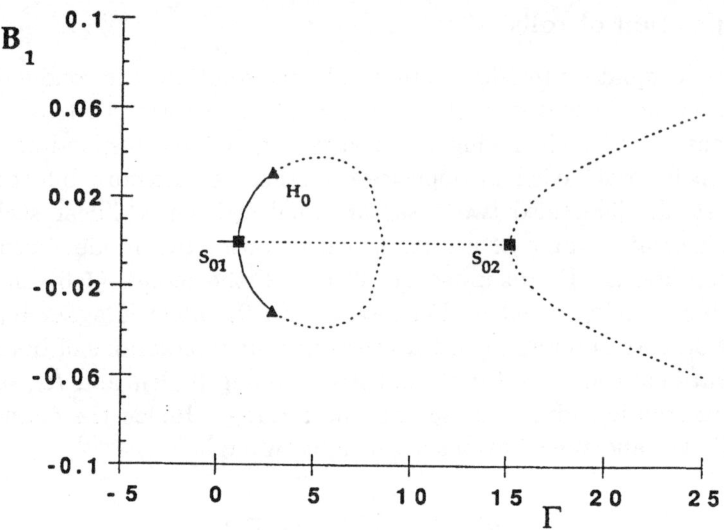

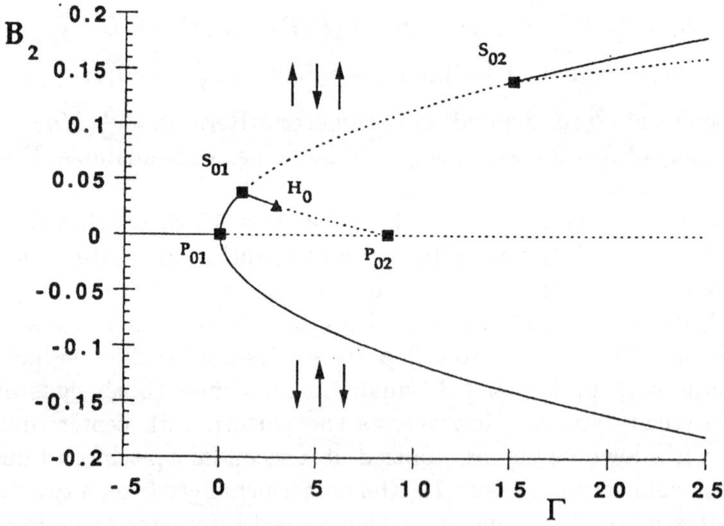

Figure 18: *Bifurcation diagram of the amplitude equations (4.13b), (after [36] and [40]). The parameter $\Gamma$ measures the distance from onset. Drawn (dotted) branches indicate stable (unstable) steady states and bifurcation points are indicated by markers. In the lower panel, the two cell solutions are indicated by the arrows along the main branches. For example, along $P_{01} - S_{02}$, the steady state consists of a two cell solution with downflow in the center.*

hexagonal pattern. This hexagonal pattern is a solution of the linear stability problem
of the unbounded case and given by

$$w = (Y \cos \frac{1}{2}\sqrt{3}\, kx \, \cos \frac{1}{2}\, ky + Z \cos ky)\, W(z)$$

$$T = (Y \cos \frac{1}{2}\sqrt{3}\, kx \, \cos\frac{1}{2}\, ky + Z \cos ky)\, \Theta(z) \qquad (4.14a)$$

If the aspect ratios $A_x$ and $A_y$ are chosen according to

$$\frac{A_x}{A_y} = \frac{m_1}{m_2\sqrt{3}} \qquad (4.14b)$$

then the modes $(m_1, m_2)$ and $(0, 2m_2)$ have the same effective wavenumber (4.4),

$$k = \frac{2m_1\pi}{\sqrt{3}A_x} = \frac{2m_2\pi}{A_y} \qquad (4.14c)$$

and correspondingly the same value of $Ma_c$. Superposition of these modes exactly leads
to the hexagonal flow pattern defined by (4.14a), while simultaneously this combination
satisfies the conditions of 'slippery' boundaries. Dauby et al. [41] applied the same
method as in Rosenblat et al. [36] to obtain amplitude equations for the amplitudes
$Y$ and $Z$ in (4.14a). One of the cases considered is $A_x = 3.49, A_y = 3.02$ such that the
(2,1) and (0,2) modes interact. For $Pr \to \infty$ and $Bi = 0$, hexagonal convection is
again stable below onset and for supercritical conditions both rolls and hexagons are
stable.

## 4.3   Rigid Sidewalls

In reality, the tangential velocity is zero at the sidewalls which is also referred to as a
rigid sidewall condition. The boundary conditions are in this case

$$x = 0, A_x : u = v = w = \frac{\partial T}{\partial x} = 0$$

$$y = 0, A_y : u = v = w = \frac{\partial T}{\partial y} = 0 \qquad (4.15)$$

Unfortunately, with these boundary conditions, the linear stability problem can no
longer be solved analytically. The two-dimensional linear stability problem with rigid
sidewalls was solved by Van de Vooren and Dijkstra [42] using finite element techniques.
The bifurcation diagrams were computed for the two-dimensional case up to aspect
ratio 4 using techniques from numerical bifurcation theory (Dijkstra, [40]).

Using the same methods (but at quite a low resolution), the three-dimensional
stability problem was tackled by Dijkstra [43-45] for 3D-square boxes just as those
used by Koschmieder and Prahl [35]. The neutral curve was determined as the value of
$Ma$ for which the first bifurcation occurred [43] and part of the bifurcation diagrams

were calculated for several aspect ratios [44]. Although similar patterns were found as those in Koschmieder and Prahl [35], it proved to be very computationally intensive to determine all possible patterns and not a detailed structure of the relevant steady states could be obtained.

This was recently done by Dauby and Lebon [46], by solving the weakly nonlinear problem with similar methods as in Dauby et al. [41], but now employing a pseudo-spectral method to represent the eigensolutions. From the amplitude equations, the bifurcation diagrams could be computed with considerable detail for aspect ratios up to 8. We consider only their results for (nearly perfect) horizontally square boxes (with $A = A_x = A_y$).

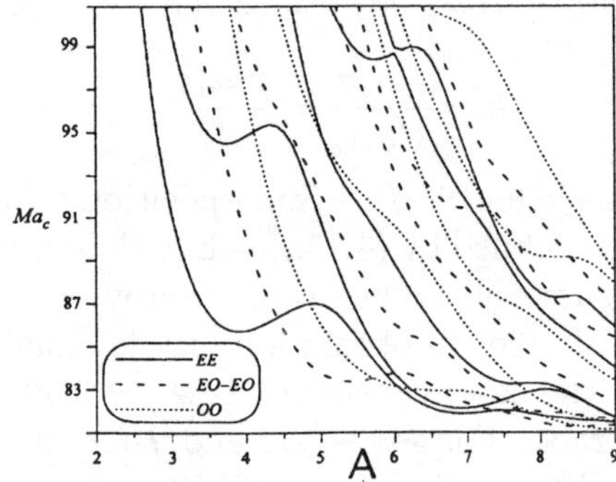


Figure 19: *Critical Marangoni numbers of particular modes as a function of aspect ratio A for horizontally square boxes (from [46]). The neutral curve is obtained by following all lower parts of the different curves.*

First, all eigensolutions were divided into classes $\{EE, EO, OE, OO\}$ according to their symmetry with respect to both symmetry axes of the problem ($x = A/2, y = A/2$). The curves at which these modes are neutrally stable are shown in Fig. 19, where each curve represent a mode from a particular class. The onset conditions for the modes of the $EO$ and $OE$ class are the same due to symmetry. The neutral curve consists of parts of the curves in Fig. 19, which have smallest $Ma$. Patterns of these neutral modes are shown in Fig. 20 and are also similar to those found in Dijkstra [43].

The bifurcation diagrams (only represented here as a plot of the amplitude of the $EE$-component of the solution versus the distance from criticality $\epsilon$) are shown in Fig. 21 for several values of the aspect ratio $A$ and $Ra = Bi = 0, Pr = 10^4$. Patterns at labelled points ((a), (b), etc.) in Fig. 21 are shown as contour plots of the vertical velocity at midheight in Fig. 22. For example, at $A = 2.4$, a 1-cell pattern (Fig. 22a) is found. This pattern still remains stable at $A = 4.7$ (Fig. 21b) for small $Ma$ (Fig. 22b) but at larger $Ma$, the diagonal pattern as in Fig. 22d is the only stable pattern. The bifurcation diagram for $A = 5.94$ (Fig. 21c) is already quite complicated, and several patterns (Fig. 4.9d-f) are stable over a certain range of $Ma$. The pattern in Fig. 4.9e

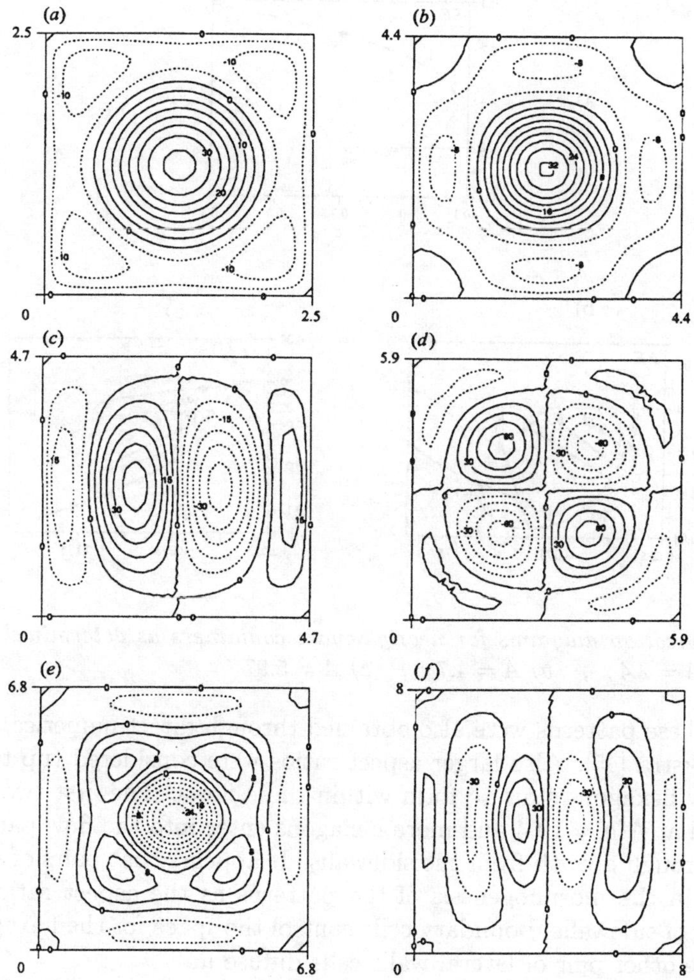

Figure 20: *Patterns of the flow at threshold for several values of the aspect ratio as shown on the axis of the plots (from [46]). A contour plot of the dimensionless vertical velocity at the midheight of the container is shown.*

is the strange three cell pattern as found in the experiments of Koschmieder and Prahl
[35] (Fig. 14c) and is indeed a stable steady state.

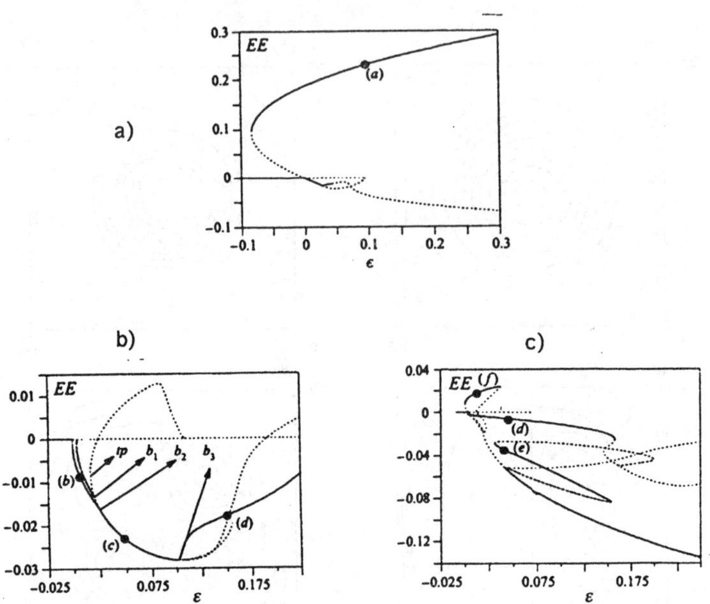



Figure 21: *Bifurcation diagrams for nearly square containers as determined by Dauby and Lebon [46].* a) $A = 2.4$ ; b) $A = 4.7$ ; c) $A = 5.97$

Several of these patterns were also obtained through direct numerical simulation of
the flow in Dijkstra [45]. Also larger aspect ratios were considered (up to 12.4) and it
was shown how hexagons start to form within these flows. The first two hexagons are
fully developed at $Ma = 10.4$ and more hexagons enter into the flow pattern at larger
aspect ratio through growth from the sidewalls. It appears that pairs of sidewalls have
different roles in the morphogenesis of the patterns as the aspect ratio is increased.
Along one pair of sidewalls, boundary cells control the space for the hexagonal pattern,
while along the other pair of lateral walls cells diffuse in.

This results for Ma = 90 and $A = 14$, $A = 16$ and $A = 19$ in the steady states
as presented in Fig. 23 [47]. In these plots, a gray-shade plot of the vertical velocity
(dark areas indicate downward flow) is overlain with a vector plot of the horizontal
velocity at a level just below the interface. For $A = 14$ (Fig. 23a) the hexagonal
pattern is well-developed. There is a center hexagon which is hexagonally surrounded
by six other hexagons. The latter are connected each to three boundary cells, which
have no perfect hexagonal shape. Note that the ordering of the cells is different at the
two pairs of lateral walls. At the east and west (controlling) wall, the boundary cells
are aligned with the wall. At the north and south (diffusing) walls, the position of the
middle cell is much more close to the wall than the other two.

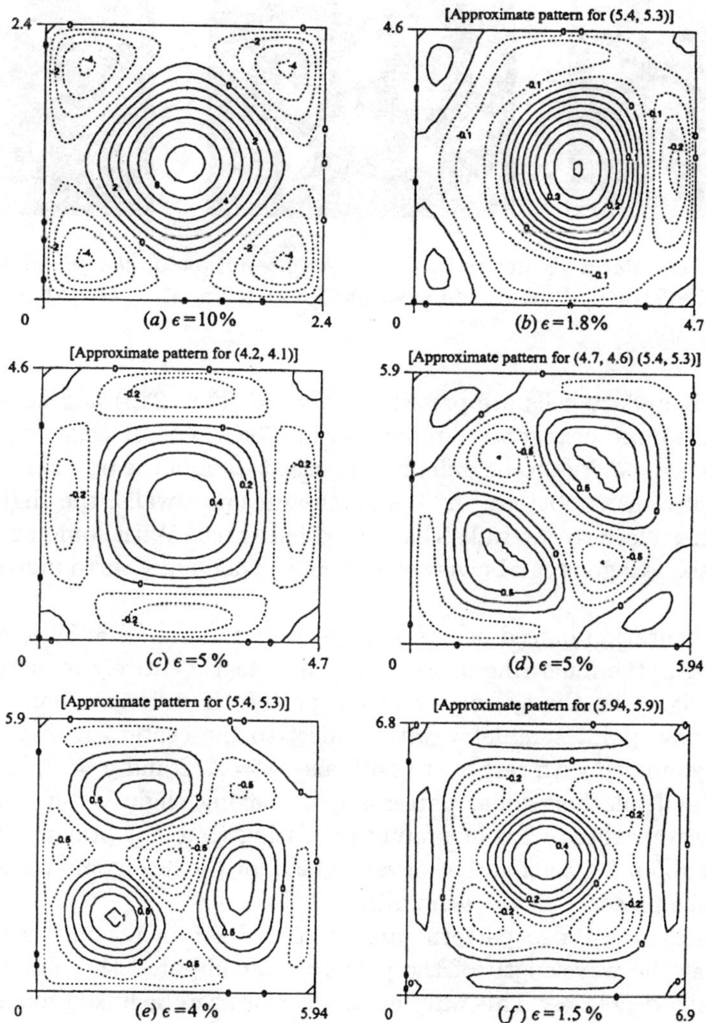


Figure 22: *Patterns at particular points on the branches of Fig. 21. Only the type of flow pattern is referred to, not the actual value of the aspect ratio.*

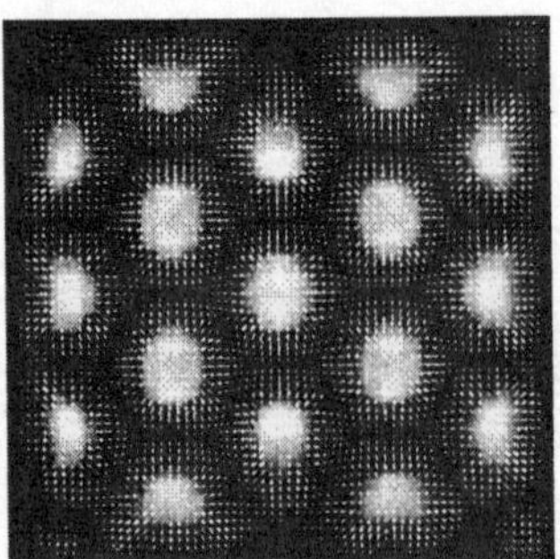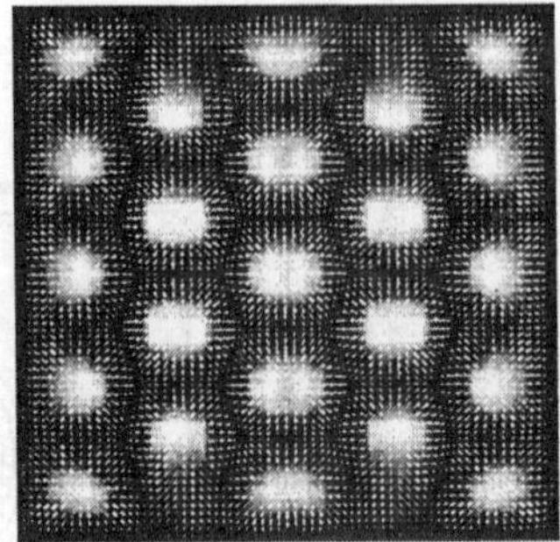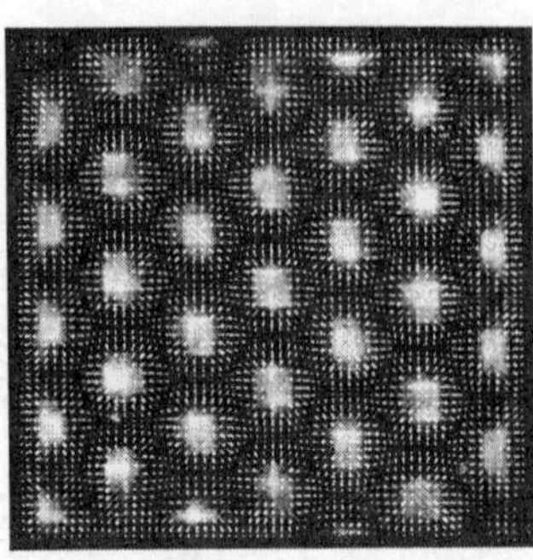

Figure 23: *Steady states as determined by time-integration of the full three-dimensional equations [47] for $Ra = 0$, $Bi = 0$, $Ma = 90$ and $Pr = 1$. (a) $A = 14$, (b), $A = 16$, (c) $A = 19$.*

Two more boundary cells are present at $A = 16$ (Fig. 23b) and consequently, the pattern contains more hexagons. Similar to Fig. 23a, a center hexagon is surrounded by six other hexagons. Again the difference in orientation of the boundary cells near the opposite pair of walls is obvious. At the north and south walls, the slightly imperfect hexagons are less close to the walls than the other three. With some imagination, the center ring of six hexagons can be viewed as being hexagonally surrounded by boundary cells.

Such a configuration indeed occurs at larger aspect ratio ($A = 19$, Fig. 23c), where the surrounding of the inner ring of six hexagons is nearly entirely by perfect hexagons. Note that in Fig. 23c, the symmetry of the pattern is different than the other two, since there is only point symmetry with respect to the center. In Fig. 23a and Fig. 23b, there is symmetry with respect to both axes $x = A/2$ and $y = A/2$. It is perfectly possible that multiple hexagonal patterns exist having slightly different symmetries as above. However, these were not found indicating (but not proving) that up to an aspect ratio of about 20, the band of wavenumbers for the hexagonal pattern is severely restricted by the presence of the sidewalls.

In [47], for $A = 14$ also patterns are computed for increasing value of $Ma$ and it is found that the wavelength of the pattern (as computed through spatial Fourier analysis, initially decreases. This wavelength decrease can be linked to a shift of energy from one mode to another mode, the latter having a larger wavenumber. This shift is accompanied by an increase in energy production of the flow. Since in steady state, energy production is balanced by dissipation, one is tempted to simply attribute the smaller cell size to induce a larger dissipation needed to balance energy production. However, because in pure Rayleigh-Bénard convection the wavelength dependence has just the opposite behavior, such a simple answer does not suffice.

## 4.4   Discussion

About hundred years after Bénard's work [3] dealing with convection in shallow fluid layers, it is about time that the intriguing problem of the preference for hexagonal patterns in Bénard-Marangoni flows is solved. This solution should not only show that hexagons appear from random initial conditions during a numerical simulation and are therefore in some sense preferred, but simultaneously it should make clear – in a physical transparant way – why other patterns are not reached, although kinematically possible. In addition, the explanation should also imply a description of the wavelength behavior of the patterns with $Ma$ or $Ra$ and an explanation of the differences.

There are many pieces of the puzzle already available. Weakly nonlinear theory for the infinite layer provides a good agreement with experiments. Lateral walls seem to restrict the band of wavenumbers of the pattern and show the importance of the presence of boundary cells. The wavelength dependence is possibly linked to integral quantities of the steady solutions for example energy production. The latter quantity is significantly different for pure surface tension driven flows, where energy is only produced at the surface, and pure buoyancy driven flows, where energy is produced over the bulk of the liquid. Although in some areas pieces of the puzzle have already been fitted together, it is still not clear what the image of the total puzzle will look like. Maybe essential pieces of the puzzle are still missing, for example, it is not clear what the role of interface deformation has on the pattern selection process. It is hoped that this text will stimulate students to continue work, either experimentally, theoretically or numerically on this fascinating issue in classical fluid mechanics.

# Acknowledgments

To make my own (very modest) contributions to this area of research I have had generous support over the years from the National Computing Facilities Foundation. Within the project SC432, I was able to perform calculations on the CRAY machines at the Academic Computing Centre (SARA), Amsterdam, the Netherlands. Use of these computing facilities was sponsored Netherlands Organization for Scientific Research. I also thank M. Jeroen Molemaker for the use of several of his CFD codes and for his interest in the subject.

# References

[1]  Koschmieder, E.L. and D.W. Switzer: The wavenumbers of supercritical surface-tension-driven Bénard convection, J. Fluid Mech., 240 (1992), 533-548.

[2]  Chandrasekhar, S.: Hydrodynamic and Hydromagnetic Stability, Clarendon Press, Oxford 1961.

[3] Bénard, H.: Les tourbillons cellulaires dans une nappe liquide transportant de la chaleur en régime permanent, Ann. Chem. Phys., 23 (1901), 62-144.

[4] Koschmieder, E.L.: Bénard cells and Taylor Vortices, Cambridge University Press 1993.

[5] Normand, C., Pomeau, Y., and M.G. Velarde: Convective instability: A physicist's approach, Rev. Mod. Phys., 49 (1977), 581-624.

[6] Rayleigh, Lord: On convection currents in a horizontal layer of fluid, when the higher temperature is on the under side, Phil. Mag., 32 (1916), 529-546.

[7] Jeffreys, H.: The surface elevation in cellular convection, Quart. J. Mech. Appl. Math., 4 (1951), 283-288.

[8] Block, M.J.: Surface tension as a cause of Bénard cells and surface deformation in a liquid film, Nature, 178 (1956), 650.

[9] Pearson, J.R.A.: On convection cells induced by surface tension, J. Fluid Mech., 4 (1958), 489-500.

[10] Marangoni, C.: Ann. Phys. Lpz., 143 (1871), 337.

[11] Thompson, J.J.: Phil. Mag., 10 (1855), 330.

[12] Pantaloni, J., Bailleux, R., Salan, J. and M.G. Velarde: Rayleigh-Bénard-Marangoni instability: new experimental results, J. Non-Equlibr. Thermodyn., 4 (1979), 201-218.

[13] Cerisier, P., Perez-Garcia, C. and R. Occelli: Evolution of induced patterns in surface-tension-driven Bénard convection, Phys. Rev.E., 47 (1993), 3316-3325.

[14] Koschmieder, E.L. and M.I. Biggerstaff: Onset of surface-tension-driven Bénard convection, J. Fluid Mech., 167 (1986), 49-64.

[15] Levich, V.G. and V.S. Krylov: 1969, Surface tension driven phenomena, Ann. Rev. Fluid Mech., 1 (1969), 293-316.

[16] Patberg, W.B., Koers, A., Steenge, W.D.E. and A.A.H. Drinkenburg: Effectiveness of mass transfer in a packed distillation column in relation to surface tension gradients, Chem. Engng. Sc., 38 (1983), 917-923.

[17] Davis, S.H.: 1987, Thermocapillary instabilities, Ann. Rev. Fluid Mech., 19 (1987), 403-435.

[18] Aris, R: Vectors, tensors, and the basic equations of fluid mechanics. Dover publications 1962.

[19] Drazin, P.G. and W.H. Reid: Hydrodynamic Stability, Cambridge University Press 1981.

[20] Joseph, D.D.: Stability of fluid motions, volumes I and II, Springer-Verlag 1976.

[21] Vidal, A. and A. Acrivos, A.: Nature of the neutral state in surface tension driven convection, Phys. Fluids, 9 (1966), 615-616.

[22] Davis, S.H.: On the principle of exchange of stabilities, Proc. Roy. Soc. A, 310 (1969), 341-358.

[23] Nield, D.A.: Surface tension and buoyancy effects in cellular convection, J. Fluid Mech., 19 (1964), 341-352.

[24] Scriven, L.E. and C.V. Sternling: On cellular convection driven by surface tension gradients: effect of mean surface tension and surface viscosity, J. Fluid Mech., 19 (1964), 321-340.

[25] Smith, K.A.: On convective instability induced by surface tension gradients, J. Fluid Mech., 24 (1966), 410-414.

[26] Takashima, M.: Surface tension driven instability in a horizontal liquid layer with a deformable surface. I. Stationary convection, J. Phys. Soc. Japan, 50 (1981), 2745-2750.

[27] Davis, S.H.: Buoyancy-surface tension instability by the method of energy, J. Fluid Mech., 39 (1969), 347-359.

[28] Davis, S.H. and G.M. Homsy: 1980, Energy stability theory for free surface problems: buoyancy-thermocapillary layers, J. Fluid Mech., 98 (1980), 527-553.

[29] Castillo, J.L. and M.G. Velarde: Buoyancy-thermocapillary instability: the role of interfacial deformation in one- and two-component fluid layers heated from below or above, J. Fluid Mech., 125 (1982), 463-474.

[30] Cloot, A. and G. Lebon: A nonlinear stability analysis of the Bénard-Marangoni problem, J. Fluid Mech., 145 (1984), 447-469.

[31] Bestehorn, M.: Phase and amplitude instabilities for Bénard-Marangoni convection in fluid layers with large aspect ratio, Phys. Rev. E., 48 (1993), 3622-3634.

[32] Schlüter, A., Lortz, D. and F.H. Busse: On the stability of steady finite amplitude convection, J. Fluid Mech., 23 (1965), 129-144.

[33] Scanlon, J.W. and L.A. Segel: Finite amplitude cellular convection induced by surface tension, J. Fluid Mech., 30 (1967), 149-162.

[34] Thess, A. and S.A. Orszag: Surface tension driven Bénard convection at infinite Prandtl number, J. Fluid Mech., 283 (1995), 201-230.

[35] Koschmieder, E.L. and S.A. Prahl: Surface tension driven convection Bénard convection in small containers, J. Fluid Mech., 215 (1990), 571-583.

[36] Rosenblat, S., Homsy, G.M. and S.H. Davis: Nonlinear Marangoni convection in bounded layers, Part 2. Rectangular cylindrical containers, J. Fluid Mech., 120 (1982), 123-138.

[37] Rosenblat, S., Homsy, G.M. and S.H. Davis: Eigenvalues of the Rayleigh-Bénard and Marangoni problems, Phys. Fluids, 24 (1991), 2115-2117.

[38] Kuznetsov, Y.A.: Elements of applied bifurcation theory, Springer Verlag 1995.

[39] Nayfeh, A.H. and B. Balachandran: Applied nonlinear dynamics, John Wiley 1995.

[40] Dijkstra, H.A.: On the structure of cellular patterns in Rayleigh-Bénard-Marangoni flows in small-aspect-ratio containers , J. Fluid Mech., 243 (1992), 73 - 102.

[41] Dauby, P.C., Lebon, G., Colinet, P. and J.C. Legros: Hexagonal Marangoni convection in a rectangular box with slippery walls, Q. J. Mech. Appl. Math., 46 (1993), 683-707.

[42] Van de Vooren, A.I. and H.A. Dijkstra, A finite-element stability analysis of the Marangoni problem in a two-dimensional container with rigid sidewalls, Computers and Fluids, 17 (1989), 467-485.

[43] Dijkstra, H.A.: Surface tension driven cellular patterns in three-dimensional boxes. I: Linear Stability, Microgravity Science and Technology, VII/4 (1995), 307-312.

[44] Dijkstra, H.A.: Surface tension driven cellular patterns in three-dimensional boxes. II: A bifurcation study, Microgravity Science and Technology, VIII/2 (1995), 70-77.

[45] Dijkstra, H.A.: Surface tension driven cellular patterns in three-dimensional boxes. III: The formation of hexagonal patterns, Microgravity Science and Technology, VIII/3, 155-162.

[46] Dauby, P.C. and G. Lebon: Bénard-Marangoni instability in rigid rectangular containers, J. Fluid Mech., 329 (1996), 25-64.

[47] Dijkstra, H.A.: Surface tension driven cellular patterns in three-dimensional boxes. IV: On the preference of hexagonal patterns, in preparation, (1998).

# THERMOCAPILLARY CONVECTION


**H.C. Kuhlmann**
**University of Bremen, Bremen, Germany**


## Abstract


The fluid motion induced by surface tension gradients due to temperature variations along liquid/gas interfaces is reviewed. Attention is focussed on the thermocapillary driven flow inside the liquid rather than on free surface deformations. The general equations for an incompressible Newtonian liquid surrounded by a passive gas are introduced followed by some basic considerations of the thermocapillary flow near the contact point. The Stokes flow in differentially heated cylindrical liquid bridges is calculated revealing the fundamental flow structures when the thermocapillary surface stresses are low. As general characteristics of thermocapillary flows the boundary layer scalings for certain limits of the Marangoni and Prandtl numbers are derived. After a brief review of hydrothermal waves in plane layers two paradigms for thermocapillary driven convection, heated cylindrical liquid bridges and rectangular cavities, are considered in more detail. Flow structures, instabilities, dynamics, and side wall effects are analyzed.

# 1 Introduction and basic equations

## 1.1 Motivation

The energetic conditions in a miscroscopically small layer between two immiscible fluids
are different from those in the bulk. Within a continuum mechanical description of
interfaces this layer can be considered infinitesimally thin and assigned a surface energy
per area, which is called the interfacial or surface tension.

The surface tension generally depends on the temperature and on other quantities
such as, e. g. the concentration of a solvant. Variations of these fields along the interface cause gradients of the surface energy and thus lead to surface forces that may
drive a significant fluid motion. The flow effects associated with this driving force are
commonly termed *Marangoni effects*. A historical review has been given by Scriven
and Sternling [1]. In case of temperature-induced surface tension gradients, the motion
is called *thermocapillary convection*.

Thermocapillary convection arises in many different processes such as the drying of
films and the motion of small droplets and bubbles. It plays a role in the mechanics of
foams and emulsions, in combustion of liquid fuels, in boiling, and in the damping of
surface waves, to name only a few areas of application.

Thermocapillary flows are also important in crystal growth from the melt [2]. Since
the initiating work of Chang and Wilcox [3, 4] the interest in thermocapillary convection in crystal growth has increased considerably. It is now clear that one major
reason for the appearance of undesired mirco-inhomogeneities in crystals grown from
the melt is the oscillatory melt flow. The desire to understand and, finally, to suppress
and control the flow motivated many studies, the number of which has increased continuously throughout the last two decades. Today, a few simple models have emerged
as paradigms for the study of thermocapillary driven flows.

This chapter provides an introduction of and an overview on the fluid flow phenomena in two of the most important crystal growth models, namely the *half-zone model*
and the *open cavity*. The former models basic aspects of the fluid dynamics in the real
*floating-zone* technique while the latter represents the *open boat* technique [2]. Models
for the Czochralski process will not be treated. We shall concentrate on the fluid flow
and pattern formation due to hydrodynamic instabilities of the bulk flow. These are
treated by analytical and numerical methods. Relevant experiments, when available,
are referred to for comparison with theory. Effects associated with dynamic surface
deformations will not be considered.

## 1.2 Basic equations

The typical flow velocities in buoyancy or surface tension driven flows close to the first
pattern forming instabilities are usually small so that the fluid motion can be described

by the Boussinesq approximation of the Navier-Stokes equations [5]

$$\partial_t \vec{u} + \vec{u} \cdot \nabla \vec{u} \;=\; -\frac{1}{\rho}\nabla p + \nu \Delta \vec{u} + g\beta T \vec{e}_z, \tag{1}$$

$$\nabla \cdot \vec{u} \;=\; 0, \tag{2}$$

$$\partial_t T + \vec{u} \cdot \nabla T \;=\; \kappa \Delta T. \tag{3}$$

The fluid is treated as incompressible with density $\rho$, kinematic viscosity $\nu$, thermal diffusivity $\kappa$, and thermal expansion coefficient $\beta = -1/\rho(\partial\rho/\partial T)_p$. We assume a homogeneous gravity field $\vec{g} = -g\vec{e}_z$ and the velocity, pressure, and temperature fields are denoted by $\vec{u}$, $p$, and $T$, respectively.

We consider the case of a liquid bounded by rigid walls as well as by a free surface with surface tension. On rigid boundaries we employ the usual no-slip, no-penetration conditions and fixed temperatures

$$\vec{u} \;=\; 0, \qquad T \;=\; const. \tag{4}$$

The forces acting on both sides of the free surface between the two fluids (1) and (2) must be the same. If the interface is flat and the surface tension constant, mechanical equilibrium requires

$$\tilde{S}^{(1)} \cdot \vec{n} \;=\; \tilde{S}^{(2)} \cdot \vec{n}, \tag{5}$$

where the pressure and viscous forces per unit surface are given by the stress tensor

$$\tilde{S}_{ij} \;=\; -p\mathbf{I} + \eta \left[ \nabla \vec{u} + (\nabla \vec{u})^T \right], \tag{6}$$

with dynamic viscosity $\eta = \rho\nu$, $\mathbf{I}$ the identity matrix, and unit normal vector $\vec{n}$ which is directed out of liquid (1) and into the ambient fluid (2). Generally the free surface is not plane and the surface tension varies along the surface. The boundary condition then reads

$$\tilde{S}^{(1)} \cdot \vec{n} + \sigma(\nabla \cdot \vec{n})\vec{n} - (\mathbf{I} - \vec{n}\vec{n}) \cdot \nabla\sigma \;=\; \tilde{S}^{(2)} \cdot \vec{n}. \tag{7}$$

The term $\sigma\left(\nabla \cdot \vec{n}\right)$ is the Laplace pressure. The curvature of the interface

$$\nabla \cdot \vec{n} \;=\; \frac{1}{R_1} + \frac{1}{R_2} \tag{8}$$

can be expressed as the sum of the inverse main radii of curvature $R_1$ and $R_2$. We take a radius of curvature positive, if the origin of the corresponding circle lies on the side of fluid (1) (the body of fluid (1) is barrel-shaped), otherwise it is negative. The second additional term describes a surface force acting tangential to the interface. It is proportional to the negative (tangential) gradient of the surface tension $\sigma$ ($\sigma$: Energy per area). The operator $\mathbf{I} - \vec{n}\vec{n}$ in (7) represents the orthogonal projection of a vector onto the tangent plane defined by $\vec{n}$. The velocity boundary conditions are completed by $\vec{n} \cdot \vec{u} = 0$.

| variable | $r,\,z$ | $t$ | $\vec{u} = (u, v, w)$ | $p$ | $T$ |
|----------|---------|-----|------------------------|-----|-----|
| scale | $d$ | $d^2/\nu$ | $\gamma\Delta T/\rho\nu$ | $\gamma\Delta T/d$ | $\Delta T$ |

Table 1: Scales used to non-dimensionalize the Oberbeck-Boussinesq equations.

In addition to the temperature-dependence of the density also the temperature-dependence of the surface tension is taken into account. For small temperature variations it suffices to expand the surface tension up to first order in $T - T_0$

$$\sigma(T) = \sigma_0(T_0) - \gamma(T - T_0) + O\left((T - T_0)^2\right). \tag{9}$$

Here $\sigma_0(T_0)$ is the surface tension at a reference temperature $T_0$, e. g. the mean temperature of the liquid, and $\gamma$ is the negative linear Taylor coefficient. For most liquid-gas interfaces $\gamma$ is positive.

On the free surface the temperature must be continuous. Since the ambient temperature distribution in the direct vicinity of the free surface is often unknown, we use Newton's law of heat transfer

$$\vec{n} \cdot (k\nabla T) = -h(T - T_a), \tag{10}$$

where $T_a$ is the ambient temperature far away from the free surface, $k$ denotes the heat conductivity of the liquid (1) and the phenomenological parameter $h$ is the heat transfer coefficient.[1]

## 1.3    Equations for the half-zone model

To be more specific, we consider a fluid volume $V$ bounded axially by two rigid parallel disks of equal radii $r = R$ at $z = \pm d/2$ a distance $d$ apart which are kept at constant temperatures $T(\pm d/2) = T_0 \pm \Delta T/2$. The radial boundary, given by a free surface of mean surface tension $\sigma_0$, is supporting the liquid. This configuration is called a *liquid bridge*. Its geometry is characterized by the aspect ratio $\Gamma = d/R$. For the particular case $V = \pi R^2 d$ and $\vec{g} = 0$ sketched in fig. 1 the fluid volume takes an upright cylindrical shape. In case of gravity we assume $\vec{g} \parallel \vec{e}_z$.

To non-dimensionalize the equations we use cylindrical coordinates $(r, \varphi, z)$ and employ the scales for length, time, velocity, pressure, and temperature given in table 1 to obtain the dimensionless equations

$$\partial_t \vec{u} + Re\,\vec{u} \cdot \nabla\vec{u} = -\nabla p + \Delta\vec{u} + Bd\,\theta\vec{e}_z, \tag{11}$$

$$\nabla \cdot \vec{u} = 0, \tag{12}$$

$$\partial_t\theta + Re\,\vec{u} \cdot \nabla\theta = \frac{1}{Pr}\Delta\theta, \tag{13}$$

---

[1]Depending on the process (the melting point of silicon is $T_m = 1410°C$) the formulation can be extended to account for radiative heat transfer; cf. [6].

 

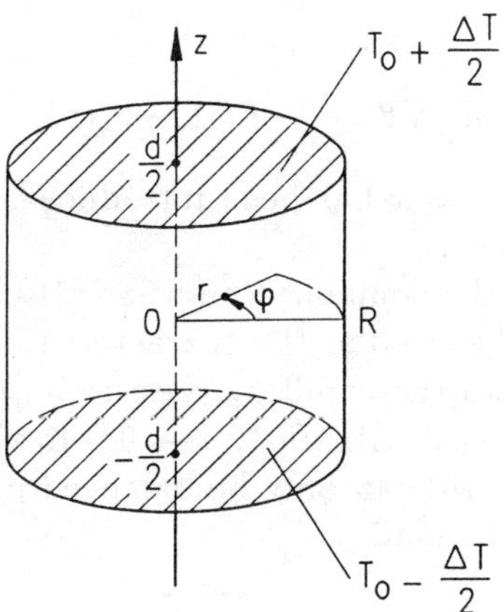


Figure 1: Geometry and coordinate system of a heated upright cylindrical liquid bridge.

where the normalized temperature $\theta = (T - T_0)/\Delta T$ has been introduced. The dimensionless groups (characteristic numbers) read

$$
\begin{aligned}
Re &= \frac{\gamma \Delta T d}{\rho \nu^2} \quad \text{(Reynolds)}, & Pr &= \frac{\nu}{\kappa} \quad \text{(Prandtl)}, \\
Gr &= \frac{g \beta \Delta T d^3}{\nu^2} \quad \text{(Grashof)}, & Bd &= \frac{Gr}{Re} \quad \text{(dynamic Bond)}.
\end{aligned}
\tag{14}
$$

The groups $Re$ and $Gr$ measure the strength of thermocapillary surface forces ($\gamma \Delta T/d$) and buoyancy forces ($\rho g \beta \Delta T d$), respectively, relative to the viscous forces per surface area ($\rho \nu^2/d^2$). Instead of $Re$ and $Gr$, the Marangoni number $Ma = RePr$ and the Rayleigh number $Ra = -GrPr$ are frequently employed.

The boundary conditions on the rigid walls at $z = \pm 1/2$ are

$$
\vec{u} = 0 \quad \text{and} \quad \theta = \pm 1/2.
\tag{15}
$$

With $r = \xi(\varphi, z)$ being the position of the free surface the dimensionless force balance on the free surface is obtained as

$$
-p\vec{n} + \left[ \nabla \vec{u} + (\nabla \vec{u})^T \right] \cdot \vec{n} + \left( \frac{1}{Ca} - \theta \right) (\nabla \cdot \vec{n}) \, \vec{n} + (\mathbf{I} - \vec{n}\vec{n}) \cdot \nabla \theta = \frac{Bo}{Ca} z \vec{n}
\tag{16}
$$

where $p$ is the pressure in the liquid bridge up to the constant ambient value $p_a(z = 0)$. Viscous shear forces in the gas phase (2) have been disregarded owing to the small dynamic viscosity of gases. The term on the right hand side of (16) is caused by the hydrostatic pressure difference. The Bond number $Bo$ and the capillary number $Ca$ are defined as[2]

$$
Bo = \frac{(\rho - \rho_a)g d^2}{\sigma_0} \quad \text{and} \quad Ca = \frac{\gamma \Delta T}{\sigma_0}.
\tag{17}
$$

---

[2]Sometimes $Ca/Re = (\rho \nu^2)/(\sigma_0 d)$ is denoted the capillary number.

These numbers measure the relative importance of hydrostatic and hydrodynamic pressure differences to the characteristic capillary pressure $\sigma_0/d$. The dimensionless thermal boundary condition reads

$$\vec{n} \cdot \nabla\theta \ = \ -Bi\left(\theta - \theta_a(z)\right), \tag{18}$$

where the Biot number $Bi = hd/k$ has been introduced. It is a measure for the heat transport through the interface.

A major simplification of the boundary condition (16) is obtained, if the hydrostatic pressure difference $(\rho - \rho_a)gd$ as well as the characteristic flow-induced normal stresses $\gamma\Delta T/d$ are small compared to the capillary pressure $\sigma_0/d$. This case formally corresponds to the limit $(Ca, Bo) \to 0$ with $Bo/Ca \to 0$.[3] In this limit, the term $Ca^{-1}\nabla \cdot \vec{n}$ in the normal component of (16) can only be balanced by the pressure $p$ in the liquid bridge. Therefore, we must demand

$$p \ = \ \frac{\nabla \cdot \vec{n}}{Ca}. \tag{19}$$

Thus the pressure in the liquid bridge is given by the capillary pressure. For $(\nabla \cdot \vec{n} \neq 0)$ it is asymptotically large compared to the hydrodynamic and the hydrostatic pressure. Equation (19) is the Young-Laplace equation for the case of weightlessness. It is second order in $z$ and $\varphi$. The solutions $\xi(\varphi, z)$ depend on the volume of the liquid bridge and on the boundary conditions at $z = \pm 1/2$ [7]. Here we assume that all contact points where liquid, gas, and solid meet are pinned and form a fixed contact line

$$\xi(\varphi, \pm\tfrac{1}{2}) \ = \ 1/\Gamma. \tag{20}$$

In the limit considered, the remaining tangential stress balances simplify to

$$\vec{t} \cdot \left[\nabla\vec{u} + (\nabla\vec{u})^T\right] \cdot \vec{n} + \vec{t} \cdot \nabla\theta \ = \ 0, \qquad \text{on} \quad r \ = \ \xi(\varphi, z), \tag{21}$$

where $\vec{t}$ stands for both linearly independent orthogonal tangent vectors. In the limit $(Bo, Ca) \to 0$ with $Bo/Ca \to 0$ the surface takes a cylindrical shape, if the volume is $V = \pi R^2 d$. Then $\xi(\varphi, z) = 1/\Gamma$ and the boundary conditions read

$$\left.\begin{aligned}
\partial_r w + \partial_z \theta &= 0 \\
r\,\partial_r\left(\tfrac{v}{r}\right) + \tfrac{1}{r}\partial_\varphi\theta &= 0 \\
u &= 0 \\
\partial_r\theta &= -Bi(\theta - z)
\end{aligned}\right\} \qquad \text{on} \quad r = \frac{1}{\Gamma}, \tag{22}$$

where a linear temperature profile $\theta_a = z$ in the gas phase has been assumed. Equations (11) – (13), (15), and (22) are the basis for our analyses.

---

[3] For zero gravity, $Bo = 0$. For gravity conditions, only static surface deformations are compatible with the Boussinesq equations.

## 1.4  Energy analysis

If the flow is evolving dynamically, the knowledge of the energy transport from the injection to the dissipation may provide useful insight into the fluid mechanics. This is particularly true for an understanding of hydrodynamic instabilities, where energy is fed from the base to the disturbance flow. For later use, we shall briefly derive the energy equations for a cylindrical liquid bridge. Generalizations are straightforward.

### 1.4.1  Reynolds-Orr equation

Consider the momentum equation (sum convention)

$$\partial_t u_i + Re\, u_j \partial_j u_i = -\partial_i p + \partial_j^2 u_i + Bd\,\theta \delta_{i3}. \tag{23}$$

Scalar multiplication by $u_i$ and integration over the volume $(< ... > = \int_V ...dV)$ yields

$$\partial_t E_{kin} = \frac{1}{2}\partial_t < u_i^2 > = -Re < u_i u_j \partial_j u_i > - < u_i \partial_i p > + < u_i \partial_j \partial_j u_i > + Bd < w\theta > . \tag{24}$$

The pressure term $< \vec{u} \cdot \nabla p >$ in this equation for the kinetic energy $E_{kin}$ vanishes upon partial integration. Now consider the nonlinear term

$$< u_i u_j \partial_j u_i > \overset{\partial_j u_j = 0}{=} < u_i \partial_j u_j u_i > \overset{Int.}{=} \underbrace{< u_i e_j u_j u_i >_S}_{\to 0} - < (\partial_j u_i) u_j u_i >, \tag{25}$$

where $< ... >_S = \int_S ...dS)$ is the integral over the surface of the volume and $e_j$ the unit normal vector. The equation has the form $a = -a$. Therefore, the nonlinear term $< u_i u_j \partial_j u_i > = 0$ vanishes identically. Consider next the diffusive term

$$< u_i \partial_j \partial_j u_i > \overset{Int.}{=} < u_i e_j \partial_j u_i >_S - < (\partial_j u_i)(\partial_j u_i) > . \tag{26}$$

The second term on the r.h.s. is just the negative of the positive rate of dissipation in the volume $< (\nabla \vec{u})^2 >$.[3] This term always contributes to a reduction of the kinetic energy. The first term is zero for all rigid surfaces. But it is non-zero on the free surface

$$< u_i e_j \partial_j u_i >_S = M_\varphi + M_z + \int_S \frac{v^2}{r} dS, \tag{27}$$

---

[3]The dissipation in the volume $< (\nabla \vec{u})^2 >$ differs from the dissipation $D = < (\partial_i u_j)^2 > + < \partial_i u_j \partial_j u_i >$ by a term that can be expressed as a surface integral. See also Raynal [8]. When the boundaries are rigid and non-rotating, $\vec{u} = 0$, this additional term vanishes. For the present case one can show

$$< (\nabla \vec{u})^2 > = < (\nabla \times \vec{u})^2 > - \int_S \frac{v^2}{r} dS.$$

where $M_\varphi$ and $M_z$ denote the work done per time by azimuthal and axial Marangoni forces, respectively. Thus we are left with the Reynolds-Orr equation

$$\partial_t E_{kin} = -D + M_\varphi + M_z + I_{Gr}, \tag{28}$$

where we have used the abbreviation $I_{Gr} = Bd < w\theta >$ for the work done per time by buoyancy forces and

$$D = < (\nabla \vec{u})^2 > - \int_S \frac{v^2}{r} dS = < (\nabla \times \vec{u})^2 > -2 \int_S \frac{v^2}{r} dS \tag{29}$$

is the total dissipation. It is reduced by the presence of a non-zero azimuthal velocity on the free surface.

### 1.4.2   Energy equations for infinitesimal perturbations

If the time-evolution of infinitesimal perturbation $(\vec{u}, p, \Theta)$ superposed to some basic nonlinear flow $(\vec{u}^{(0)}, p^{(0)}, \theta^{(0)})$ is investigated, we must consider the linearized momentum equation

$$\partial_t u_i + Re \left( u_j^{(0)} \partial_j u_i + u_j \partial_j u_i^{(0)} \right) = -\partial_i p + \partial_j^2 u_i + Bd\, \Theta \delta_{i3}, \tag{30}$$

and the energy balance for the disturbances must thus be supplemented on the right hand side by

$$\partial_t E_{kin} = \ldots - Re \left( < u_i u_j^{(0)} \partial_j u_i > + < u_i u_j \partial_j u_i^{(0)} > \right). \tag{31}$$

The first term vanishes due to the same arguments as for the nonlinear term (25) above. The second term describes the change of energy of the perturbation flow component $u_i$ by transport of basic state stress $\partial_j u_i^{(0)}$ by convection $u_j$ due to to the disturbance flow. If the base state is axisymmetric $\vec{u}^{(0)} = u_0 \vec{e}_r + w_0 \vec{e}_z$ five contributions remain

$$-Re < u_i u_j \partial_j u_i^{(0)} > = -Re < u_i (u\partial_r + \frac{v}{r}\partial_\varphi + w\partial_z)u_i^{(0)} > \tag{32}$$

$$= \sum_{i=1}^{5} I_{vi} = -Re \int_V \left( v^2\frac{u_0}{r} + u^2\frac{\partial u_0}{\partial r} + uw\frac{\partial u_0}{\partial z} + wu\frac{\partial w_0}{\partial r} + w^2\frac{\partial w_0}{\partial z} \right) dV,$$

and we obtain the rate of change of kinetic energy of the disturbance flow

$$\partial_t E_{kin} = -D + M_\varphi + M_z + I_{Gr} + \sum_{i=1}^{5} I_{vi}. \tag{33}$$

The terms $D, M_\varphi, M_z$, and $T_{Gr}$ have the same form as those for the basic state. A similar procedure leads to the balance for the *thermal energy*[4] of the disturbance flow, defined as $E_T = \frac{1}{2}\int_V \Theta^2\, dV$,

$$\partial_t E_T = -D_T - H + \sum_{i=1}^{3} I_{Ti}, \tag{34}$$

---

[4] $E_T$ must not be confused with the thermal energy in thermodynamics.

where

$$D_T \;=\; \frac{1}{Pr} < (\nabla\Theta)^2 >, \qquad\qquad H \;=\; \frac{Bi}{Pr}\int_S \Theta^2 \, dS, \tag{35}$$

$$\sum_{i=1}^{3} I_{Ti} \;=\; -Re \int_V \Theta \left( u\frac{\partial\Theta_0}{\partial r} + w\frac{\partial\Theta_0}{\partial z} + w \right) dV, \tag{36}$$

and $\Theta_0 = \theta_0 - z$ is the deviation of the temperature from the conductive profile $z$ (heating from above).

## 2    Stokes flow

### 2.1    On the flow in thermocapillary corners

Consider, for the moment, the local flow in the vicinity of a cold corner with contact angle $\alpha$.[5] We assume that the velocities are small, the interface is locally plane, and the temperature varies linearly with the distance from the contact line. Using polar coordinates $(\rho, \theta)$ in a plane $\varphi = const.$ the stationary the flow is governed by (curl of (11))

$$\nabla^4\psi \;=\; \left[\frac{1}{\rho}\partial_\rho\rho\partial_\rho + \frac{1}{\rho^2}\partial_\theta^2\right]^2 \psi(\rho,\theta) \;=\; 0, \tag{37}$$

where the stream function $\psi$ has been introduced by

$$u \;=\; \frac{1}{\rho}\partial_\theta\psi, \qquad\qquad v \;=\; -\partial_\rho\psi, \tag{38}$$

and $u, v$ are the radial and azimuthal velocities with respect to the coordinates $\rho$ and $\theta$. The linear temperature variation along the free surface leads to a constant shear stress. It can be normalized to unity by an appropriate length and temperature scale. The boundary conditions for the stream function then read

$$\psi(\theta = 0) \;=\; \partial_\theta\psi(\theta = 0) \;=\; \psi(\theta = \alpha) \;=\; \frac{1}{\rho^2}\partial_\theta^2\psi(\theta = \alpha) + 1 \;=\; 0. \tag{39}$$

It is tempting to look for separable similarity solutions of (37) that are compatible with the boundary condition (39). Such a solution must have the form

$$\psi \;=\; \rho^2 f(\theta), \qquad \text{with} \quad f(0) = f'(0) = f(\alpha) = f''(\alpha) + 1 = 0. \tag{40}$$

The resulting fourth order differential equation for $f$ is easily solved to give

$$f(\theta) \;=\; A\cos 2\theta + B\sin 2\theta + C\theta + D, \tag{41}$$

---

[5]The behavior in a hot corner is analogous.

where the four integration constants

$$A = \frac{1}{4}\frac{\sin 2\alpha - 2\alpha}{\sin 2\alpha - 2\alpha\cos 2\alpha}, \qquad C = -2B,$$
$$B = \frac{1}{4}\frac{1 - \cos 2\alpha}{\sin 2\alpha - 2\alpha\cos 2\alpha}, \qquad D = -A, \tag{42}$$

are determined by the boundary conditions. This solution has also been given by Canright [9]. For the important case of $\alpha = \pi/2$ the solution is

$$\psi(\rho,\theta,\alpha = \pi/2) = \frac{\rho^2}{4}\left[1 - \cos 2\theta + \frac{2}{\pi}(\sin 2\theta - 2\theta)\right]. \tag{43}$$

Typical streamline patterns for $\alpha = \pi/4$, $\pi/2$, $3\pi/4$, and $\pi$ are shown in fig. 2. For the vorticity one obtains

$$\vec{\omega} = \nabla \times \vec{u} = -\nabla^2\psi\vec{e} = -(4f + f'')\vec{e} = -4(C\theta + D)\vec{e}, \tag{44}$$

where $\vec{e}$ is the unit vector parallel to the contact line. The surface on which the vorticity vanishes (the angle $\theta_0$ is shown as a dashed line in fig. 2) is given by

$$\theta(\omega = 0) = \theta_0 = -\frac{D}{C} = \frac{1}{2}\frac{2\alpha - \sin 2\alpha}{1 - \cos 2\alpha}. \tag{45}$$

Since the vorticity depends on $\theta$ only, it takes different values when the origin is approached along different paths. Therefore, the vorticity is always singular at the origin $\rho = 0$. The same applies to other derivatives of the velocity field. As can be seen from (41,42) the solution becomes singular everywhere, if $\alpha$ is a root of

$$\tan 2\alpha - 2\alpha = 0. \tag{46}$$

The only non-trivial root in the interval $[0, \pi]$ is $\alpha_0 = 0.715\pi = 2.2467 = 128.7°$. At this contact angle the surface $\theta = \theta_0$ on which the vorticity vanishes (45) coincides with the free surface. The boundary conditions, however, impose a constant non-zero vorticity $\omega = 1$ on the free surface. Therefore, the vorticity field must exhibit strong gradients all along the free surface when $\alpha \approx \alpha_0$. As a result high velocity gradients arise and the streamlines become asymptotically dense for $\alpha \to \alpha_0$. For $\alpha < \alpha_0$ the flow on the free surface is directed towards the cold corner ($u < 0$), while it is directed away from it ($u > 0$) for $\alpha > \alpha_0$.

The validity of the Stokes flow solution (40) depends on Reynolds number (which depends on the temperature and length scales) in the known fashion [5]. If $f = O(1)$, then $\psi = O(\rho^2)$ and the velocities are $O(\rho)$. The condition under which inertia terms can be neglected is readily obtained as

$$Re|\vec{u}\cdot\nabla\vec{u}| = O(\rho Re) \ll O\left(\frac{1}{\rho}\right) = |\nabla^2\vec{u}|, \tag{47}$$

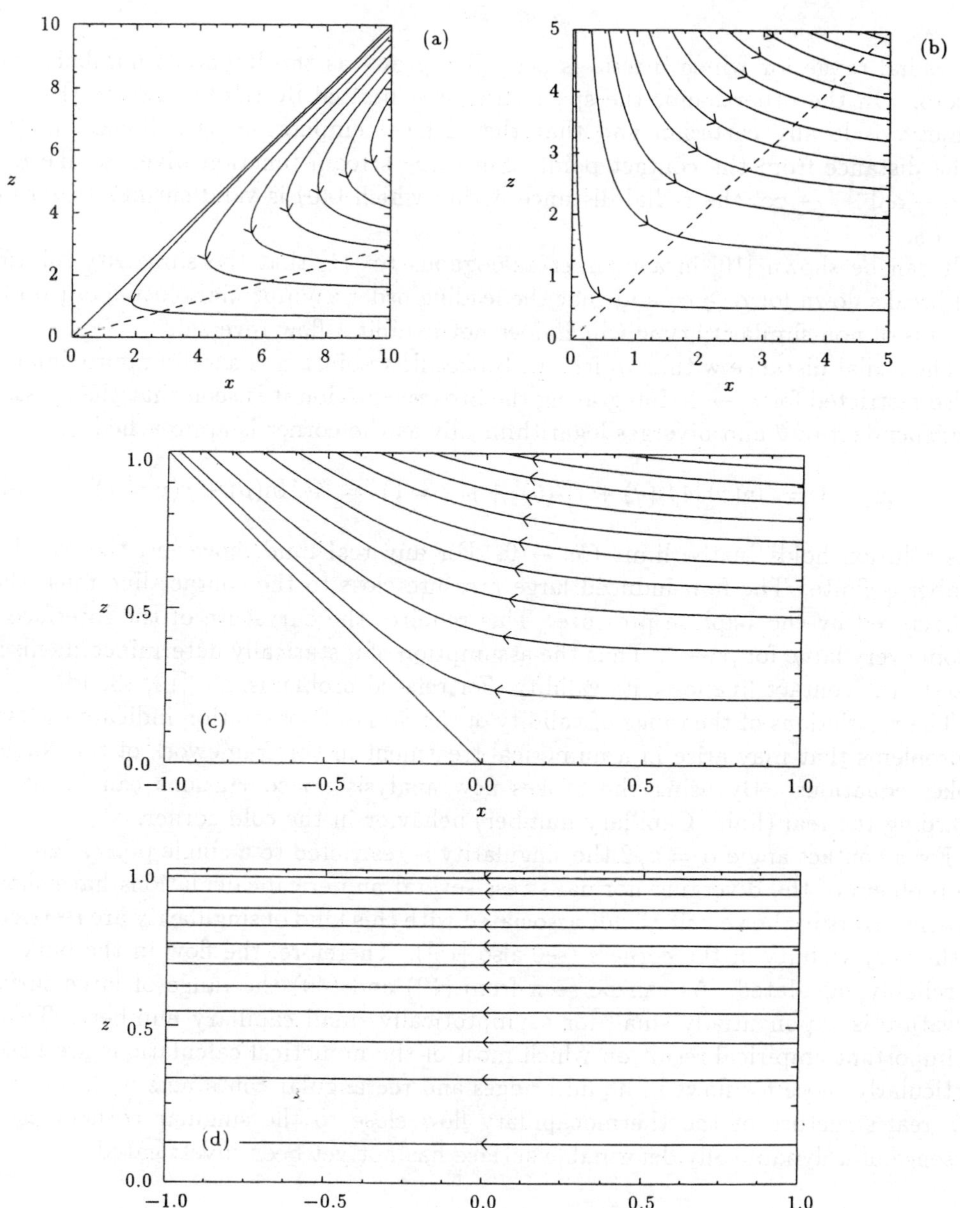

Figure 2: Streamlines in the cold corner for $\alpha = \pi/4$ (a), $\pi/2$ (b), $3\pi/2$ (c), and $\alpha = \pi$ (d). On the dashed line $\omega = 0$.

from which follows

$$\rho \ll Re^{-\frac{1}{2}}. \tag{48}$$

The radial range for which inertia is negligible grows as the Reynolds number tends to zero. On the other hand, the spatial range is limited by the conditions that the geometry is locally cartesian and that the surface temperature is a linear function of the distance from the contact point. Since the stream function diverges like $\psi = O(|\alpha - \alpha_0|^{-1}) \to \infty$, the radial distance within which (40) is valid shrinks to zero as $\alpha \to \alpha_0$.

It can be shown [10] in a manner analoguous to [11] that the similarity solution (41) breaks down for $\alpha \geq \alpha_c$ and that the leading order asymptotic solution of (37) for $\rho \to 0$ is of non-similarity type which does not exhibit a flow reversal.

The radial distance within which the Stokes flow solution is a valid approximation is also restricted for $\rho \to 0$. Integrating the Stokes equation it is seen that the pressure is independent of $\theta$ and diverges logarithmically as the corner is approached

$$p(\rho,\theta) \;=\; \ln(\rho)\left[4f'(\theta) + f'''(\theta)\right] + p(\rho = 1) \;=\; 4C\ln(\rho) + p(\rho = 1). \tag{49}$$

This solution holds in the limit $Ca \to 0$. For any real fluid, however, the capillary number is finite. The flow-induced large pressure close to the contact line must then be balanced by the Laplace pressure. This requires the curvature of the interface to become very large for $\rho \to 0$. Thus the assumption of a statically determined meniscus close to the contact line loses its validity. For related problems, see [12, 13, 14].

The restrictions of the range of validity of the Stokes flow solution indicate the type of problems that may arise in a numerical treatment in the framework of the Navier-Stokes equations. By using the Stokes flow analysis no conclusions can be drawn regarding the real (finite Capillary number) behavior in the cold corner.

For a contact angle $\alpha = \pi/2$ the singularity is restricted to a single point. Ignoring the problem of the diverging normal stress several numerical calculations have shown that the errors in the velocity fields associated with this kind of singularity are restricted to the very vicinity of the corners (see also [15]). Therefore, the flow in the bulk can be reliably calculated. As can be seen from (19) and (49) the range of large surface curvature is exponentially small for asymptotically small capillary numbers. This is an important empirical result on which most of the numerical calculations are based, particularly those for flows in liquid bridges and rectangular containers with $\alpha = \pi/2$. The real structure of the thermocapillary flow close to the singular corners in the presence of a dynamically deformable surface has not yet been investigated.

## 2.2  Creeping flow in liquid bridges

If the thermocapillary driving forces are small, the flow is weak and nonlinear effects can be neglected throughout the whole volume $V$. For stationary conditions the resulting creeping flow can be obtained as a solution of a biharmonic equation with inhomogeneous boundary conditions. It can be given analytically in form of an infinite series. The method will be outlined here very briefly. It can be traced back to Smith [16] and has been generalized by Joseph [17] and Joseph and Sturges [18]. A further simplification is obtained in the limit of zero Prandtl number $Pr \to 0$. Then the temperature field is purely diffusive and decouples from the velocity field. The latter assumption is not too restrictive for most applications in crystal growth (low Prandtl numbers). To avoid difficulties associated with a deformation of the free surface we consider, moreover, $Bo = 0$ and the limit $Ca \to 0$.

For the calculation of creeping flows it is appropriate to use the viscous scales $\nu/d$ and $\rho\nu^2/d^2$ for the velocity and the pressure instead of those listed in table 1. The equations are re-scaled by replacing $\vec{u}$, $\psi$ and $p$ by $(\vec{u}, \psi, p)/Re$. In the limit $(Re, Pr, Ca) \to 0$ the temperature is given by the conductive solution $\Theta = 0$, where we have denoted $\Theta = \theta - z$ the deviation of the temperature from the conductive profile. It can be shown by a systematic expansion [19] that the stationary flow is described by the two-dimensional biharmonic equation for the streamfunction. In cylindrical coordinates it reads

$$\left[ DD_* + \partial_z^2 \right]^2 \psi(r, z) = 0, \tag{50}$$

where the stream function[6] $\psi$ is given by

$$u = \partial_z \psi, \qquad w = -D_* \psi, \tag{51}$$

with $D_* = D + 1/r = \partial_r + 1/r$. In the above limit and for a given volume $V = \pi R^2 d$ the fluid domain is cylindrical. Then the symmetry conditions at $r = 0$ for axisymmetrical flows and the boundary conditions (15) and (22) read

$$
\begin{aligned}
\psi &= & DD_* \psi &= 0 & \text{on} \quad r &= 0, \\
\psi &= DD_* \psi - 1 &= 0 & \text{on} \quad r &= 1/\Gamma, \\
\psi &= & \partial_z \psi &= 0 & \text{on} \quad z &= \pm 1/2.
\end{aligned}
\tag{52}
$$

The separable solution of (50) may be written as

$$\psi \propto I_1(2\lambda r)\, \phi(z, \lambda), \tag{53}$$

where $I_1$ denotes the Bessel function of first order that is regular at $r = 0$. If this form of $\psi$ is inserted into (50), the following differential equation for $\phi$ is obtained

$$\left[\partial_z^2 + 4\lambda^2\right]^2 \phi = \left[\partial_z^4 + 2(4\lambda^2)\partial_z^2 + (4\lambda^2)^2\right] \phi = 0. \tag{54}$$

The solution is a superposition of harmonics and products of harmonics with $z$. Thus

$$\psi = I_1(2\lambda r)\left[A\cos(2\lambda z) + B\sin(2\lambda z) + 2Cz\cos(2\lambda z) + 2Dz\sin(2\lambda z)\right]. \tag{55}$$

---

[6] The *streamfunction* $\psi$ used here differs from the Stokes streamfunction. The Stokes streamfunction which contour lines coincide with the streamlines is obtained as $\psi_S = r\psi$.

The four integration constants $A$, $B$, $C$, and $D$ are determined by the boundary conditions. The solvability condition of the resulting algebraic fourth order system yields the characteristic roots $\{\lambda_n\}$ as solutions of the transcendental equation[7]

$$(\lambda - \sin\lambda\cos\lambda)(\lambda + \sin\lambda\cos\lambda) = 0. \tag{56}$$

Owing to the symmetry of the differential equation, the solutions of (54) separate into even and odd functions of $z$. Since the boundary conditions (52) for $\psi$ are symmetrical in $z$, only the even functions are required. The odd functions can be obtained in an analogous way [17]. The characteristic roots belonging to the even functions are solutions of

$$\lambda + \sin\lambda\cos\lambda = 0. \tag{57}$$

They must be determined numerically. If $\lambda$ is a root then also $-\lambda$ and $\lambda^*$ are roots of (57). We define $\lambda_{-n} := \lambda_n^*$ and order the index $n$ according to the real parts of $\lambda_n$. The even functions are obtained as

$$\phi^{(n)}(z) = \phi(\lambda_n, z) = \lambda_n \left(\sin\lambda_n \cos(2\lambda_n z) - 2z\cos\lambda_n \sin(2\lambda_n z)\right). \tag{58}$$

They are called Papkovich-Fadle functions [22, 23]. The general symmetric solution can be written as a superposition of all modes

$$\psi(r, z) = \sum_{n=-\infty}^{\infty} A_n I_1(2\lambda_n r)\, \phi^{(n)}(z). \tag{59}$$

We have not yet made use of the boundary conditions (52) on the free surface. These boundary conditions are now used to determine the unknown coefficients $A_n$. Since two boundary conditions must be satisfied simultaneously, it is of advantage to write the fourth order equation (54) as two second order equations. The coefficients $A_n$ will then be obtained by projecting the equations onto suitably constructed orthonormal modes. In this formulation the modes consist of two-component vector functions. We define

$$\phi_1 := \phi, \qquad \phi_2 := \frac{1}{4\lambda^2}\phi_1''. \tag{60}$$

The differential equation (54) for $\phi$ may then be written as

$$L\vec{\Phi} = 0, \tag{61}$$

where $\vec{\Phi} = (\phi_1, \phi_2)^T$, $L := [\partial_z^2 + 4\lambda^2 \mathbf{A}]$, and

$$\mathbf{A} = \begin{pmatrix} 0 & -1 \\ 1 & 2 \end{pmatrix}. \tag{62}$$

---

[7]Hillman & Salzer [20] and Robbins & Smith [21] were the first to calculate the roots.

It is possible to define adjoint functions $\vec{\Psi}_m(z)$ with a scalar product leading to the orthogonality condition

$$\int_{-1/2}^{1/2} \vec{\Psi}^{(m)} \cdot \mathbf{A} \cdot \vec{\Phi}^{(n)} \, \mathrm{d}z \; = \; K_m \delta_{nm}, \qquad \text{with} \quad K_n \; = \; -2\cos^4 \lambda_n. \tag{63}$$

Equation (63) is the required *biorthogonality relation* for the two-component vector functions $\vec{\Phi}^{(n)}$. The unknown amplitudes $A_n$ can now be determined by projection of the boundary conditions (52) at $r = 1/\Gamma$ which can be written as

$$\begin{pmatrix} 1 \\ 0 \end{pmatrix} = \begin{pmatrix} DD_*\psi \\ \partial_z^2\psi \end{pmatrix}_{r=1/\Gamma} = \sum_{n=-\infty}^{\infty} 4\lambda_n^2 A_n \mathrm{I}_1(2\lambda_n/\Gamma) \begin{pmatrix} \phi_1^{(n)} \\ \phi_2^{(n)} \end{pmatrix}. \tag{64}$$

Multipying this equation with $\vec{\Psi}^{(m)} \cdot \mathbf{A}$, integrating over $[-\frac{1}{2}, \frac{1}{2}]$, and making use of the biorthogonality condition (63) the amplitudes $A_n$ are determined leading to

$$\psi(r,z) \; \approx \; \sum_{n=-N}^{N} \frac{-\mathrm{I}_1(2\lambda_n r)}{4\lambda_n^2 \cos^4 \lambda_n \mathrm{I}_1(2\lambda_n/\Gamma)} \, \phi_1^{(n)}(z), \tag{65}$$

where the infinite sum has been truncated at some finite value $N$. It turns out that the sequence of finite series converges rapidly for increasing $N$. Once the leading order flow field is determined, the leading order temperature field can be calculated. This is, however, omitted here. The solution for $Gr = 0$ in the limit $Ca \to 0$ has the following asymptotic form

$$\psi(r,z) \; = \; Re\,\psi_{100}(r,z) + O(Re^2), \qquad \Theta(r,z) \; = \; RePr\,\Theta_{110}(r,z) + O(Re^2), \tag{66}$$

where $\psi_{100}$ is given by (65) above. The subscripts $\{n,m,l\}$ in (66) indicate the order $O(Re^n Pr^m Ca^l)$ of the respective term in an expansion for small $Re$, $Pr$, and $Ca$. The first non-vanishing contribution to the temperature field arises in $O(Re^1 Pr^1 Ca^0)$.

In a next step, the flow induced surface deformation at leading order $O(Ca^1)$, $\xi_{001}$, may be calculated by evaluating the normal stress balance (see [19]).

### 2.2.1  Flow patterns

As an example, the streamlines (contour lines of $\psi_S = r\psi_{100}$) are shown in fig. 3a. The flow pattern consists of a single toroidal vortex with a stream function minimum $\psi_{S,min} = -0.0108$ at $r = 0.80$.[8] In fig. 3b the associated temperature field $\Theta_{110}$ is shown. As expected for small Prandtl numbers, the regions of elevated and lowered temperatures are caused by the axial flow components perpendicular to the conductive temperature profile.

Characteristic profiles of the surface deformation $\xi_{001}$ are presented in fig. 4 for different values of $\Gamma$. The constriction near the hot endwall of the liquid bridge ($z = 0.5$) and the bulging near the cold wall ($z = -0.5$) are typical for all aspect ratios.

---

[8]The minimum of $\psi_{100}$ is $\psi_{100,min} = -0.0140$ and is located at $r = 0.75$

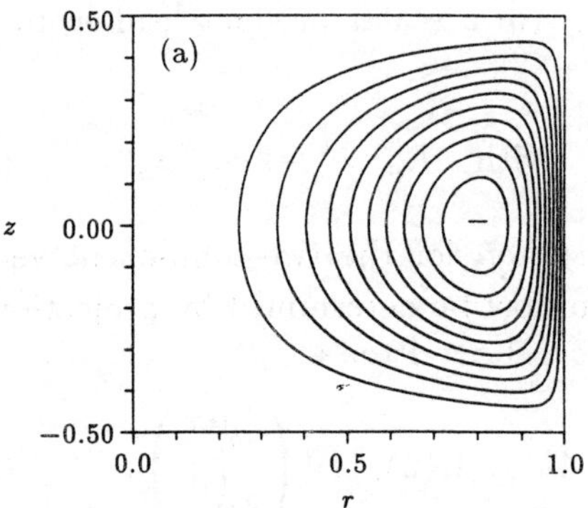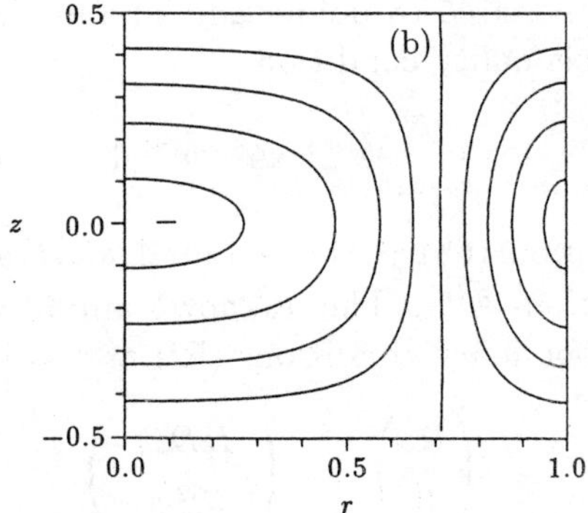

Figure 3: Streamlines (a) of the Stokes flow ($\psi_S = r\psi_{100}$) (level-lines at $-0.001 \times n$ with $n \in [1,10]$) and isotherms of $\Theta_{110}$ (b) in a small Prandtl number thermocapillary liquid bridge for $\Gamma = 1$; $N = 20$.

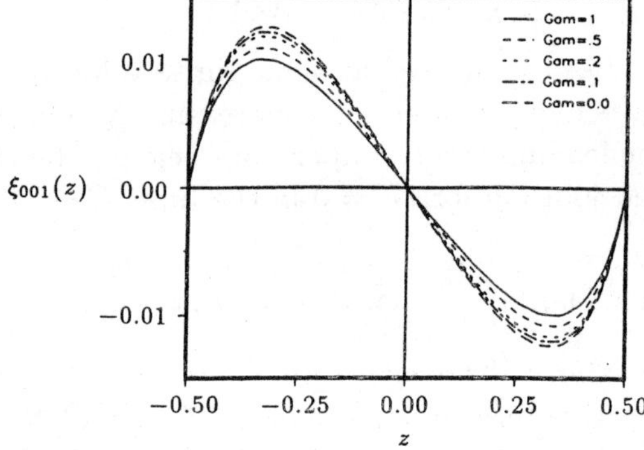


Figure 4: Surface deflection $\xi_{001}$ for $\Gamma = 0$, 0.1, 0.2, 0.5, and 1.0; $N = 20$.

The influence of a nonlinear temperature distribution in the ambient atmosphere ($\theta_a \neq z$) can easily be investigated. In order that such a modified temperature distribution $\theta_a$ influences the flow the Biot number must be non-zero. If $\theta_a$ has contributions symmetric with respect to $z$, also the antisymmetric basis functions have to be taken into account in (59) (see also [24, 25]).

### 2.2.2   Aspect ratio limits

$\Gamma \to 0$

In the limit of a very shallow liquid bridge ($\Gamma \to 0$) the expression for the stream function far away from the lateral boundaries at $r = 0$ and $r = 1/\Gamma$ can be simplified. For $1 \ll r \ll 1/\Gamma$ an expansion of the Bessel functions for large arguments in (59) yields

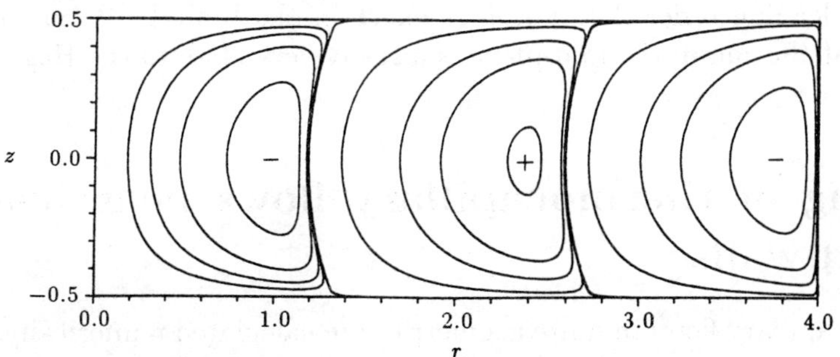

Figure 5: Creeping thermocapillary flow in a liquid bridge with $\Gamma = 0.25$; $N = 20$. Streamlines $\psi_S$ are drawn for $(-0.03, -0.01, -3 \times 10^{-3}, -3 \times 10^{-4})$, $(10^{-4}, 3 \times 10^{-5}, 10^{-5}, 10^{-6})$ and $(-10^{-7}, -3 \times 10^{-8}, -10^{-8}, -3 \times 10^{-9})$.

a nearly exponential dependence on $r$. Owing to the rapid decay of these functions for $r \to 0$ only the mode $n = 1$ contributes substantially to the flow, resulting in the approximation

$$\psi(1 \ll r \ll 1/\Gamma, z) \cong \frac{-1}{4\lambda_1^2 \cos^4 \lambda_1} \frac{\phi_1^{(1)}(z)}{\sqrt{\Gamma r}} \exp\left\{ 2\lambda_1 \left( r - \frac{1}{\Gamma} \right) \right\} + \text{c.c.} \qquad (67)$$

As can be seen from (67), the flow consists of a sequence of radially nested vortices having a self-similar shape. The flow amplitude decays nearly exponentially for $r \to 0$. The radial diameter of the vortices is $\pi/2\Im(\lambda_1) = 1.396$. In fig. 5 the streamlines for the complete stream function $\psi_S = r\psi_{100}$ according to (65) are shown for $\Gamma = 0.25$. The nested vortices have the same diameter as the rectilinear viscous Moffatt eddies between two semi-infinite parallel rigid planes [26, 27].

In the limit $\Gamma \to 0$ the surface deflection takes the form [19]

$$\xi_{001}(z, \Gamma \to 0) = \sum_{n=-\infty}^{\infty} \left( \frac{\sin(2\lambda_n z) - 2z \sin \lambda_n}{2 \cos^5 \lambda_n} - \frac{z \cos(2\lambda_n z) - z \cos \lambda_n}{\lambda_n \cos^3 \lambda_n} \right). \qquad (68)$$

The deformation has a sinusoidal shape with extrema $|\xi_{001}(z = \pm 0.33)| = 0.0125$. The curve is nearly identical with the one for $\Gamma = 0.1$ in fig. 4.

$$\Gamma \to \infty$$

An isothermal liquid bridge breaks due to the Rayleigh instability if $\Gamma > \Gamma_c = 2\pi$ [28, 29]. The critical aspect ratio $\Gamma_c$ changes only slightly in the presence of weak thermocapillary flow. Considering solution (65) at midplane $z = 0$ and expanding the Bessel functions into ascending series [30] for small $1/\Gamma$ one obtains, as expected, the well-known Hagen-Poiseuille profile

$$w_{100}(r, \Gamma \to \infty) = \frac{\Gamma}{2} \left( \frac{1}{2\Gamma^2} - r^2 \right) + O\left( \Gamma^{-3} \right). \qquad (69)$$

Note that the leading order flow becomes exact in the limit $\Gamma \to \infty$. For $\Gamma = 5$, the velocity distribution in the mid-plane is already very close to the Hagen-Poiseuille profil.

# 3 Scaling of thermocapillary flows near heated or cooled walls

Before thermocapillary flows in finite geometries are calculated numerically, it is useful to consider some scaling properties of the flow. The driving of the motion is provided by the temperature-induced surface tension gradients on the free surface. Thus the regions of high surface temperature gradients are particularly important for the flow. We consider the flow near rectangular corners which are pertinent to rectangular cavities but also apply to liquid bridges.

If the Prandtl number is low and the flow is creeping, the temperature gradient will be $O(1)$ everywhere on the free surface. As the flow becomes stronger, the isotherms on the free surface become convectively compressed towards the cold boundary and the driving forces become more localized. The raised temperature gradient near a cold wall increases the local surface stress leading to a further isotherm crowding, until the process is balanced by viscous effects and thermal diffusion. Thus the main driving is located close to the cold corner for low Prandtl number flows.

For high Prandlt number fluids, thermal boundary layers develop on the hot wall. These lead to large surface temperature gradients near the hot wall which provide the main driving forces in this case. The locally enhanced driving results in a suction of fluid from the bulk and an acceleration along the free surface away from the hot corner. Due to this suction effect the isotherms being dense along the hot wall are also dense near the free surface. Except for both corner regions with strong thermal gradients the temperature attains a nearly constant intermediate value over the remainder of the free surface.

To derive scaling laws, we must consider asymptotic limits for large Marangoni or Reynolds numbers. Depending on the particular limit, one of the corner regions will dominate in driving the flow and the forcing provided by the respective other corner will be asymptotically small. It is then sufficient to consider only a single corner with suitable assumptions about the bulk flow.

Consider the thermocapillary flow in a rectangular domain depicted in fig. 6. The above nondimensionalization (table 1) is used and the governing steady state equations ($Gr = 0$) read

$$Re\,\vec{u} \cdot \nabla\vec{u} = -\nabla p + \nabla^2\vec{u}, \tag{70}$$

$$Ma\,\vec{u} \cdot \nabla\theta = \nabla^2\theta, \tag{71}$$

$$\nabla \cdot \vec{u} = 0, \tag{72}$$

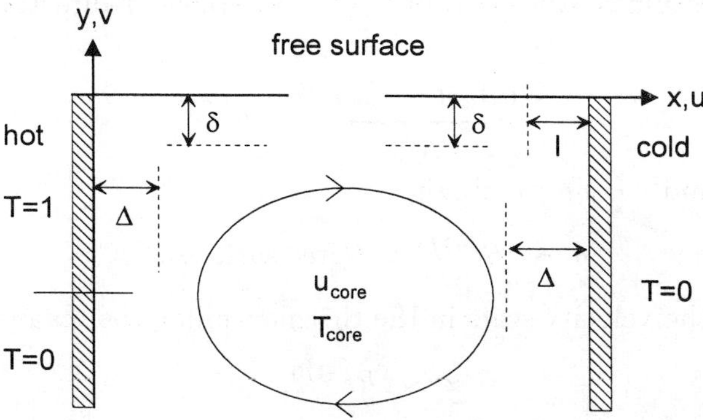


Figure 6: Structure and notation used for the discussion of the boundary layer scalings.

together with the nontrivial stress boundary condition

$$\partial_y u + \partial_x \theta = 0. \tag{73}$$

Since the inverse of the Reynolds and the Marangoni number appear in front of the highest derivatives ($\nabla^2$), the limit of large parameters is singular and a boundary layer character is expected at large Reynolds and Marangoni numbers. The diffusive terms are thus important only on decreasingly small scales close to the boundaries.[9]

First, we consider, however, the case when the temperature gradient is $O(1)$ everywhere on the free surface.

## 3.1   Low Prandtl number inertial flow

As we have seen in §2.2, the velocity scale is constant $\sim O(1)$ and boundary layers are absent when both the Prandtl and the Reynolds number tend to zero (creeping flow).

### 3.1.1   Free surface layer

In the limit Pr $\to 0$ thermal boundary layers are absent and the temperature on the free surface varies on an $O(1)$ length scale. Here we are primarily interested in the boundary layer on the free surface. If we denote the shear layer thickness by $\delta$, the thermocapillary stress condition (73) yields $\partial_y u = O(1)$, from which we get $u \sim \delta$. From the continuity equation one obtains

$$\partial_x u + \partial_y v \sim \frac{u}{1} + \frac{v}{\delta} \sim 0 \quad \Rightarrow \quad v \sim \delta u. \tag{74}$$

---

[9] For an order of magnitude analysis we can disregard the pressure term, since it is equivalent to a quadratic nonlinearity ($\nabla \cdot (70)$). Alternatively, it may be eliminated by taking the curl of (70).

The primary momentum balance parallel to the interface ($x$-direction) is determined by

$$\underbrace{Re(u\partial_x u + v\partial_y u)}_{O(\delta^2)} \sim \frac{1}{\delta^2}u. \tag{75}$$

Noting that $u \sim \delta$ and $v \sim \delta u$ we obtain

$$\delta \sim Re^{-1/3} \qquad \text{(free surface)}, \tag{76}$$

from which we get the velocity scale in the thermocapillary boundary layer

$$u \sim Re^{-1/3}. \tag{77}$$

This is the most important scaling law and has been derived by [31] and [32].

### 3.1.2 Conventional rigid wall layers

On the rigid hot and cold wall we expect conventional viscous boundary layers. This is shown in the usual way. Assume, for the moment, that the streamwise velocity is $v \sim O(1)$ outside the viscous boundary layer. From continuity we get

$$\partial_x u + \partial_y v = 0 \quad \Rightarrow \quad \frac{u}{\Delta} + O(1) \sim 0, \tag{78}$$

where $\Delta$ denotes the viscous boundary layer thickness. Therefore, $u \sim \Delta$. The momentum balance in the streamwise ($y$) direction then yields (balance between convective and leading diffusive terms)

$$Re(\underbrace{u\frac{v}{\Delta}}_{O(1)} + \underbrace{v\partial_y v}_{O(1)}) \sim \frac{1}{\Delta^2}v, \tag{79}$$

from which we get

$$\Delta \sim Re^{-1/2} \qquad \text{and} \qquad u \sim Re^{-1/2} \qquad \text{(solid wall)}. \tag{80}$$

### 3.1.3 Rigid wall layers in thermocapillary scaling

In the present case the free-stream velocity $v$ is not $O(1)$, but weaker, $v \sim Re^{-1/3}$ (cf. (77)). Continuity, therefore, requires

$$\frac{u}{\Delta} + \frac{Re^{-1/3}}{1} \sim 0, \tag{81}$$

thus $u \sim \Delta Re^{-1/3}$. The streamwise momentum balance then yields

$$\underbrace{Re(u\frac{v}{\Delta} + v\partial_y v)}_{\sim Re^{-2/3}} \sim \frac{1}{\Delta^2}v. \tag{82}$$

We obtain

$$\Delta \sim Re^{-1/3} \qquad \text{(solid wall)}, \tag{83}$$

and within the particular velocity scale (table 1)

$$u \sim Re^{-2/3} \qquad \text{(solid wall)}. \tag{84}$$

Thus the viscous boundary layer thickness on the rigid walls scales like the free surface boundary layer thickness $\sim Re^{-1/3}$. Note that the relative cross-stream velocity $u/v \sim Re^{-1/3}$ scales like the boundary layer thickness (as must be for viscous wall boundary layers). The boundary layer thickness could have also be obtained by considering the velocity field on a scale $Re^{2/3}$ instead of the present $Re$-scale, such that $v = O(1)$. In that case $Re^{2/3}$ instead of $Re$ would appear in front of the convective term.[10]

## 3.2    Scaling of the flow near the cold corner

Different from the pure inertial flow, we now investigate the high Prandtl number flow near the cold corner. Then the temperature is no longer $O(1)$ over the full length of the free surface but rather varies only on a length scale $l$ close to the corner. We use the length $l$ and the corresponding temperature drop to define the Marangoni number here. For an insulating free surface it is reasonable to assume that the temperature field varies only weakly in $y$-direction perpendicular to the free surface. Using these assumptions, Canright [9] derived several scaling limits for $\{Re, Ma\} \to \{0, \infty\}$. As an example we consider the convective viscous limit $Ma \to \infty$, $Re \to 0$.

Since all the local driving is located within the length $l$ from the corner, the shear stress condition of the free surface yields

$$\partial_y u + \partial_x \theta \ \longrightarrow \ \frac{u}{\delta} + \frac{(\theta \sim 1)}{l} \ \sim 0 \ \longrightarrow \ u \sim \frac{\delta}{l}, \tag{85}$$

where $\delta$ is the free surface layer thickness. From the continuity equation we get

$$\partial_x u + \partial_y v \ \longrightarrow \ \frac{u}{l} + \frac{v}{\delta} \ \sim 0 \ \longrightarrow \ v \sim u\frac{\delta}{l}. \tag{86}$$

The viscous dominated ($Re \to 0$) momentum balance requires $\delta^{-2} + l^{-2} \sim 0$, from which we conclude $\delta \sim l$. Thus $u \sim v \sim 1$. The energy equation in the limit $Ma \to \infty$ now yields

$$Ma\left[1 + \frac{1}{l}\right] \ \sim 1 + \frac{1}{l^2}, \tag{87}$$

where we have used that $\partial_y \theta \sim O(1)$. From this we obtain in the scaling

$$l \ \sim Ma^{-1}. \tag{88}$$

---

[10]Since the fluid is ejected along the cold wall from a relatively small region near the corner, it might be reasonable to interpret the local flow in terms of a wall jet [33] with its characteristic scaling.

| Type | limit | $l$ | $\delta$ | $w$ |
|---|---|---|---|---|
| conductive-viscous | $Ma \to 0$, $Re \to 0$ | 1 | 1 | 1 |
| conductive-inertial | $Ma \ll Re^{\frac{1}{3}}$, $Re \to \infty$ | 1 | $Re^{-\frac{1}{3}}$ | $Re^{-\frac{1}{3}}$ |
| convective-viscous | $Ma \to \infty$, $Re \ll Ma$ | $Ma^{-1}$ | $Ma^{-1}$ | 1 |
| convective-inertial | $Ma^3 \gg Re$, $Re \to \infty$ | $Ma^{-1}Pr^{-\frac{1}{2}}$ | $Ma^{-1}$ | $Pr^{\frac{1}{2}}$ |

Table 2: Scaling of the boundary layers near the thermocapillary cold corner with contact angle $\alpha = \pi/2$; after [9]. $\rho_0$ is the distance from the cold corner within which the flow is creeping.

Using these results, the momentum equation may be reconsidered to see that the limit $Re \to 0$ may be relaxed to $Re \ll Ma$.

Canright [9] has derived several other scaling limits for the flow in the cold corner which are summarized in table 2.

## 3.3   The scaling of high Prandtl number thermocapillary flow in a hot corner (viscous convective limit)

Cowley and Davis [34] considered the viscous flow of a high Prandtl number fluid in a hot corner ($Re = O(1), Ma \to \infty$). We shall not repeat the calculations here but rather briefly present the result.

The analysis starts with assuming a scaling of the form

$$|\vec{u}_{core}| = O(Ma^{-b}), \qquad \Delta = O(Ma^{-f}), \qquad \delta = O(Ma^{-d}), \tag{89}$$

where $\vec{u}_{core}$ is the velocity in the core far away from the boundaries, $\Delta$ the thermal boundary layer thickness on the hot wall, and $\delta$ the thickness of the free surface layer. By considering the order of magnitude relations obtained for the momentum and heat transport in the boundary layers on the rigid hot wall and on the free surface, and by using the thermocapillary boundary condition as well as the global heat conservation, one can derive the following scaling

$$\begin{aligned} |\vec{u}_{core}| &\sim Ma^{-1/7}, \qquad \Delta \sim Ma^{-2/7} \quad \text{(hot wall)}, \qquad Nu \sim Ma^{2/7}, \\ \theta_{surface} &\sim Ma^{-1/7}, \qquad \delta \sim Ma^{-3/7} \quad \text{(free surface)}, \end{aligned} \tag{90}$$

where $Nu$ denotes the Nusselt number on the rigid wall.

The question arises, which of the two regions of localized forcing, the hot or the cold corner, dominates the flow in a finite size geometry. Numerical evidence suggests that, for high Prandtl numbers and Reynolds numbers not too large, the hot corner determines the scaling of the surface temperature, bulk velocity, and Nusselt number.

The reason probably is that the high velocity near the cold corner arises within a region $O(Ma^{-1})$ which tends to zero as $Ma$ increases. Since the fluid is accelerated towards the wall, the locally high velocity decays rapidly away from the corner. On the other hand, the fluid near the hot corner is accelerated away from the rigid wall within a region of $O(Ma^{-2/7})$, which is much larger.[11]

# 4  Hydrothermal waves in plane thermocapillary layers

Before thermocapillary flows in finite geometries are investigated in detail it is useful to consider the much simpler problem of thermocapillary plane layers. A comparison with realistic systems shows that the convective instability mechanisms in plane layers are also operative in finite size systems,[12] at least for high Prandtl numbers.

Extended liquid layers have been studied by [35, 36, 37, 38, 39, 40]. A review is due to Davis [41]. The physical mechanisms of the convective instabilities in plane thermocapillary layers have been discussed by Smith & Davis [42].

We consider a liquid layer extended infinitely in $(x, y)$-direction and having a depth $d$. Let the liquid be bounded from below by a rigid wall at $z = -d/2$ and from above by a free surface at $z = d/2$. If the medium above the liquid layer is a passive gas at a temperature depending linearly on the $x$-coordinate $T = T_0 - bx$ with $T_0 = const.$ a constant surface stress in $x$-direction is induced due to the thermocapillary effect. We select the scales $d$, $d^2/\nu$, $\gamma bd/\rho\nu$, $\gamma b$, and $bd$ for length, time, velocity, pressure, and temperature $(\theta = (T - T_0)/bd)$, and the volume equations (11) – (13) are considered for $Gr = 0$. The boundary conditions for the basic state $(\vec{u}_0, \theta_0)$ are

$$\vec{u}_0 = \partial_z\theta_0 \;=\; 0, \qquad \text{on} \quad z = -1/2, \tag{91}$$

$$w_0 = \partial_z v_0 = \partial_z u_0 - 1 = \partial_z\theta_0 \;=\; 0, \qquad \text{on} \quad z = 1/2, \tag{92}$$

where the free surface has been assumed thermally insulating. The basic equations allow for a set of solutions satisfying

$$\vec{u}_0 \;=\; -g''(z)\,\vec{e}_x, \qquad\qquad \theta_0 \;=\; -x + RePr\,g(z), \tag{93}$$

where $g$ is a fourth order polynomial and the Reynolds number is given by

$$Re \;=\; \frac{\gamma bd^2}{\rho\nu^2}. \tag{94}$$

---

[11]This argument requires that the extended region of constant temperature between the heated walls remains at a temperature sufficiently different from both wall temperatures.

[12]In finite length systems also other types of instabilities may arise, particularly those that are just caused by the presence of no-penetration boundaries.

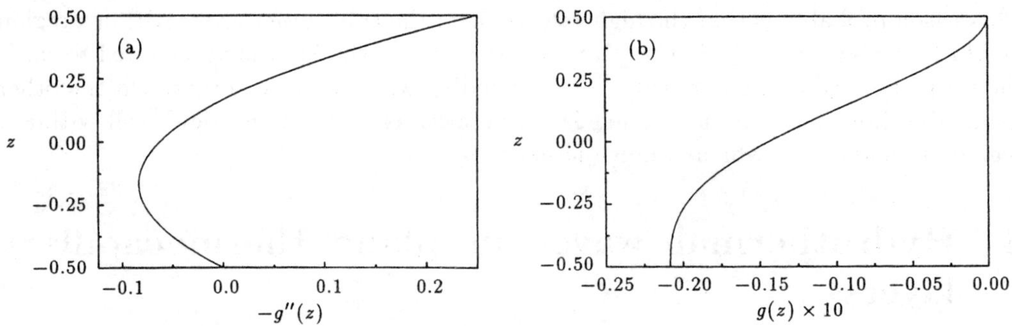

Figure 7: Profiles of the basic state velocity $-g''(z)$ (a) and temperature $g(z)$ (b) in an adiabatic plane thermocapillary layer without through flow.

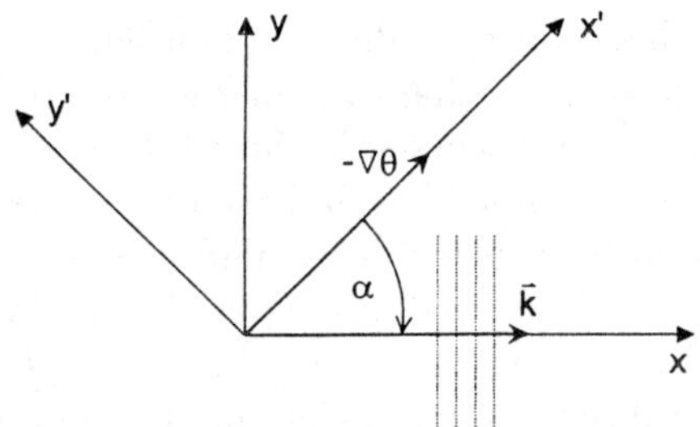


Figure 8: Orientation of $-\nabla\theta_a$ and $\vec{k}$ relative to the coordinate system.

Since most systems of interest are not subject to a through flow, the solution with vanishing lateral through flow ($\int_{-1/2}^{1/2} u_0 dz = 0$) is of particular interest. This flow is called *return flow*.[13] The solution is easily found to be

$$g(z) = -\frac{1}{64}\left(4z^4 + \frac{8}{3}z^3 - 2z^2 - 2z + \frac{11}{12}\right).$$

(95)

The base state velocity and temperature profiles are shown in fig. 7.

We are interested in the stability of this simple solution. Since the system is laterally unbounded, the solutions of the linearized disturbance equations can be written as normal modes in the $x$- and $y$-directions. These modes are plane waves propagating under an angle $\alpha$ with respect to the negative temperature gradient, i. e. the direction of the basic flow at the free surface. In a coordinate system in which the $x$-axis is parallel to the $\vec{k}$-vector (see fig. 8) the basic state reads

$$\begin{aligned}
u_0 &= -g''(z)\cos\alpha, & w_0 &= 0, \\
v_0 &= -g''(z)\sin\alpha, & \theta_0 &= -(x\cos\alpha + y\sin\alpha) + RePr\,g(z).
\end{aligned}$$

(96)

---

[13]For the stability of the thermocapillary plane Couette flow, see [38]. See [37] for the stability of the return flow with conducting bottom boundary.

After eliminating the pressure and introducing the stream function $\psi$

$$u = \partial_z \psi, \qquad w = -\partial_x \psi, \tag{97}$$

we obtain the nonlinear equations governing the deviations $(\psi, v, \theta)$ from the basic state

$$(\partial_t - \Delta)\Delta\psi = -Re\left[(\psi_z\partial_x - \psi_x\partial_z - cg''\partial_x)\Delta\psi + cg''''\psi_x\right], \tag{98}$$

$$(\partial_t - \Delta)v = -Re\left[(\psi_z\partial_x - \psi_x\partial_z - cg''\partial_x)v + sg'''\psi_x\right], \tag{99}$$

$$\left(\partial_t - \frac{\Delta}{Pr}\right)\theta = -Re\left[(\psi_z\partial_x - \psi_x\partial_z - cg''\partial_x)\theta - c\psi_z - sv\right] + Re^2 Pr g'\psi_x, \tag{100}$$

together with the corresponding boundary conditions. We have used the abbreviations $c = \cos\alpha$ and $s = \sin\alpha$. By choice of the coordinate system the unknowns $\phi \in \{\psi, v, \theta\}$ do no longer depend on $y$. Therefore, the solution of the linearized problem can be written as

$$\phi(x, z, t) = \tilde{\phi}(z)e^{ikx}e^{(\sigma - i\omega)t} + \text{c.c.} \tag{101}$$

Due to the symmetry of the linearized perturbation equations only the quadrant $\alpha \in [0, \pi/2]$ needs to be considered.

The linearized system of equations has to be solved numerically. One can employ a Chebyshev-$\tau$ method [43]. To that end the variable $z \longrightarrow 2z$ is stretched and all quantities $\tilde{\phi}(z)$ are expanded into Chebyshev polynomials $T_n(z) = \cos[n\arccos(z)]$ with $z \in [-1, 1]$

$$\left\{\tilde{\psi}, \tilde{v}, \tilde{\theta}\right\}(z) = \sum_{n=0}^{N} \left\{\hat{\psi}_n, \hat{v}_n, \hat{\theta}_n\right\} T_n(z), \tag{102}$$

where the infinite sum has been truncated at order $N$. This ansatz is inserted into the linearized version of (98) – (100) and the resulting equations are projected onto the orthogonal polynomials $T_m(z)$. This way one obtains a linear algebraic system of equations of the form

$$\begin{bmatrix} \mathbf{A}^{(\Re)} & -\mathbf{A}^{(\Im)} \\ \mathbf{A}^{(\Im)} & \mathbf{A}^{(\Re)} \end{bmatrix} \begin{pmatrix} \vec{\psi}^{(\Re)} \\ \vec{\psi}^{(\Im)} \end{pmatrix} = 0. \tag{103}$$

Here, $\vec{\psi} = \vec{\psi}^{(\Re)} + i\vec{\psi}^{(\Im)} = (\hat{\psi}_n, \hat{v}_n, \hat{\theta}_n)$ denotes the $3(N+1)$-dimensional vector of the complex field amplitudes $((\vec{\psi}^{(\Re)}, \vec{\psi}^{(\Im)}) \in \mathbb{R}^{6(N+1)})$. Considering neutral stability $(\sigma = 0)$, the complex coefficient matrix $\mathbf{A} = \mathbf{A}^{(\Re)} + i\mathbf{A}^{(\Im)}$ depends on the parameters $\alpha$, $k$, $\omega$, $Pr$, $Re$, and eventually on $Bi$. For the existence of a non-trivial solution, the coefficient determinant in (103) must vanish. Owing to the structure of $\mathbf{A}$ it is easy to show that $\det(\mathbf{A}) \geq 0$. Thus the zero $(\omega, Re)$ of $\det(\mathbf{A})$ is also a minimum. The zero may be calculated numerically by a Newton method for given parameters $(Pr, Bi, \alpha, k)$. After that, $Re$ can be minimized with respect to $\alpha$ and $k$ by help of a gradient method [44] to find the critical values $Re_c$, $\omega_c$, $k_c$, and $\alpha_c$.

The critical mode consists of a pair of plane waves that propagate under the angles $\pm|\alpha|$ with respect to the negative temperature gradient. For these waves, Smith &

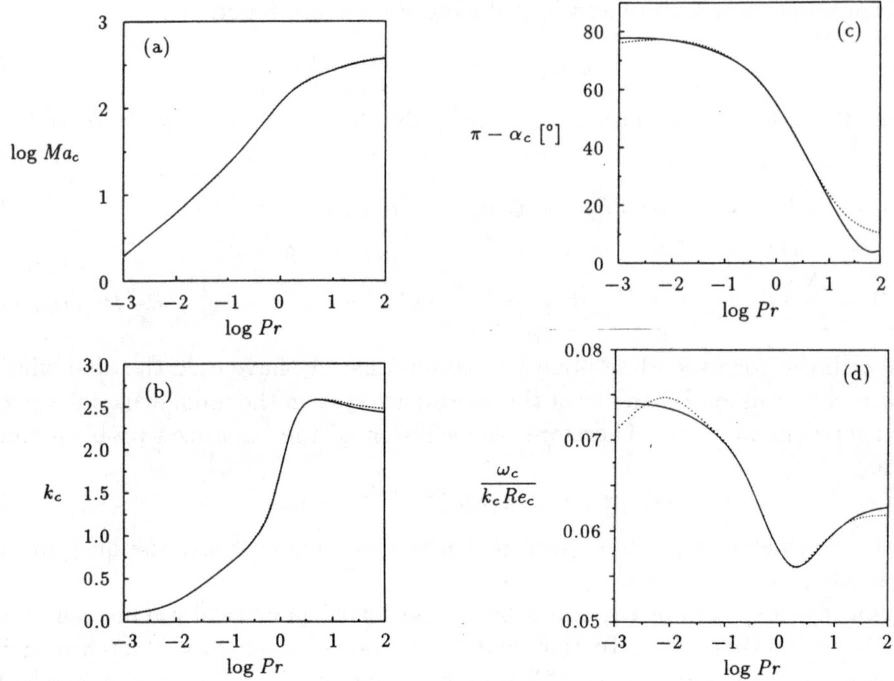

Figure 9: Critical Marangoni number $Ma_c$ (a), critical wavenumber $k_c$ (b), critical angle of propagation $\alpha_c$ (c), and critical phase velocity $\omega_c/k_c$ (d) as a function of the Prandtl number for the *return flow* at $Gr = Bi = 0$ (full lines). For comparison, the results of [38] are shown as dashed lines. The mode truncation is $N = 10$.

Davis [38] have coined the name *hydrothermal waves*. The critical Marangoni number $Ma_c$, the wavenumber $k_c$, the propagation angle $\alpha_c$, and the phase velocity $\omega_c/k_c$ are shown in fig. 9 for $N = 10$ as functions of the Prandtl number for $Bi = 0$ (full curves). Up to the lines' thickness the curves for $N = 10$ agree with those for $N = 20$. Fig. 10 shows examples of the critical modes for low ($Pr = 0.01$) and high ($Pr = 100$) Prandtl numbers.

For high Prandtl numbers the hydrothermal waves propagate with a wavenumber $k \approx 2.6$ nearly parallel to the temperature gradient and opposite to the surface flow, whereas, for small Prandtl numbers, they propagate nearly perpendicular to the applied temperature gradient having a very small wavenumber. Further calculations show that a non-zero Biot number does not change the character of the instability. It can be shown that the basic flow (93) does not depend on the Biot number. The only term depending on $Bi$ is the stabilizing energy term $H$ (cf. (35)). Therefore, the critical Reynolds numbers for $Bi > 0$ are monotonically shifted to higher values (see also [38]).

Using an energy analysis, Smith [42, 45] showed that the hydrothermal waves at small Prandt numbers obtain their energy from the velocity field. Due to the classical

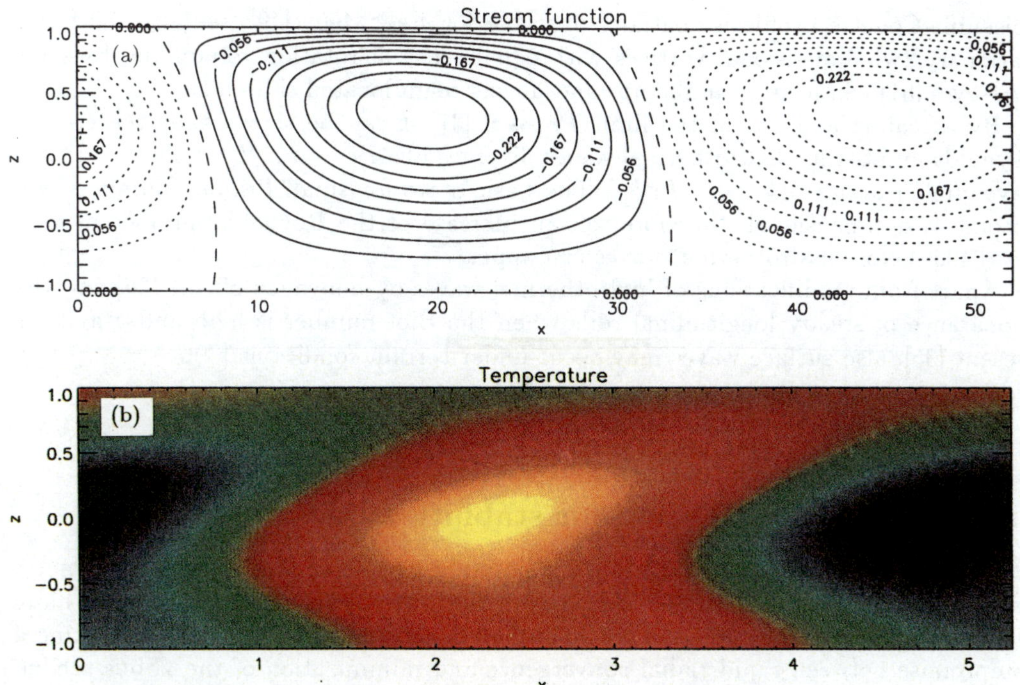

Figure 10: Streamlines of the critical mode for $Pr = 0.01$ (a) and critical temperature field for $Pr = 100$ (b) in a cut parallel to the direction of propagation which is to the left.

Marangoni effect [46] $x$-momentum of the disturbance field is amplified through a supply of $x$-momentum from the basic flow.

At high Prandtl numbers the hydrothermal waves exhibit strong temperature extrema in the bulk of the fluid layer (see fig. 10). Convective transport of the basic state temperature field owing to the disturbance flow maintains the internal temperature extrema. The pattern propagates, since the associated temperature extrema on the free surface are phase shifted with respect to the internal ones. A similar mechanism can also be found in thermocapillary liquid bridges and will be discussed in §5.3.2.

By a perturbation technique, Sen & Davis [47] have shown that the thermocapillary flow in finite length containers with $L \gg d$ far away from the lateral endwalls can be well described by the above Couette-Poiseuille profile.[14] This has been confirmed by Saedeleer *et al.* [48] for small Marangoni numbers. In view of the simplifications inherent to the model it is difficult, however, to experimentally observe the ideal hydrothermal waves in cavities with insulating bottom. For instance, the temperature distribution on the free surface cannot simply be enforced [48]. Moreover, the

---

[14]In order that no dynamic surface deformations occur the relation $Ca < O(Re(d/L)^4)$ must, however, be satisfied in the limit $d/L \to 0$.

Poiseuille-Couette profile is spatially unstable for $Re \gtrsim 400$ ([49], see also [50]) and stationary two-dimensional vortices with equal sense of rotation resembling Kelvin's cat's-eyes [51] can form near the up- and downstream located sidewalls.[15]

By a scaling analysis in the limit $Pr \to 0$ [37] it can be shown that the critical Reynolds number for hydrothermal waves diverges like $Re_c \sim Pr^{-1/2}$ while the critical wavenumber scales like $k_c \sim Pr^{1/2}$. The basic state for small Prandtl numbers will become two-dimensional, therefore, on an increase of the Reynolds number beyond $Re \approx 400$, before hydrothermal waves can appear.

Apart from modifications of hydrothermal waves by buoyancy effects [36] and the appearance of steady longitudinal rolls when the Biot number is high and gravity is present [35], also surface waves may occur under certain conditions [39].

## 5   Thermocapillary liquid bridges

### 5.1   A numerical method for stability analyses

For a theoretical treatment of the flows in liquid bridges one relies on numerical methods. Several techniques can be employed. Here we use a mixed method of finite differences in axial direction and a Chebyshev collocation method radially making a compromise between rapid radial convergence and minimization of the Gibbs oscillations due to the discontinuities of the boundary conditions. Details of the method can be found in [54]. The advantages of mixed methods have been demonstrated by application to the Taylor-Couette problem [55].

Starting point for the analysis is the stationary axisymmetric base state. A stream function - vorticity formulation $(\psi_0, \omega_0)$ is employed, where $\psi_0$ and $\omega_0 = \tilde{L}\psi_0 = (DD_* + \partial_z^2)\psi_0$ are chosen as in (51). The bulk equations for the base state are obtained from (11,13)

$$\tilde{L}^2\psi_0 \;=\; GrD\Theta_0 - \left[\frac{2}{r}(\partial_z\psi_0) + (D_*\psi_0)\partial_z - (\partial_z\psi_0)D_*\right]\tilde{L}\psi_0, \qquad (104)$$

$$\Delta\Theta_0 \;=\; -Pr\{[(D_*\psi_0)\partial_z - (\partial_z\psi_0)D]\Theta_0 + D_*\psi_0\}. \qquad (105)$$

The unknowns $(\psi_0, \omega_0, \Theta_0)$ are calculated on a grid consisting of $M+1$ equidistant finite difference points in axial direction and of $N+1$ radial Gauß-Lobatto points [43], i. e. the roots of $T'_N(x_i) = 0$,

$$\{\psi_0, \omega_0, \Theta_0\}(x, z) \;\longrightarrow\; \{\psi_0, \omega_0, \Theta_0\}_{ij}, \qquad \text{where} \qquad (106)$$

$$x_i \;=\; \cos(i\pi/N), \quad i \in [0, N], \qquad \text{and} \qquad z_j = -1/2 + j/M, \quad j \in [0, M].$$

Here the radial coordinate has been transformed as $x = 2\Gamma r - 1$. The radial derivatives are obatined by multiplying the vector of unknowns at the Gauss-Lobatto points

---

with the Chebyshev collocation derivative matrix (see, e. g. [43]). The so discretized nonlinear equations are solved with a Newton-Raphson method.

For the stability analysis of the discretized base state small perturbations are written as normal modes

$$\{u, v, w, p, \Theta\}(r, z, \varphi, t) = \{\hat{u}, \hat{v}, \hat{w}, \hat{p}, \hat{\Theta}\}(r, z)\, e^{\gamma t + im\varphi} + \text{c.c.}, \qquad (107)$$

where $m$ is an azimuthal wavenumber and $\gamma$ a complex growth rate. For a solution of the resulting linear stability equations, the azimuthal velocity is eliminated by use of the continuity equation[16]

$$\hat{v} = -\frac{r}{im}\left(D_*\hat{u} + \partial_z\hat{w}\right). \qquad (108)$$

The equations to be solved consist of radial and axial momentum balances, the temperature equation, and the Poisson equation. The linearized equations read

$$\gamma\hat{u} + Re\left(\vec{u}_0 \cdot \nabla\hat{u} + \hat{\vec{u}} \cdot \nabla u_0\right) = -D\hat{p} + \left(\Delta - \frac{1}{r^2}\right)\hat{u} + \frac{2}{r}\left(D_*\hat{u} + \partial_z\hat{w}\right), \quad (109)$$

$$\gamma\hat{w} + Re\left(\vec{u}_0 \cdot \nabla\hat{w} + \hat{\vec{u}} \cdot \nabla w_0\right) = -\partial_z\hat{p} + \Delta\hat{w} + Bd\,\hat{\Theta}, \qquad (110)$$

$$\gamma\hat{\Theta} + Re\left(\vec{u}_0 \cdot \nabla\hat{\Theta} + \hat{\vec{u}} \cdot \nabla\Theta_0 + \hat{w}\right) = \frac{1}{Pr}\Delta\hat{\Theta}, \qquad (111)$$

$$-2Re\nabla \cdot \left(\hat{\vec{u}} \cdot \nabla\vec{u}_0\right) = \Delta\hat{p}, \qquad (112)$$

where $\partial_\varphi$ is to be replaced by $im$ in the differential operators. The boundary conditions require

$$\hat{u} = \hat{w} = \partial_z\hat{w} = \hat{\Theta} = 0, \qquad \text{on} \qquad z = \pm 1/2, \qquad (113)$$

as well as

$$\left.\begin{array}{r} \hat{u} = 0 \\ D\hat{w} + \partial_z\hat{\Theta} = 0 \\ r^2(D_*D\hat{u} + \partial_z D\hat{w}) + m^2\hat{\Theta} = 0 \\ (D + Bi)\hat{\Theta} = 0 \end{array}\right\} \quad \text{on} \quad r = \frac{1}{\Gamma}, \qquad (114)$$

$$\left.\begin{array}{rl} \hat{u} = D\hat{w} = D\hat{p} = D\hat{\Theta} = 0, & \text{if} \quad m = 0 \\ D\hat{u} = \hat{w} = \hat{p} = \hat{\Theta} = 0, & \text{if} \quad m = 1 \\ \hat{u} = \hat{w} = \hat{p} = \hat{\Theta} = 0, & \text{if} \quad m > 1 \end{array}\right\} \quad \text{on} \quad r = 0. \qquad (115)$$

The discretization of the perturbation equations is carried out in analogy to the base state discretization leading to the generalized eigenvalue problem

$$\mathbf{A}\vec{x} = \gamma\mathbf{B}\vec{x}, \qquad (116)$$

---

[16]For $m = 0$, however, a stream function - vorticity formulation as for the base state has been used.

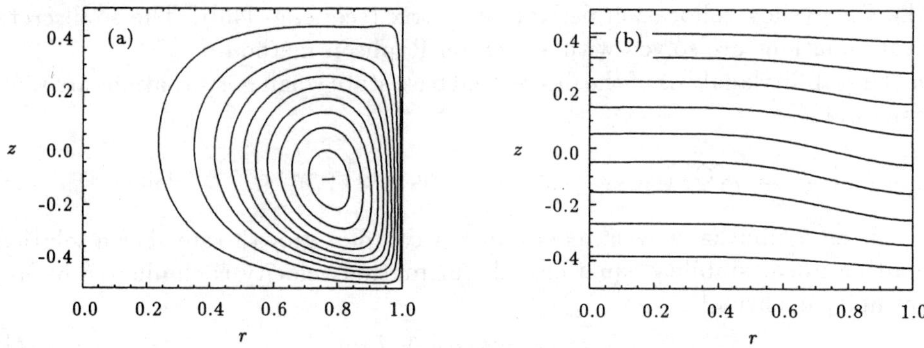

Figure 11: Stream function (a) and isotherms (b) in a liquid bridge for $\Gamma = 1$, $Pr = 0.02$, $Re = 2000$, and $Bi = Gr = 0$.

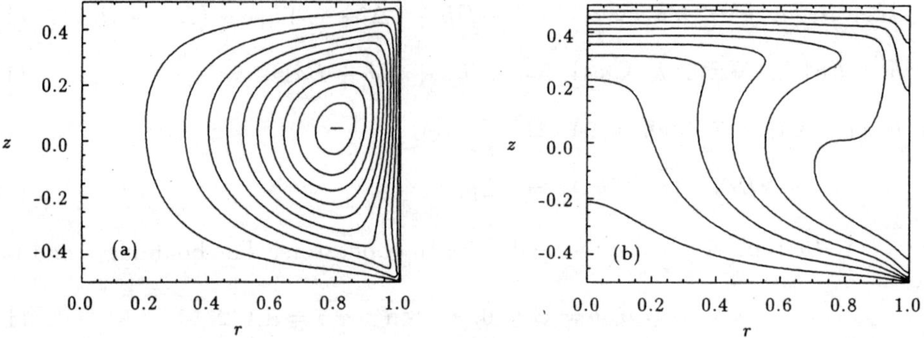

Figure 12: Stream function (a) and isotherms (b) in a liquid bridge for $\Gamma = 1$, $Pr = 2$, $Re = 2000$, and $Bi = Gr = 0$.

where $\vec{x} = (\hat{u}, \hat{w}, \hat{p}, \hat{\Theta})_{ij}$. The matrix $\mathbf{A}$ has a block diagonal structure owing to the particular method of discretization, if the index $i$ of the Chebyshev polynomials is selected as the fast (inner) index. $\mathbf{B}$ is diagonal and singular. The system (116) was either solved for all eigenvalues and eigenvectors or for the eigenvector and eigenvalue with the largest real part using inverse iteration.

## 5.2   Two-dimensional flows

Semi-conductor melts, e. g. in the floating-zone method, have a low Prandtl number ($Pr(\mathrm{Si}) \approx 0.02$). For low Prandtl number and not too high Reynolds number convective effects will be weak and the flow will scale $\sim Re^{-1/3}$ (cf. (77)). Since the inertia terms violate the mirror symmetry with respect to $z = 0$ in (11) – (13) and due to slight convective effects the streamlines become increasingly asymmetric for increasing Reynolds number and the center of the vortex is shifted towards the cold corner. For

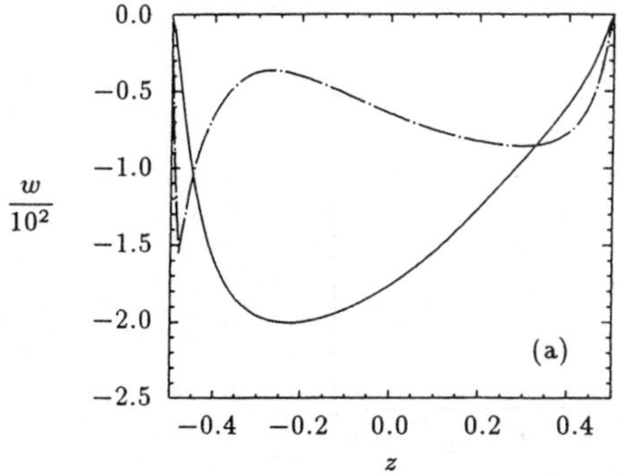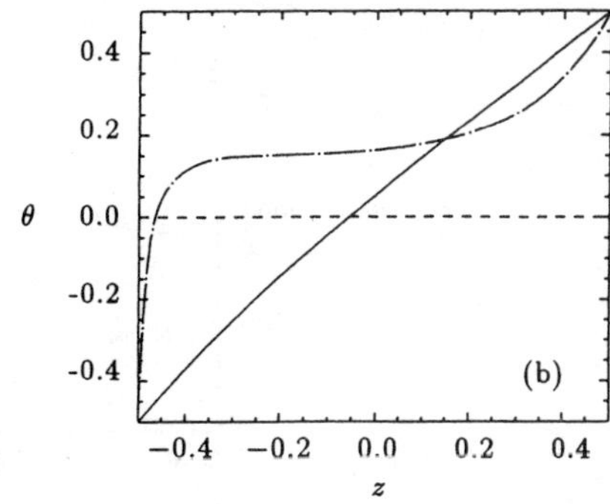

Figure 13: Axial velocity $w(z)$ in units of $\nu/d$ (a) and temperature $\theta(z)$ (b) on the free surface at $r = 1/\Gamma$ for $\Gamma = 1$, $Re = 2000$, and $Bi = Gr = 0$. ( —— ): $Pr = 0.02$, $(- \cdot - \cdot -)$: $Pr = 2$.

$Pr = 0.02$ and $Re = 2000$ this is illustrated in fig. 11. When the Reynolds number is further increased, the vortex develops a nearly inviscid core (Prandtl-Batchelor theorem, see e. g. [56, 57]) in which the vorticity $\omega$ varies linearly with $r$ ($\omega/r = const.$, [58]) and which is surrounded by boundary layers.

Contrary to liquid metals, convective effects are more pronounced in transparent liquids which are commonly used for experimental modeling. Here, the Prandtl numbers range from $Pr \approx 1$ for KCl melts [59] up to $Pr > 100$ for some silicone oils. As an example the two-dimensional flow for $Pr = 2$ and $Re = 2000$ is shown in fig. 12. The (not yet fully developed) thermal boundary layer near $z = 0.5$ is clearly visible as well as the compression of the isotherms near the cold corner ($z = -0.5$, $r = 1$) together with the associated crowding of streamlines. The center of the vortex is located closer to the hot corner indicating its importance for the driving of the flow. The axial velocity and the temperature along the free surface are shown in fig. 13. The high gradients near $z \approx -0.5$ (for higher Marangoni numbers also near $z \approx 0.5$) make numerical computations for high Marangoni numbers increasingly costly. The strong localization of the axial thermal gradient close to the cold corner leads to a heat transfer to the solid wall at $z = -0.5$ that is mainly concentrated to a narrow annular region.

The above properties of the two-dimensional thermocapillary convection essentially agree with the experimental findings [60]. Observed deviations from the results of model calculations are mainly caused by non-cylindrical surface deformations, additional buoyancy forces, contamination of the free surface, temperature-dependent material parameters, and momentum and heat transfer to the ambient gas. The influence of these effects is largely unknown and cannot not be discussed here.

When buoyancy is more important than thermocapillarity, multiple two-dimensional flows are possible. These flows and their stability have been investigated by [61].

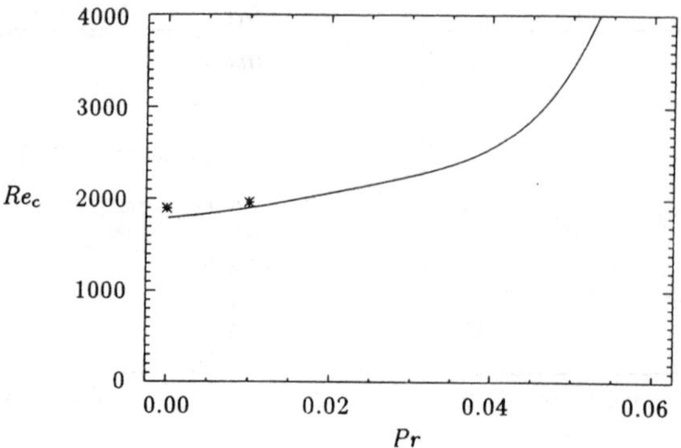

Figure 14: Critical Reynolds numbers $Re_c$ for $\Gamma = 1$, $Bi = 0$ and small Prandtl numbers; (*): [62].

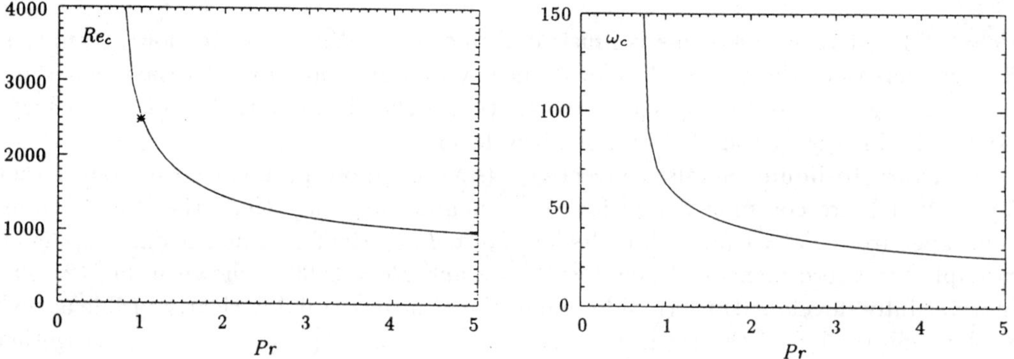

Figure 15: Critical Reynolds numbers $Re_c$ (a) and oscillation frequencies (b) for $\Gamma = 1$, $Bi = 0$ and large Prandtl numbers; (*): [63].

## 5.3  Linear stability

The stability of the basic two-dimensional flow depends strongly on the Prandtl number. The typical behavior of the critical Reynolds number $Re_c$ as function of $Pr$ can be demonstrated for the aspect ratio $\Gamma = 1$ and for an insulating free surface ($Bi = 0$) (figs. 14 and 15(a)). In this case the critical azimuthal wavenumber is $m_c = 2$ independent of $Pr$. For very small Prandtl numbers the critical Reynolds number increases moderately from the limiting value of $Re_c = 1793$ at $Pr \approx 0$. For Prandlt numbers larger than $Pr \approx 0.05$ the critical Reynolds number increases, however, very rapidly and the instability boundary has not been followed to higher Prandtl numbers, because the numerical error in the calculated critical values becomes too large. In the range of intermediate Prandtl numbers ($0.05 \lesssim Pr \lesssim 0.7$) the basic flow is linearly stable for $Re \lesssim 5000$, not only for $m = 2$ but also for other wavenumbers. The comparatively

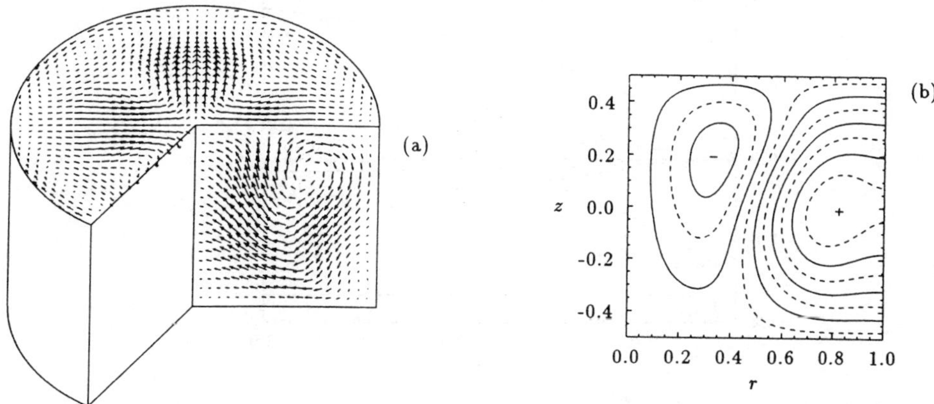

Figure 16: Critical mode for $Pr = 0.02$ at $Re_c = 2062$ ($m = 2$, $\Gamma = 1$, $Bi = 0$):
(a) velocity field at $\varphi = 0$ and projection of the velocity field onto $z = -0.25$, (b)
temperature field $\Theta$ at $\varphi = 0$.

stable range is followed by a range of Prandtl numbers for which the first instability is
oscillatory. The oscillation frequencies are shown in fig. 15(b). Since the character of
the critical mode does not change along each of the two individual critical curves, the
instabilities are analyzed in more detail for two representative, high and low Prandtl
numbers.

### 5.3.1  Small Prandtl numbers: $Pr = 0.02$

The critical mode for $Pr = 0.02$ ($Re_c = 2062$) is shown in fig. 16. The pattern consists
of four symmetric convection cells each occupying a radial section $\Delta\varphi = \pi/2$. The
$m = 2$ symmetry is readily identified by the regions of alternating radial in- and
outflow visible in a horizontal cut shown in fig. 16a. The azimuthal velocity vanishes
on the vertical cell boundaries at constant $\varphi$. Similar as the basic state the critical
velocity field at $\varphi = 0$ has a single vortical structure. Owing to the small Prandtl
number the convective temperature transport is comparatively weak. Therefore, the
temperature field of the neutral mode is mainly determined by the convective transport
($w\partial_z$) of the conductive base state temperature profile ($\theta \approx z$) leading to the balance
$\Delta\Theta \approx RePrw$.

To understand the mechanics of this mode it is useful to analyze its energetics. The
normalized rates of change of kinetic energy are shown in fig. 17. For small Reynolds
numbers the dissipation $D$ is dominating and the mode is strongly damped. With
increasing Reynolds number the interaction term $I_{v4}$ increases an finally dominates the
balance for $Re \gtrsim 2500$. On the stability threshold $D$ and $I_{v4}$ nearly compensate each
other. All other contributions to the kinetic energy balance are much smaller, even
locally.

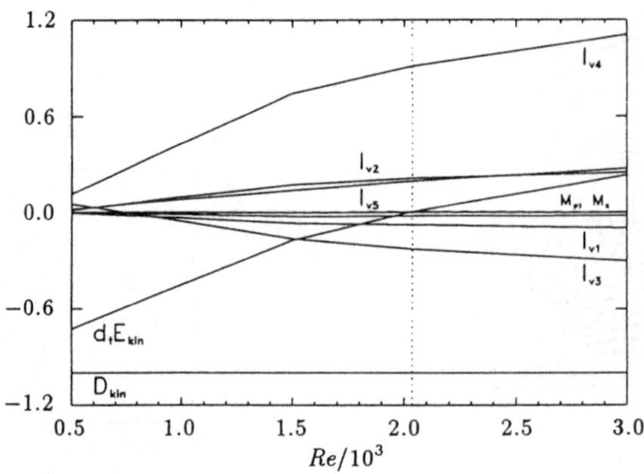

Figure 17: Kinetic energy balance as function of the Reynolds number for $Pr = 0.02$, $\Gamma = 1$, and $Bi = 0$.

The intergral $I_{v4} = -Re \int w\, u\, \partial_r w_0\, dV$ is a measure for the transport of axial basic state momentum $(w_0)$ to the disturbance $(w)$ by means of the radial velocity field of the perturbation $(u)$. The efficiency of this process is proportional to the magnitude of the radial gradient of the axial base state velocity $(\partial_r w_0)$, i. e. proportional to the shear rate of the axial base flow. The high absolute value of the shear gradient $(\partial_r w_0 < 0)$ of the basic flow near the free surface (see fig. 11a; the Reynolds number is close to the critical one $(Re_c = 2062)$) is caused by the driving Marangoni forces. For $\varphi = 0$, the radial velocity of the neutral mode close to the free surface is essentially negative (fig. 16a) and it transports negative axial momentum of the base flow from the free surface into the interior of the liquid bridge. Since the axial velocity $w$ of the neutral mode is also negative for $r \gtrsim 0.7$ $(\varphi = 0)$, it is enhanced there. On the other hand, the vertical velocity of the basic flow is positive for radial distances $r \lesssim 0.7$ and it has a weak positive radial gradient for $r \lesssim 0.5$ (cf. fig. 11a). Since the axial component of the perturbation velocity field is also positive there, the perturbation is similarly amplified by the radial inward flow of the perturbation. This explanation is confirmed by the two positive maxima of the local energy transfer rate $-Re\, w\, u\, \partial_r w_0$ for $\varphi = 0$ which are shown in fig. 18 as function of $r$ and $z$. As can be seen the peak close to the free surface makes by far the largest contribution to the energy gain.

For these reasons, the thermocapillary instability for small Prandtl numbers is an inertial instability, the critical mode receiving its energy from the axial shear flow close the free surface. Although the shear flow is driven by the thermocapillary effect, the rates of change of energy due to Marangoni forces induced by the the neutral temperature field, $M_\varphi$ and $M_z$, do not play a role for the instability mechanism (fig. 17). Since the critical perturbations for $m = 2$ are $\pi$-periodic, the same energy transfer mechanisms apply to $\varphi = \pi$. The explanation is equally valid at $\varphi = \pi/2$ and $\varphi = 3\pi/2$, albeit with negative signs of $u$ and $w$.

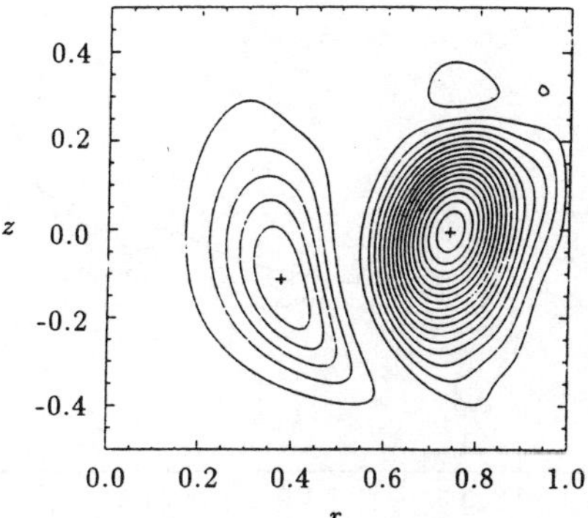

Figure 18: Local transfer rate of kinetic energy (integrand of $I_{v4}$) at $\varphi = 0$ for $Pr = 0.02$, $Re_c = 2062$, $m = 2$, $\Gamma = 1$, and $Bi = 0$.

In order that the discussed amplification mechanism leads to instability a feedback from the amplified velocity $w$ to $u$ is required. Here the feedback is provided by the rigid boundaries at $z = \pm 1/2$ together with the continuity of the flow ($\nabla \cdot \vec{u} = 0$) which enables a transfer from the amplified axial momentum to the radial disturbance momentum. Bounding rigid walls are not present in the model system of an infinitely long liquid bridge [64]. Therefore, a feedback is lacking and the stationary instability is not present for small Prandtl numbers $Pr \to 0$ and $\Gamma \to \infty$.[17]

A similar explanation was proposed by Levenstam & Amberg [62]. The basic axial shear flow near the free surface contains a straining flow that makes an angle of 45° with respect to the $z$-axis. It is this strain which ultimately leads to the instability shifting the originally toroidal vortex core into the main strain directions leading to a saddle-shaped vortex. Note that the vorticity is not constant for a toroidal inertial vortex ($\omega \sim r$, [58]) and thus a toroidal curved vortex is also subject to a self-induced strain (see also [65]). This picture is consistent with the tendency of the stability boundary to increase for $\Gamma \to 0$ (fig. 19).

The question arises why the critical Reynolds number increases for larger Prandtl numbers. By artificially neglecting the azimuthal Marangoni forces in the boundary condition for the neutral mode Prange [66] has shown that azimuthal Marangoni effects are the dominant cause for the stabilization. With increasing Prandtl number the radial outward disturbance flow creates ever colder surface spots, since the interior of the liquid bridge is becoming increasingly colder than the free surface. Azimuthal Marangoni forces are thus induced opposing the azimuthal disturbance surface flows that arise due to continuity.

The influence of the aspect ratio on the instability for $Pr = 0.02$ is shown in fig. 19. With increasing normalized radius $1/\Gamma$ the critical azimuthal wavenumber increases due to a crossing of the neutral curves for different values of $m$. The structure of

---

[17]Instead, the first instability in infinitely long liquid bridges is oscillatory.

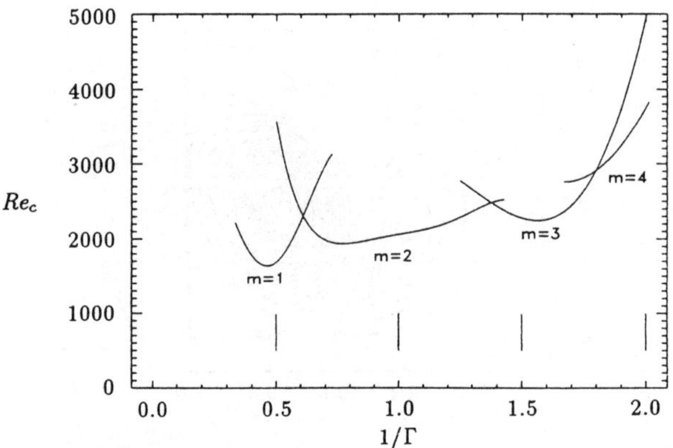

Figure 19: Critical Reynolds number as a function of the aspect ratio for $Pr = 0.02$ and $Bi = 0$. The vertical marks indicate the aspect ratios for which $m = 2/\Gamma$.

the critical modes within a single azimuthal convection cell $(0 \leq \varphi \leq \pi/m)$ is similar and the same instability mechanism is operative for all $m$. The wavenumbers are well ordered in the range of $\Gamma$ considered. It is found that the azimuthal wavelength on the free surface, $2\pi R/m \approx \pi d$, scales with the height of the liquid bridge.[18]

### 5.3.2   Large Prandtl numbers: $Pr = 4$

The instability mechanism at high Prandtl numbers differs from the one at small $Pr$. At the critical threshold $Re_c$ (fig. 15a) a pair of azimuthally travelling waves with frequencies $\pm\omega_c$ become unstable through a Hopf bifurcation [69]. Within the interval $0.9 < Pr < 5$ the critical wavenumber is given by $m_c = 2$. The oscillation frequency $\omega_c$ (fig. 15b) has a similar dependence on $Pr$ as $Re_c$, and both decrease with increasing Prandtl number.

For $Pr = 4$, $\Gamma = 1$, and $Bi = 0$ the critical Reynolds number is $Re_c = 1047$ with Hopf frequency $\omega_c = 27.9$. In what follows the instability mechanism is considered for the mode with $\omega_c > 0$.[19]

The azimuthal velocity components on the free surface are by far larger than the vertical ones. This is consistent with the energy analysis. Those terms in (34) that include axial variations are comparatively small. An evaluation of the kinetic energy balance shows that the flow of the neutral mode is mainly driven by the azimuthal Marangoni effect induced by the temperature field of the neutral mode. Inspecting the thermal energy balance in fig. 20 it is seen that $I_{T3}$ is always stabilizing and

---

[18]The scaling of the wavelengths of symmetry breaking convection patterns with the macroscopic characteristic length is observed in many systems (see, e. g. [67] and [68]).

[19]The behavior of the mode with $\omega_c < 0$ is analogous. At $Re_c$ both modes become unstable simultaneously and the neutral solution is a superposition of both waves. In the framework of a linear stability analysis the modal amplitudes for $Re > Re_c$ cannot be determined.

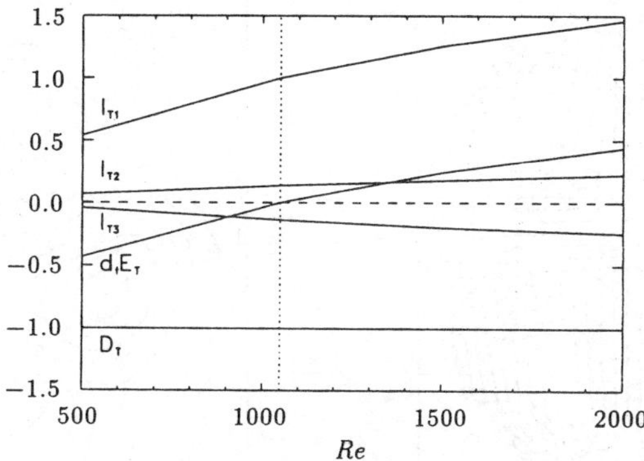

Figure 20: Thermal energy balance as function of the Reynolds number for $Pr = 4$, $\Gamma = 1$, $m = 2$, and $Bi = 0$.

nearly compensating $I_{T2}$. The main energy input is due to the process described by $I_{T1} = -Re \int \Theta u \partial_r \Theta_0 dV$. It is the rate of change of thermal energy due to radial transport ($u \partial_r$) of basic state temperature ($\Theta_0$) to the temperature field ($\Theta$) of the perturbation. This process is now explained in some more detail.

Since the axial terms in the energy equations are very small, the energetics can be discussed by considering the fields in a horizontal cut at $z = 0$. The temperature field of the neutral mode has well pronounced extrema in the interior of the liquid bridge (fig. 21). An internal disturbance temperature maximum is heating the free surface conductively and induces Marangoni flows that are directed from the interior to the hot spot on the free surface and along the free surface away from the hot surface spot. In the present case the interior disturbance temperature extrema are only infinitesimal perturbations of the basic temperature which is lower in the interior than on the free surface. Therefore, the hot free surface spot is cooled by the radial outward convection. This situation is stable in the sense of the classical Marangoni effect. It establishes, however, a frustration owing to competing processes: Conductive heating and convective cooling of the free surface hot spot. Clearly, the convective cooling becomes more efficient with increasing Reynolds number.

The frustration is resolved, if both competing processes are phase-shifted in azimuthal direction. This happens, in fact, as can be seen from fig. 21b where the phase shift between the internal disturbance temperature extrema and those on the free surface is clearly visible. The flow at the free surface is directed azimuthally away from the surface temperature maxima at $\varphi \approx 0$ and $\pi$ and it is directed towards the corresponding minima at $\varphi \approx \pi/2$ and $3\pi/2$. Due to continuity, the azimuthal Marangoni flows are connected with radial flows that transport colder fluid (base state) from the interior to the free surface. This way the internal temperature minima are amplified. The location of maximum amplification, however, is slightly phase-shifted azimuthally. Similarly, the internal temperature maxima are amplified by the corresponding hot

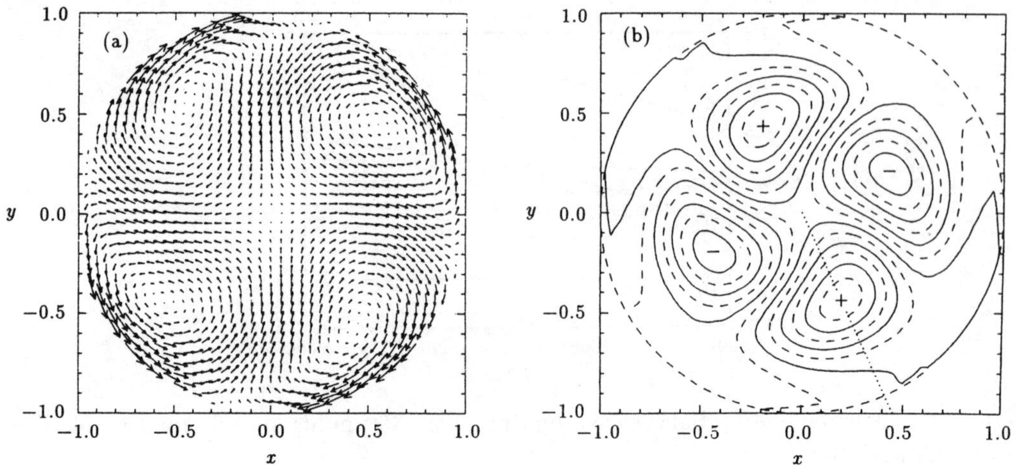

Figure 21: Flow (a) and temperature field (b) of the critical mode for $Pr = 4$, $Re_c = 1047$, $m = 2$, $\Gamma = 1$, and $Bi = 0$ in a horizontal cut at $z = 0$. The dotted line indicates the instantaneous angle of a temperature maximum (cf. fig. 22). The pattern rotates clockwise.

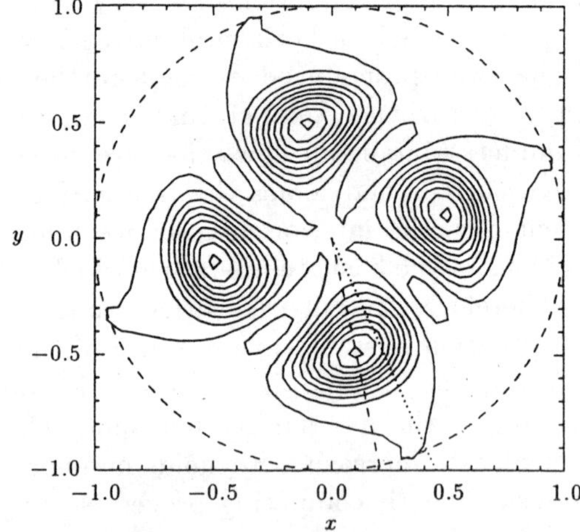

Figure 22: Isolines of the local transfer rate of thermal energy $-Re\,\Theta\,u\,\partial_r\Theta_0$ (integrand of $I_{T1}$) at $z = 0$ for $Pr = 4$, $Re_c = 1047$, $m = 2$, $\Gamma = 1$, and $Bi = 0$. The dashed line indicates the angle under which the local amplification rate takes a maximum value. The dotted line corresponds to the angle for the associated temperature maximum (compare fig. 21).

inward flows. The phase-shift between the internal temperature extrema and the locations of their maximum amplification leads to a rotation of the pattern as a whole

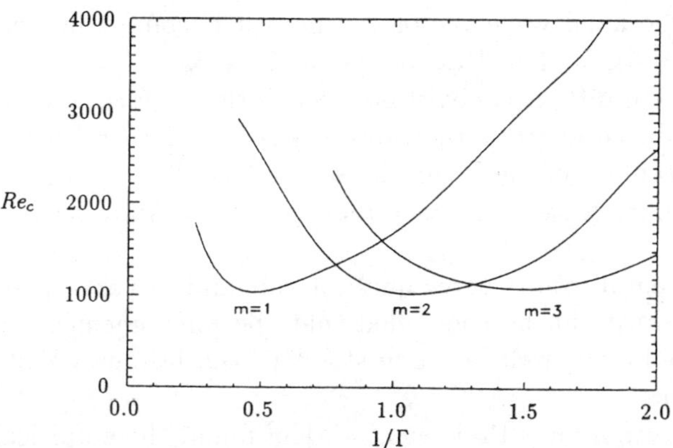



Figure 23: Neutral stability boundaries as function of the inverse aspect ratio $1/\Gamma$ for $Pr = 4$ and $Bi = 0$.

in negative azimuthal direction. The spatial distribution of the local amplification rate confirms this explanation. The integrand $-Re\,\Theta\,u\,\partial_r\Theta_0$ of $I_{T1}$ is shown in fig. 22 at $z = 0$. It can be seen that the locations of maximum amplification are advancing those of the corresponding temperature extema in negative azimuthal direction.

The dependence of the linear stability boundary on the aspect ratio $\Gamma$ is shown in fig. 23 for $Pr = 4$. Similar as for small Prandtl numbers an approximate scaling law $m_c \approx 2/\Gamma$ is valid. The instability mechanism and the modal structure (within one azimuthal wavelength) do not change qualitatively in the considered range of $\Gamma$.

### 5.3.3   Additional remarks

The analyses carried out in the previous sections have shown that there exist at least two different types of instabilities in thermocapillary liquid bridges. For the range of medium Prandtl numbers $Pr < 0.5$ and for $\Gamma = 1.2$, [70] predicted the first instability of the axisymmetric flow to be stationary. This is in contrast to recent results of [71] who found oscillatory instabilities with $m = 3$ for $0.1 \leq Pr \leq 0.8$ at very high Reynolds numbers ($\Gamma = 1$). Owing to the high Reynolds numbers these results must, however, be considered with care.

The instability mechanism for high Prandtl number liquid bridges is very similar to the one for plane thermocapillary liquid layers (cf. [38, 42]). Therefore, the critical modes in high Prandtl number liquid bridges may also be termed *hydrothermal waves*. Xu & Davis [64] have investigated the stability of a long thermocapillary liquid bridge in the limit $\Gamma \to \infty$. They found a critical mode with $m = 1$ which is consistent with the present analysis. Notwithstanding the different azimuthal wavenumbers for finite zones, the critical modes are quite similar in both systems. Both modes are slightly corkscrew twisted (see also [72]). This corresponds to a small axial component of the wave vector which is directed streamwise of the free surface flow in both cases. The relative strength of the axial wave vector component becomes smaller for decreasing Prandtl number ($Pr < 5$), for $\Gamma = 1$ as well as for $\Gamma \to \infty$.

For $Pr < 1$ significant differences exist between both systems. The critical Reynolds number for inertia-induced hydrothermal waves diverges like $Pr^{-1}$ [64]. This is consistent with the linear stability of the Hagen-Poiseuille flow [28]. On the contrary, a finite critical Reynolds number for a stationary mode exists in finite length liquid bridges (see figs. 14 and 19).

Since the experimental results agree qualitatively and in parts quantitatively with the numerical analyses, it can be concluded that the phenomena observable close to the instability threshold can well be understood in the framework of the instability mechanisms discussed.

For the critical wavenumbers Preisser *et al.* [60] found the empirical correlation

$$m_c \approx \frac{2.2}{\Gamma}. \tag{117}$$

This scaling of the azimuthal wavenumber with the height $d$ of the liquid bridge agrees well with $m_c \approx 2/\Gamma$ found numerically for *both* high and low Prandtl numbers. The corresponding aspect ratios are marked in fig. 19.

Velten *et al.* [59] confirmed that the amplitude of the temperature oscillations grow with the square-root of the distance from the critical point $\propto (Re - Re_c)^{1/2}$. Thus the instability is a supercritical Hopf bifurcation. Far above the threshold and depending on $\Gamma$ a rich variety of different spatio-temporal patterns exists.

For an understanding of the supercritical flow, it must be taken into account that two hydrothermal waves become unstable propagating in opposite azimuthal directions. The relative amplitudes are not known *a priori*. A further complication arises due to the wave fronts of the hydrothermal waves not being planes $\varphi = const$. The local wave vector has components in $r$- and $z$-directions, since the phase $G(r,z)$ is a function of both $r$ and $z$ (the amplitudes in (107) are complex). The neutrally stable hydrothermal waves travelling in positive ($F_+$) and negative ($F_-$) azimuthal direction have the form

$$F_{\pm} = A(r,z)\mathrm{e}^{i(\pm m\varphi - \omega t + G(r,z))} + \mathrm{c.\,c.}, \qquad G, A \in \mathbb{R}, \tag{118}$$

where $F$ is any scalar perturbation field. It can be shown quite generally [73] that for slightly supercritical driving only pure travelling or pure standing waves are nonlinear solutions. Stable bifurcating solutions are possible only if both solutions bifurcate supercritically. For a standing wave,

$$F_{standing} = 4A(r,z)\cos(m\varphi - \varphi_0)\cos(G(r,z) - \omega t), \tag{119}$$

and the azimuthal nodal planes are determined by the arbitrary phase $\varphi_0$. Note, however, that the radial and axial nodal surfaces are given by $G(r,z) = \omega t + \frac{2n+1}{2}\pi$ and thus are not generally plane. Muehlner *et al.* [72] observed a pure travelling hydrothermal wave with $m = 1$ at $Ma \approx 13,000$ for $Pr = 35$, $\Gamma = 1$. The total azimuthal twist of the phase front was $\approx 95°$ between both heated walls. Thus the wave had a pronounced spiral character.

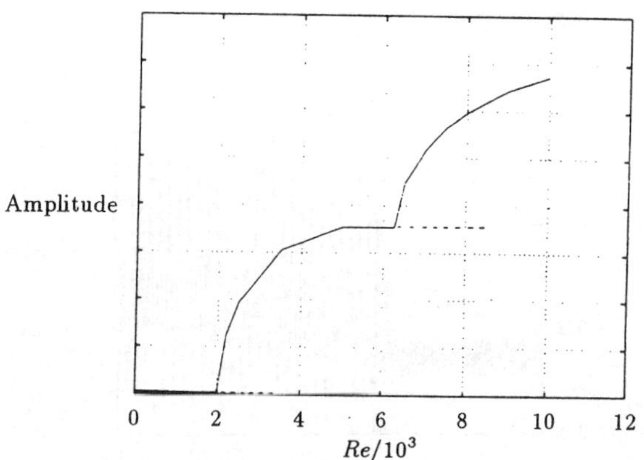

Figure 24: Bifurcation diagram for $Pr = 0.01$, $\Gamma = 1$, $Bi = 0$, and $Gr = 0$ according to [62]. The amplitude shown is a normalized azimuthal velocity.

## 5.4   Three-dimensional simulations

The first three-dimensional simulation of thermocapillary flows in liquid bridges was due to Rupp *et al.* [70]. Recently, Levenstam & Amberg [62] simulated the flow for $Pr = 0.01$, $\Gamma = 1$, and $Bi = Gr = 0$. In good agreement with the linear stability analysis (fig. 14) they found a supercritical bifurcation out of the basic flow to a stationary mode with fundamental wavenumber $m = 2$ for $Re_c = 1960$. Beyond the threshold, the flow is essentially toroidal with a saddle-shaped deformed vortex core. This flow becomes oscillatory at $Re_{c2} = 6250$ [62]. Since the three-dimensional base flow for the secondary bifurcation is not axisymmetric, the oscillatory perturbations cannot be azimuthal normal modes. The oscillations thus do not appear as harmonic waves propagating azimuthally nor do they occur in form of a rotation of the basic three-dimensional flow as a whole.

Since the bifurcation scenario is the same in the limit $Pr \to 0$, the temperature field, which is unimportant for the primary instability, cannot be of decisive importance for the secondary instability either. The bifurcation diagram is shown in fig. 24. The stationary as well as the oscillatory amplitudes grow with the square-root of the distance from the critical point, i. e. $\sim a(Re - Re_c)^{1/2}$ and $\sim [b + c(Re - Re_{c2})^{1/2}]$, respectively, where $a, b$, and $c$ are constants.

Three-dimensional simulations of the flow in cylindrical liquid bridges of high Prandtl number have been carried out by [74]. The extrapolated critical Reynolds numbers for the onset of oscillations agree well with the linear stability boundaries. For a comparison with the data measured by [59] the numerical calculations [74] have been performed in the range $Re_c < Re < 2Re_c$ for $Pr = 7$ and various aspect ratios taking into account buoyancy $(Bd = 0.13(d/\text{mm})^2)$ and free surface heat transfer $(Bi = 6.4)$. When the unstable axisymmetric basic state is used as the initial condition for supercritical driving hydrothermal waves propagating either clockwise or counter-

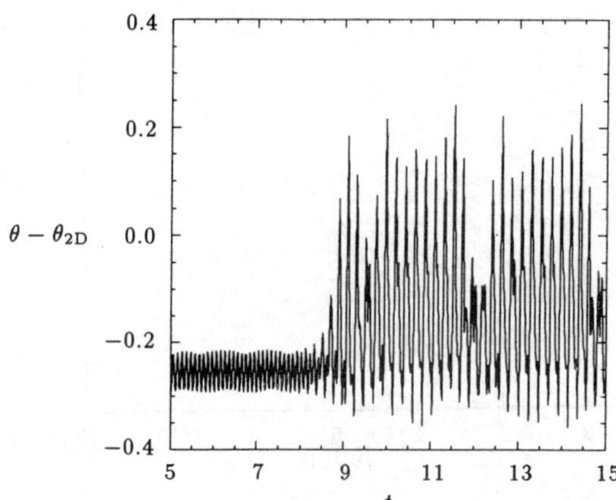

Figure 25: Time-dependence of the three-dimensional part of the temperature field $\theta - \theta_{2D}$ at $(r = 0.2, z = 0)$ for $\Gamma = 1$, $Pr = 4$, $Re = 1300$, and $Bi = Gr = 0$.

clockwise grew out of the *numerical noise* (amplitude $\simeq O(10^{-14})$) initially. These waves are composed of counter-propagating fundamental modes with significantly different amplitudes ($F_+ \neq F_-$, cf. (118)). At later times, all these flows developed into pure azimuthally propagating waves. In all cases investigated, the azimuthal wavenumbers of the fundamental mode agreed well with the predictions of the linear stability theory and with the experimental results[20] of [59].

A different transient behavior is obtained, when finite amplitude initial conditions satisfying the symmetry of a standing wave are used, e. g. a finite amplitude perturbation of the temperature field in form of $\Theta = f(r, z)\cos(m\varphi)$ with $f \in \mathbb{R}$, $|f|_{max} = 5 \times 10^{-3}$ (in units of $\Delta T$). In such cases standing hydrothermal waves are growing. If $\epsilon$ is sufficiently large, they can reach a saturated amplitude on a short time scale (slightly less than the thermal diffusion time). Disturbances that break the chiral (left/right) symmetry grow exponentially at a rate that depends on the distance $\epsilon = (Re - Re_c)/Re_c$ from the critical point. Finally, the standing waves decay to pure travelling waves. An example for such a transition process is clearly visible in the time-dependence of the temperature signal at $(r = 0.2, z = 0)$ shown in fig. 25. This scenario is consistent with recent numerical simulations of Savino & Monti [75] who investigated the flow for $Pr = 30$ and 74 and aspect ratio $\Gamma = 2$ when the Reynolds number is increased at a constant rate.

In agreement with the general theory these calculations indicate the existence of stationary standing hydrothermal wave states.[21] They are not stable, however, for the parameters investigated to date.

---

[20]Except for the *mixed modes* for which [59] did not specify a wavenumber.

[21]Recently, Xu & Zebib [76] found stable hydrothermal waves travelling streamwise while standing spanwise in a rectangular cavity.

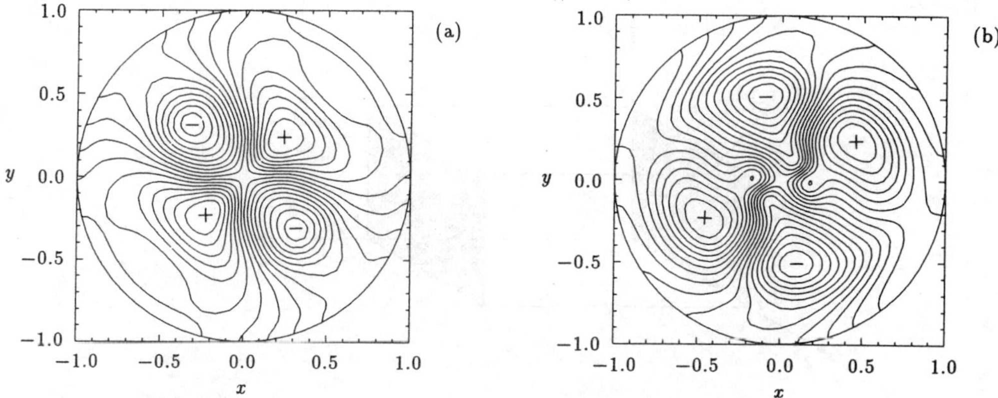

Figure 26: Instantaneous temperature field $\theta - \theta_{2\mathrm{D}}$ (the axisymmetric contribution has been subtracted) of a standing wave (a) and a wave propagating counter-clockwise (b) with $m = 2$ for $Pr = 4$, $\Gamma = 1$, $Re = 1300$, and $Bi = Gr = 0$. Shown is a cut at $z = 0$.

In fig. 26 the instantaneous temperature fields of an unstable standing hydrothermal wave (a) and of a wave that propagates counter-clockwise (b) are shown at $z = 0$. The plotted temperature deviations from the axisymmetric part of the field exhibit the typical pronounced internal extrema. In the plane $(z = 0)$ shown, the extrema of the standing wave propagate towards the center and vanish there. During this process new temperature extrema of opposite sign develop further radially outward. This propagation toward the center $(r = 0)$ is consistent with the mathematical description ($r$-dependence of the phase $G$ of the fundamental mode $m = 2$, cf. (118)). The pattern of the travelling wave (fig. 26b) still contains a weak counter-propagating component.

# 6 Thermocapillary convection in rectangular containers

Thermocapillary convection in rectangular containers is a model for the *open boat* crystal growth method [2]. Two-dimensional flows in rectangular geometries are similar to those in cylindrical liquid bridges. There are, however, differences that will be addressed in the following.

## 6.1 Formulation

For the mathematical description we consider a rectangular volume of liquid with lengths $L_x$, $L_y$, and $L_z$ in $x$, $y$, and $z$-directions. The sidewalls at $x = 0$ and $L_x$ are kept at constant temperatures $T_0 + \Delta T$ and $T_0$, respectively. Let the free surface be at $y = L_y$ (see fig. 27) and take the depth of the liquid layer $L_y$ as the length scale of the

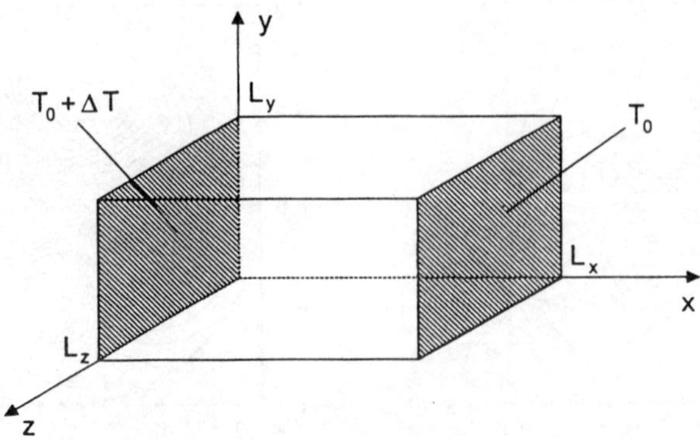


Figure 27: Geometry and coordinate system for the rectangular *open boat* model.

problem. Two aspect ratios arise, defined by $\Gamma = L_x/L_y$ and $\Gamma_z = L_z/L_y$. Employing the scalings given in table 3 and defining the Reynolds and Grashof numbers

$$Re = \frac{\gamma \Delta T L_y}{\rho \nu^2}\frac{1}{\Gamma}, \qquad\qquad Gr = \frac{g\beta \Delta T L_y^3}{\nu^2}\frac{1}{\Gamma}, \qquad (120)$$

the volume equations (11) – (13) remain unchanged. Here the acceleration of gravity is $\vec{g} = -g\vec{e}_y$. On the rigid walls at $x = (0,\Gamma)$, $y = 0$, and $z = (0,\Gamma_z)$ the velocity field must vanish ($\vec{u} = 0$). Neglecting surface deformations[22] ($Ca \to 0$) the boundary conditions for the velocity field on the free surface are

$$v = \partial_y u + \partial_x \theta = \partial_y w + \partial_z \theta = 0, \qquad \text{on} \quad y = 0. \qquad (121)$$

Using the reduced temperature $\theta = \Gamma(T - T_0)/\Delta T$ the thermal boundary conditions are given by

$$\theta(x = 0) = \Gamma \qquad \text{and} \qquad \theta(x = \Gamma) = 0. \qquad (122)$$

The ambient temperature field is assumed to be conducting with $\theta_a = \Gamma - x$. If not noted otherwise, all other boundaries are taken as adiabatic ($Bi = 0$)

$$\partial_y \theta = 0, \quad \text{on} \quad y = (0,1) \qquad \text{and} \qquad \partial_z \theta = 0, \quad \text{on} \quad z = (0,\Gamma_z). \qquad (123)$$

---

[22]As shown by [77] for $\Gamma = \Gamma_z = 1$, $Pr = 100$ (silicone oil), and $Re = 400$ surface deformations are of relative order of magnitude $O(10^{-3} - 10^{-2})$ even for $Gr = 0$.

| variable | $x, y, z$ | $t$ | $\vec{u} = (u, v, w)$ | $p$ | $T$ |
|---|---|---|---|---|---|
| scale | $L_y$ | $L_y^2/\nu$ | $\gamma \Delta T L_y/\rho\nu L_x$ | $\gamma \Delta T/L_x$ | $\Delta T L_y/L_x$ |

Table 3: Scales for the non-dimensionalization of the Navier-Stokes equations in case of rectangular containers.

## 6.2  Two-dimensional flows

### 6.2.1  Short containers ($\Gamma \leq 1$)

In the limit of small Prandtl and Reynolds numbers $((Re, Pr) \to 0)$ the flow and temperature field can be written in terms of sums over even Papkovich-Fadle functions as for liquid bridges (see §2.2). All flow properties derived there apply at least qualitatively to the rectangular geometry. In particular, for $\Gamma \lesssim 1/1.4$ the creeping flow consists of a sequence of counter-rotating Moffatt eddies [25, 26, 78]. The extension in $y$-direction of the corresponding vortices is $\approx 1.396 \times \Gamma$ as in liquid bridges, they have a similar structure, and their strength decays exponentially in negative $y$-direction $\sim e^{\approx 4.21 y/\Gamma}$.

The stationary nonlinear $(Re \gg 0)$ two-dimensional thermocapillary convection in a square container $(\Gamma = 1)$ has been calculated numerically by Zebib $et\ al.$ [79] for different Prandtl and Reynolds numbers up to $Re = 5 \times 10^4$. The flow consists of a main vortex and two small corner vortices at $y = 0$ and $x = (0, 1)$. For high Reynolds numbers boundary layers develop along all boundaries. Using the assumption that the free surface temperature gradient remains $O(1)$ it was shown that the thermocapillary boundary layer thickness scales like $Re^{-1/3}$ (cf. §3).

Extending the work of [79], Carpenter & Homsy [57] numerically showed that a further secondary vortex separates on the rigid wall close to the hot corner $(x = 0, y = 1)$ if $Re > 4 \times 10^4$ for $Pr = 10^{-3}$ or $Re > 1.3 \times 10^5$ for $Pr = 1$. They found that the assumption $\partial_x \theta|_{y=1} = O(1)$ does not always hold for high Prandtl numbers, since the free surface becomes essentially isothermal apart from small corner regions. This observation led to the conclusion that the asymptotic regime with $\delta \sim Re^{-1/3}$ is only reached for Reynolds numbers which become increasingly large the higher the Prandtl number is.[23]

In another work Carpenter & Homsy (1989) investigated the additional influence of buoyancy for $Pr = 1$ and aspect ratio $\Gamma = 1$. Their numerical calculations for constant dynamic Bond number $(Bd = Gr/Re = const.)$ showed a change of the scaling of the free surface boundary layer thickness from $\delta \sim Re^{-1/2}$ to $\delta \sim Re^{-1/3}$ within a narrow range of Reynolds numbers. Supported by a scaling analysis, the numerically observed behavior led to the hypothesis that the combined flow is thermocapillary dominated for every fixed value of the dynamic Bond number $Bd$, if only $Re$ is sufficiently high.

### 6.2.2  Stationary flow in long containers ($\Gamma \gg 1$)

For shallow liquid layers with $\Gamma \gg 1$ and $Gr = 0$ the main vortex becomes laterally elongated. In the limit $\Gamma \to \infty$ and for small Reynolds numbers $Re = O(\Gamma^{-1})$ (cf. (120)) one obtains the asymptotic surface temperature $\theta = \Gamma - x$. The velocity field is

---

[23]The main driving for $Re \to \infty$ and $Pr = const.$ is due to the cold corner.

then given by the plane Poiseuille-Couette profile [47, 80, 81]

$$u = \left( \frac{3}{4} y^2 - \frac{1}{2} y \right), \tag{124}$$

(cf. (93) and (95) with $z \to y - 1/2$). The temperature field in the bulk of the liquid depends on the thermal boundary conditions. For an ideal heat conducting boundary at $y = 0$ one obtains, different from (93),

$$\theta = \Gamma - x - \frac{1}{48} RePr \left( 3y^4 - 4y^3 + y \right). \tag{125}$$

For $\Gamma > 4$ the streamlines in the middle of the container are already nearly horizontal [81] and (124,125) represent good approximations to $u$ and $\theta$ in that region.

Two-dimensional thermocapillary flows in long containers up to $\Gamma = 25$ have been calculated by Ben Hadid & Roux [82] for $Gr = 0$ and $Pr = 0.015$. For heat conducting boundary conditions at the bottom ($y = 0$) and for Reynolds numbers not too large, the temperature field depends on the velocity field only in the vicinity of $x = (0, \Gamma)$. For Prandtl number $Pr = 0.015$ and up to $Re = O(10^4)$ it is nearly conductive in the interior and almost decouples from the flow.

When the Reynolds number is small the flow consists of the asymptotic core flow (124,125) with symmetric turning zones at $x = 0$ and $\Gamma$ (fig. 28a). On an increase of the Reynolds number three distinct zones develop. Close to the cold wall ($x = \Gamma$) a number of co-rotating vortices is created (for $\Gamma = 12.5$ two vortices appear for moderate Reynolds numbers). The flow close to the hot wall ($x = 0$) behaves like the entrance flow of a channel [83]. Ben Hadid & Roux [82] showed that the interior flow ($x \approx \Gamma/2$) can well be approximated by the Poiseuille-Couette profile (124) as long as $Re < 20\Gamma$. For higher Reynolds numbers the end-zones grow into the interior and the Poiseuille-Couette-like flow is destroyed.

For $Re \lesssim 200\Gamma$ the classical boundary layer scaling for shear-induced flows holds in the entrance region ($0 < x < \Gamma/2$). The dynamic equation ($\nabla \times$ (11)) in two dimensions yields

$$Re\,\vec{u} \cdot \nabla \vec{\omega} \sim \nabla^2 \vec{\omega}. \tag{126}$$

Approximating $\partial_y \sim \delta^{-1}$ in (126), where $\delta$ is the boundary layer thickness, one obtains the order of magnitude relation

$$Re \frac{u}{x} \sim \frac{1}{\delta^2}. \tag{127}$$

Together with the thermocapillary boundary condition, $1 = \partial_y u \sim u/\delta$, the scaling of the horizontal velocity within the entrance region is obtained as

$$u \sim \left( \frac{x}{Re} \right)^{1/3}, \tag{128}$$

where $x$ denotes the distance from the hot side ($x = 0$) of the container.

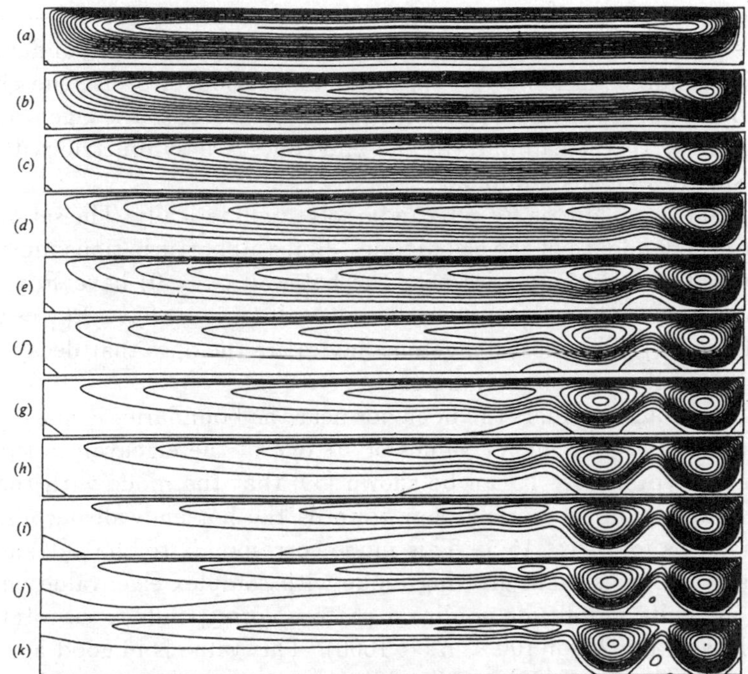

Figure 28: Streamlines in a long open rectangular container with $\Gamma = 12.5$ and $Pr = 0.015$, $Gr = 0$, and $Bi = \infty$. The conductive temperature profile is imposed on the free surface. The Reynolds numbers are: (a) $Re = 66.7$, (b) $Re = 333$, (c) $Re = 667$, (d) $Re = 1330$, (e) $Re = 2000$, (f) $Re = 3330$, (g) $Re = 5000$, (h) $Re = 6670$, (i) $Re = 10^4$, (j) $Re = 1.33 \times 10^4$, (k) $Re = 2 \times 10^4$. After [82].

The flow structures found by [82] close to the endwalls have been explained by Laure *et al.* [49]. They considered the limit $\Gamma \to \infty$ with $Pr \to 0$ and investigated the spatial stability of the Poiseuille-Couette flow (*return flow*) with heat conducting boundary conditions at the bottom. Using the stream function $\psi_0$ for the basic state the linearized equation governing small two-dimensional stationary perturbations $\psi$ is

$$\nabla^4 \psi \;=\; Re \left\{ (\partial_x \psi)\, \partial_{yyy} \psi_0 - (\partial_y \psi_0)\, \nabla^2 \partial_x \psi \right\}. \tag{129}$$

This equation can be written as an evolution equation in $x$-direction

$$\frac{\partial}{\partial x} \vec{Z}(x, y) \;=\; \mathbf{L}(Re) \cdot \vec{Z}(x, y), \tag{130}$$

where $\mathbf{L}$ is a linear differential operator and $\vec{Z} = (\psi, \psi_x, \psi_{xx}, \psi_{xxx})^T$. Using the separation ansatz $\psi = f(y) \mathrm{e}^{\gamma x}$ one obtains

$$\gamma^4 f + 2\gamma^2 f'' + f'''' \;=\; -\gamma Re \left( \psi_0' f'' + \gamma^2 \psi_0' f - \psi_0''' f \right), \tag{131}$$

where the prime denotes differentiation with respect to $y$. For $Re \to 0$ it follows $f'''' + 2\gamma^2 f'' + \gamma^4 f = 0$. Inserting the general solution (55) into the boundary conditions yields the eigenvalue equation $2\gamma = \sin 2\gamma$. The solutions are just the characteristic roots of the odd Papkovich-Fadle functions (cf. (56)). Because the eigenvalue equation is symmetric in $\gamma$, the streamlines in the end-zones are symmetric with respect to $x = \Gamma/2$ for small Reynolds numbers.

The eigenvalue problem for $Re \neq 0$ can be solved numerically. The set of eigenvalues describes the $x$-dependence of the eigenmodes. Since all perturbations must decay from the boundaries, the relevant modes near the hot wall ($x = 0$) have $\Re(\gamma < 0)$ (decay in positive $x$-direction) and those near the hot wall ($x = \gamma$) have $\Re(\gamma > 0)$ (decay in negative $x$-direction). The most dangerous modes are the ones that decay slowest, i. e. those with the smallest absolute value of $\gamma$.

For small Reynolds numbers the behavior near the boundaries is quasi-symmetrical as long as $Re < 15.5$. For higher values of $Re$ one of the eigenvalues turns real and the behavior is asymmetric. It can be shown [49] that the mode with the real eigenvalue growing upstream the surface flow towards the hot endwall becomes saturated by nonlinear terms for $Re > 15.5$. This mode corresponds to the entrance flow. For $Re > 400$ also the downstream growing mode with complex eigenvalue becomes saturated. Here the finite amplitude nonlinear pattern corresponds to co-rotating vortices with a wavelength $\lambda \approx 2$ (for $100 < Re < 1000$). This value is in good agreement with the numerical simulation of [82] (see fig. 28).

### 6.2.3    Two-dimensional oscillatory flows

$Pr \ll 1$

Smith & Davis [38] considered the linear stability of (93) (see §4). For small Prandtl numbers, the most dangerous modes of the Poiseuille-Couette flow are long waves propagating nearly in $z$-direction, perpendicular to the applied temperature gradient. The critical Marangoni number tends to zero like $Ma_c \sim Pr^{1/2} \to 0$. The first instability in infinite non-deformable thermocapillary liquid layers with $Gr = 0$ thus appears in form of *three-dimensional* hydrothermal waves. This is consistent with the results of Carpenter & Homsy [57] for square systems and the ones of Ben Hadid & Roux [82] for systems of finite extent who did not find two-dimensional instabilities.[24]

Buoyancy forces can act destabilizing on the otherwise stable two-dimensional thermocapillary flow. Ben Hadid & Roux [85] considered the mixed convection in long containers for $Pr = 0.015$. The critical Rayleigh number for the onset of two-dimensional oscillations was found to increase (stabilization), if the Reynolds number is increased ($Re > 0$). In this case both driving mechanisms support the same sense of rotation of the vortex flow. For opposing driving ($Re < 0$), a destabilization occurs. The corresponding stability diagram for $\Gamma = 4$ and $Pr = 0.015$ has been completed by [86]. It is

---

[24]See, however, [84]

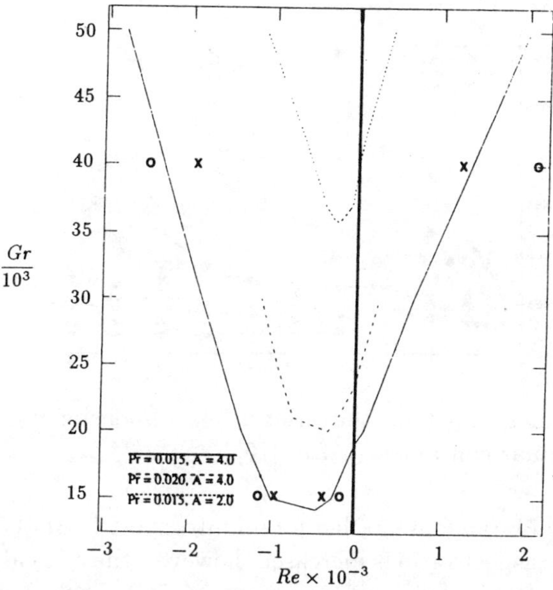



Figure 29: Critical Grashof number $Gr_c$ as function of the Reynolds number $Re$ for the onset of two-dimensional oscillations in a rectangular system with adiabatic boundary conditions; after [86]. ( —— ): $Pr = 0.015$, $\Gamma = 4$; ( - - - ): $Pr = 0.02$, $\Gamma = 4$; ( ..... ): $Pr = 0.015$, $\Gamma = 2$. The points marked as $\times$ (oscillatory) and $\circ$ (stationary) have been calculated by [85].

shown in fig. 29. For $Re < 0$ a maximum destabilization occurs (minimum in fig. 29). A further reduction of the Reynolds number leads to an increase of the critical Rayleigh number. In this parameter range, the flow is thermocapillary dominated. For a discussion of the instability mechanisms the reader is referred to [86]. Three-dimensional buoyant-thermocapillary instabilities in layers have been treated by [35] and [36].

## $Pr > 1$

Contrary to small Prandtl numbers, the hydrothermal wave instability is nearly two-dimensional for high Prandtl numbers. Peltier & Biringen [87] simulated the two-dimensional flow for $Pr = 6.8$ and $Gr = Bi = 0$. They found a supercritical oscillatory bifurcation when the aspect ratio is larger than $\Gamma > 2.3$. These results were confirmed by Xu & Zebib [76] and extended to $Pr = 1$, 4.4, 6.78, 10, and 13.9. They even found multiple regions of instability at higher values of $\Gamma$. A stability diagram is shown in fig. 30.

A comparison with hydrothermal waves is in order, since they are nearly two-dimensional for high Prandtl numbers. The typical critical Marangoni numbers at small aspect ratios are $Ma_c(\Gamma = 2.5) \approx 3200$ and $Ma_c(\Gamma = 4) \approx 2500$ which is much

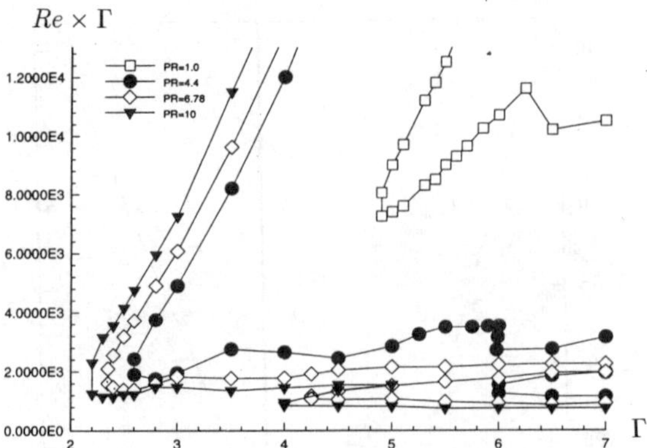


Figure 30: Critical values $Re \times \Gamma$ for the onset of two-dimensional thermocapillary flow oscillations in rectangular containers. After [76].

larger than the critical Marangoni number for infinite layers[25] ($Ma_c(Pr = 6.8) = O(200)$ at $\lambda \approx 2.5$). When the aspect ratio is increased, however, the critical threshold value for two-dimensional hydrothermal waves is approached [76]. Thus the lateral boundaries act strongly stabilizing for low aspect ratios and even suppress the instability for $\Gamma <$ 2.3. Moreover, the instability mechanism for low aspect ratios is different from that of hydrothermal waves. For $\Gamma = 2.6$ Peltier & Biringen [87] explained the oscillations in terms of a vortex that is periodically expanding and contracting in $y$-direction. This process is associated with a periodic appearance of a cold spot in the vicinity of the hot wall. The cold spot is caused by convective transport of cold fluid from the cold wall along the adiabatic bottom boundary. This behavior was also confirmed by [76]. It is clear that a finite geometry ($L_y$) is required for this process. It is interesting to note that the minimum of the critical curve in fig. 30 occurs at $\Gamma \approx 2.5$ which is about the wavelength of the most dangerous hydrothermal wave in plane layers. For an experimental work on thermocapillary convection in finite size containers ($Pr = 4$) the reader is referred to [50].

## 6.3 Three-dimensional flows

### 6.3.1 Large aspect ratios

Three-dimensional numerical simulations have been carried out by Xu and Zebib [76] for selected high Prandtl numbers in the range $Pr = 1 - 13.9$. When the aspect ratio $\Gamma_z$ is $O(1)$, the oscillations are nearly two-dimensional except for a small region near $z = (0, \Gamma_z)$. On an increase of the spanwise length $\Gamma_z$ the hydrothermal wave still propagates upstream the basic surface flow. In addition, also spanwise oscillations (in

---

[25]It must be taken into account, however, that the slope of the surface temperature in the bulk is weaker than suggested by the Marangoni number, because of significant temperature variations near the endwalls.

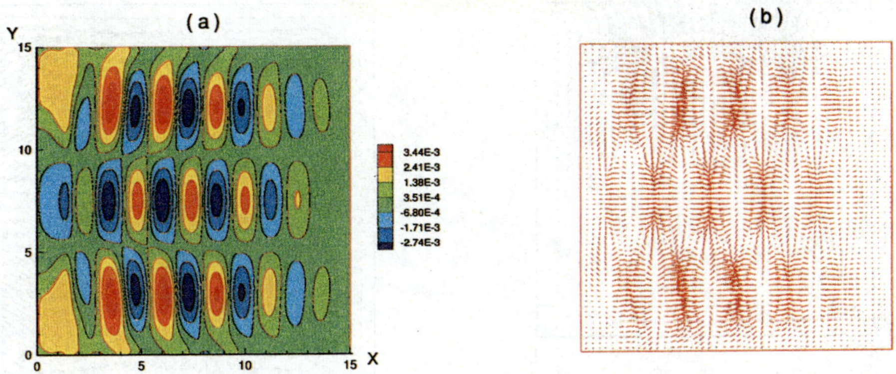

Figure 31: Hydrothermal wave in a cavity of aspect ratio $\Gamma = \Gamma_z = 15$ for $Pr = 13.9$ at $Re = 710/\Gamma = 47.3$. Shown are instantaneous surface values of the fluctuations of the temperature (a) and the velocity field (b). From [76].

$z$-direction) in form of a standing wave arise. This behavior may be explained in terms of two plane hydrothermal waves of equal amplitudes propagating at angles $\pm\alpha$ with respect to the applied temperature gradient. For increasing Prandtl numbers, ever higher aspect ratios $\Gamma_z$ are required for the appearance of spanwise standing waves. This may be due to the decrease of the critical angle of the hydrothermal waves (see fig. 9) with $Pr$. The wave in a cavity of aspect ratios $\Gamma = \Gamma_z = 15$ is shown in fig. 31. For this shallow cavity the critical Reynolds number obtained from the slope of the basic state surface temperature in the bulk and the angle of propagation $\alpha$ are already in good agreement with the respective data for plane layers.

## 6.3.2   Sidewall effects

When the aspect ratio is of $O(1)$ boundary effects become significant. To elucidate the three-dimensional flows induced by the lateral sidewalls we consider stationary three-dimensional thermocapillary convection ($Gr = 0$) in a cube ($\Gamma = \Gamma_z = 1$) with insulating boundary conditions [88].

As in two dimensions, secondary vortices driven by the primary thermocapillary convection exist in the corners at $y = 0$ and $x = (0,1)$. These vortices are most pronounced at midplane $z \approx 0.5$, they develop out of viscous Moffatt eddies [26], and their size increases with increasing Reynolds number. In the cube the secondary vortices are much smaller than in the two-dimensional square cavity.

In addition to these structures, two symmetric secondary vortices with vorticity components primarily in $x$-direction appear directly below the free surface at $y = 1$ and at $z = (0,1)$ (see fig. 32a). These vortices are caused by thermocapillary effects.

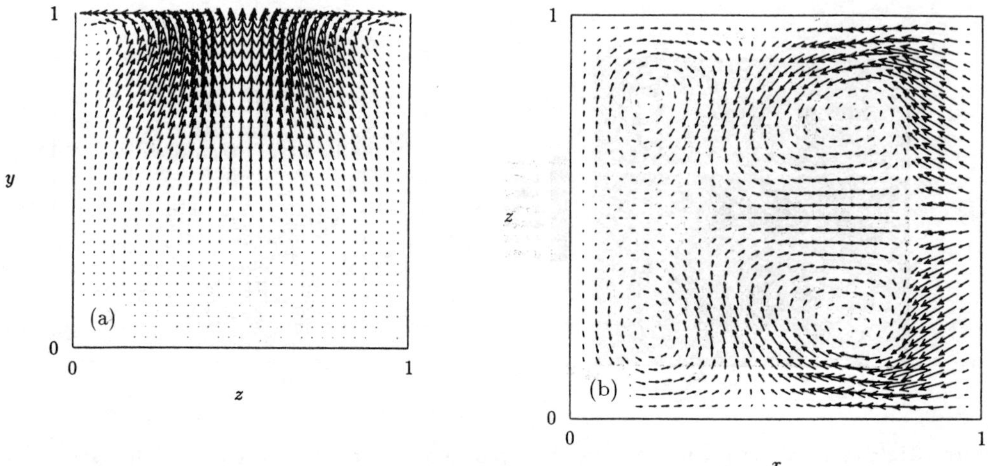

Figure 32: Projections of the velocity field for $Pr = 1$ and $Bi = Gr = 0$. (a): $x = 0.5$, $Ma = 100$. (b): $y = 0.625$, $Ma = 5 \times 10^4$.

Since the primary vortex flow is weaker close to the rigid sidewalls than in the center of the cavity, the surface temperature is nearly conducting near the sidewalls whereas it is strongly convective in the middle of the container. The resulting spanwise temperature gradients drive the observed secondary vortices near the free surface. For high Marangoni numbers the secondary flows may reach sizable amplitudes. The ratio $|w_{max}|/|u_{max}|$, e. g., is 25% for $Ma = 5 \times 10^4$ and $Pr = 1$.

Since the main vorticity is perpendicular to the sidewalls, secondary Bödewadt vortices [89] are induced at high Reynolds numbers. They extend in a curved fashion along each sidewall and can be well distinguished in the lower part of the container. In the upper part of the cavity they are not visible, since thermocapillary forces act opposing to the Bödewadt vortices. A cross-section through the vortices at $y = 0.625$ is shown in fig. 32b. Owing to the superposed primary flow field in positive (negative) $x$-direction at $y \approx 1$ ($y \approx 0$) the secondary vortex centers in the $(x, z)$-plane close to the cold wall appear in the *corners* of the cold wall for $y \approx 1$, whereas they appear in the middle ($x = 0.5$, $z = 0.5$) for $y \approx 0$. The apparent position of the Bödewadt vortices in the vicinity of the hot wall are just opposite.

Within the range of parameters investigated by [88] no multiple two- or three-dimensional solutions were found. The numerical calculations were based on the asumption of steady flow. For that reason, no conclusion can be drawn regarding the stability of the calculated flows with respect to time-dependent perturbations.

### 6.3.3  Steady spatially periodic three-dimensional flows

Apart from the secondary three-dimensional flows caused by the rigid sidewalls and the slightly three-dimensional hydrothermal waves in high Prandtl number layers, the

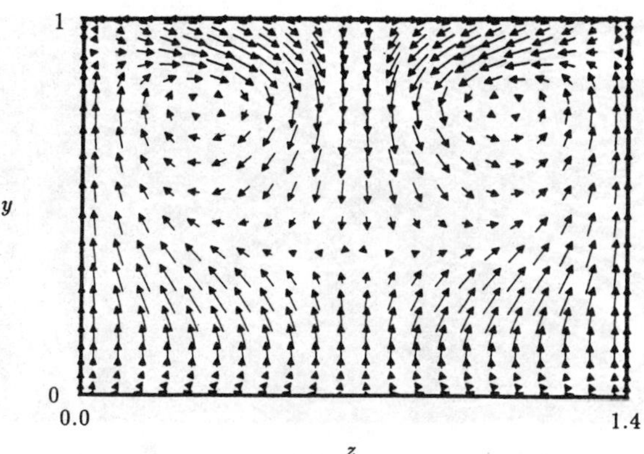

Figure 33: Velocity field in a thermocapillary cavity with $\Gamma = 1.47$ projected onto the plane $x = 1.10$ after [93]. The parameters are $Re = 1.95 \times 10^5$, $Bd = 5.76$, $Pr = 8.4$, and $Bi = 0$. The wavelength in $z$-direction is $\lambda = 1.4$ in units of $L_y$.

two-dimensional base state for aspect ratios $\Gamma = O(1)$ and $\Gamma_z \gg 1$ can become three-dimensional through a stationary symmetry breaking bifurcation.

To investigate the mechanisms and the conditions under which these instabilities appear it is useful to consider volumes that are infinitely extended in spanwise direction ($\Gamma_z \to \infty$). Then sidewall effects are absent.[26]

The first work along these lines was due to Gillon & Homsy [90].[27] They experimentally investigated the combined buoyant thermocapillary convection in a liquid volume with $\Gamma = 1.47$, $\Gamma_z = 3.8$, $Pr = 8.4$, and dynamic Bond number $Bd = Gr/Re = \rho g \beta L_y^2 / \gamma = 2.5$. It was observed that the basic flow which was two-dimensional except for sidewall effects became unstable for high Marangoni numbers $Ma = O(10^5)$ giving way to stationary three-dimensional convection cells with a spanwise wavelength of $O(1)$. Similar experimental results have been obtained by Daviaud & Vince [92] for a certain range of aspect ratios $\Gamma = O(1)$.

In order to simulate the experiment of [90] numerically, Mundrane & Zebib [93] calculated the three-dimensional flow using the same parameters (but $Bd = 5.76$) and $Bi = 0$. They employed periodic boundary conditions in $z$-direction corresponding to the experimentally determined value $\lambda = 1.4$. For $Ma = 2.93 \times 10^3$ the flow was found to be two-dimensional and stable. For $Ma = 1.95 \times 10^5$ the stationary calculation yielded a three-dimensional flow pattern. A projection of this flow onto the $(y, z)$-plane at $x = 1.10$ is shown in fig. 33. The calculated flow structure agrees qualitatively with the projection of the streamlines observed by [90].

Similarly, Schneider [53] also observed stationary convection cells in a radially heated annular system with an upper free surface and filled with ethanol (radii

---

[26][52] avoided sidewall effects by selecting an annular system for their experiments.

[27]The aspect ratio $\Gamma_z = 1$ used by Schwabe & Metzger [91] obviously was too small to observe the patterns.

Figure 34: The cat's eye flow near the free surface in a deep cavity (heated from the right side). The flow is driven by combined buoyant and thermocapillary forces. After [91].

$R_i = 20$mm, $R_o = 40$mm; $\Gamma = (R_a - R_i)/L_y < 4.5)$. The observed wavelength in azimuthal direction normalized by the layer's height $L_y$ agrees well with the one found by [90] in the rectangular container.

A satisfactory physical explanation of the observed and calculated stationary instabilities has not yet been given. The analysis is complicated by the presence of both thermocapillary and buoyancy forces. Since the vertical velocity at the cell boundaries in fig. 33 transports relatively cold fluid from the bulk to the free surface, the Marangoni effect is suppressing the secondary flow in the $(y, z)$-plane. The instability should therefore be caused by another mechanism. It is important to note that the basic flow is subject to high strain rates before the instability sets in. The base flow in the annular gap was observed to consist of an elongated stretched vortex just before the onset of instability [53]. Similarly, Gillon & Homsy [90] noticed a stretched thermocapillary driven vortex immediately below the free surface with two smaller co-rotating eddies embedded in the main circulation. An example of such a flow structure is shown in fig. 34.

A stationary three-dimensional instability of vortices with elliptical streamlines was recently observed in a two-sided lid-driven cavity by [94, 95]. This experiment could be a key to the understanding of the three-dimensional buoyant-thermocapillary instability. The quasi-two-dimensional flow structure just before the onset of a three-dimensional steady flow (which is shown in fig. 35) has a typical cat's eye structure (fig. 36). By a linear stability analysis Kuhlmann *et al.* [94] showed that the instability is due to the elliptic instability mechanism [96, 97, 98] and that it is determined mainly by the flow properties in the center between both regions of closed streamlines. In view of the topological equivalence of both types of flows, it is very likely that also the combined buoyant-thermocapillary instability is due to this mechanism. The cellular instability arises only for non-unit aspect ratios $\Gamma \neq 1$. It relies on a strong straining

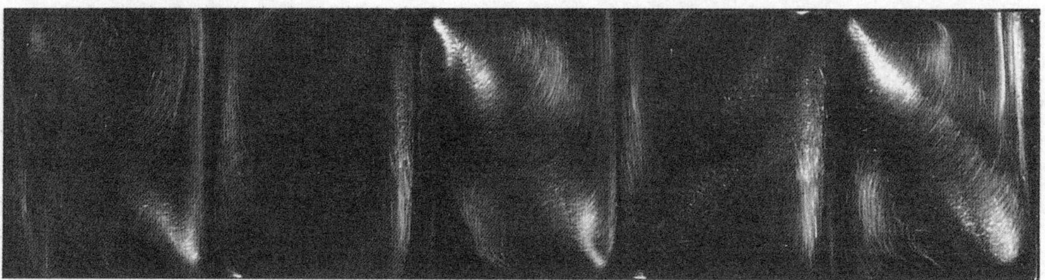

Figure 35: Typical pattern as a result of the stationary elliptic instability in a two-sided lid-driven cavity [94]. The moving walls are at the top and bottom and move with equal velocities into and out of, respectively, the plane shown.

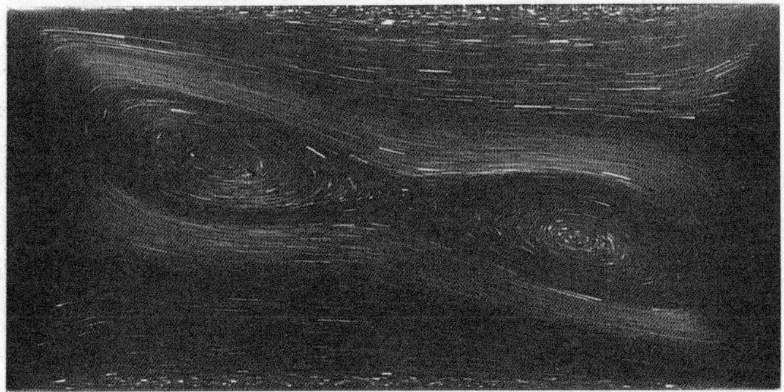

Figure 36: Basic cat's eye flow just before the onset of three-dimensional instability in a two-sided lid-driven cavity [94]. The walls on the left and the right side move with equal absolute values of the velocity up and down, respectively.

motion in the center of the vortex. The straining motion typically appears in form of elliptical streamlines or, for high strain rates, as a free hyperbolic stagnation point. Such a hyperbolic stagnation point is clearly visible in fig. 34 and fig. 36. It was also observed by [90] just before the onset of the instability.

The steady three-dimensional cellular flows can become time-dependent. Braunsfurth & Homsy [99] considered the cavity flow for $Pr = 4.4$ and equal aspect ratios $\Gamma = \Gamma_z$ in the range of $1 - 8$. It was observed that the fully developed steady cellular flow is reversed near the cold wall and directed opposite to the thermocapillary surface force. The resulting stagnation line on the free surface may be even located close to the hot wall where a distinct small but intense eddy is induced. The surface temperature gradient is high close to the hot corner, while it is low along the remainder of the surface. For Marangoni numbers above a second critical value ($Ma_{c2} \approx 5.5 \times 10^4$) the corner eddy was observed to periodically oscillate in strength and diameter. The instability mechanism has been discussed in terms of a positive feedback due to a temperature fluctuation at the stagnation line: If the temperature decreases at the stagnation line due to a perturbation, the temperature gradient near the hot corner is increased and the driving of the corner eddy is enhanced. Simultaneously, the reversed flow towards the hot corner, which tends to suppress the hot corner eddy, feels less opposition due to a reduced surface stress on the cold side of the cavity. This effect also contributes to the amplification of the strength of the hot corner vortex, yielding a positive feedback. Note that the presence of a strong three-dimensional steady cellular flow must be present to provide the flow reversal near the cold wall in order that this type of instability can occur.

## 7  Summary

The incompressible fluid flow driven by the thermocapillary effect in differentially heated cylindrical liquid bridges and rectangular open cavities has been reviewed. After an introduction of the governing equations and a derivation of the energy equations the Stokes flow is considered in corners made up of an isothermal rigid plane and a flat free surface on which the temperature varies linearly from the line of intersection. The creeping flow in the whole volume of cylindrical liquid bridges is then calculated by use of the biorthogonal series method and the flow patterns are discussed. Boundary layer scalings are derived for some representative limiting cases of the Reynolds and Prandtl numbers. As a paradigm for hydrothermal waves the linear instability of the return flow (plane Poiseuille-Couette flow) is presented. The mechanics of hydrothermal waves is discussed for the instability of the toroidal axisymmetric flow in high Prandtl number liquid bridges. For low Prandtl numbers, the axisymmetric flow in differentially heated liquid bridges is due to an inertial instability which is not present in plane layers nor in the infinite cylinder model. The modal structure and the time-dependence of the supercritical flow in liquid bridges is briefly discussed. The last section explains the thermocapillary flows in differentially heated rectangular cavities. The two-dimensional flow structures depend on the aspect ratio. For high Reynolds numbers time-dependence sets in. The oscillatory flow corresponds to hydrothermal waves if the aspect ratio is sufficiently large. If the aspect ratio is $O(1)$ but the extent perpendicular to the applied temperature gradient large a steady three-dimensional flow can bifurcate from the two-dimensional base flow.

The area of thermocapillary and buoyancy driven flows in systems with free surfaces is currently evolving rapidly. One reason for this interest is the importance of heat and mass transfer in crystal growth from the melt. The geometries and the experimental conditions in technical processes are more complicated than the models considered here. The simple models presented allow, however, a basic understanding of the underlying fluid physics. Based on the knowledge gained refined models can be studied more efficiently in the future.

# Acknowledgments

The author is very grateful to his co-workers of the *Hydrodynamic Stability* group at ZARM for many discussions, the good cooperation, and the motivating atmosphere. This work has been supported in parts by Deutsche Forschungsgemeinschaft and Deutsche Agentur für Raumfahrtangelegenheiten.

# References

[1] L. E. Scriven and C. V. Sternling. The Marangoni effects. *Nature* **187**, 186 (1960).

[2] D. T. J. Hurle, editor. *Handbook of Crystal Growth*. North Holland 1994.

[3] Ch. E. Chang and W. R. Wilcox. Inhomogeneities due to thermocapillary flow in floating zone melting. *J. Crystal Growth* **28**, 8 (1975).

[4] C. E. Chang and W. R. Wilcox. Analysis of surface tension driven flow in floating zone melting. *Int. J. Heat Mass Transfer* **19**, 355 (1976).

[5] L. D. Landau and E. M. Lifshitz. *Hydrodynamik* volume VI of *Lehrbuch der Theoretischen Physik*. Akademie Verlag 1986.

[6] F. Dupret, P. Nicodème, Y. Ryckmans, P. Wouters, and M. J. Crochet. Global modelling of heat transfer in crystal growth furnices. *Int. J. Heat Mass Transfer* **33**, 1849 (1990).

[7] A. D. Myshkis, V. G. Babskii, N. D. Kopachevskii, and L. A. Slobozhanin. *Low-Gravity Fluid Mechanics*. Springer 1987.

[8] F. Raynal. Exact relation between spatial mean enstrophy and dissipation in confined incompressible flows. *Phys. Fluids* **8**, 2242 (1996).

[9] D. Canright. Thermocapillary flow near a cold wall. *Phys. Fluids* **6**, 1415 (1994).

[10] H. C. Kuhlmann, C. Nienhüser, and H. J. Rath. The local flow in a wedge between a rigid wall and a surface of constant shear stress. *J. Eng. Math.* (1998. submitted).

[11] H. K. Moffatt and B. R. Duffy. Local similarity solutions and their limitations. *J. Fluid Mech.* **96**, 299 (1980).

[12] W. W. Schultz and C. Gervasio. A study of the singularity of the die swell problem. *Q. J. Appl. Math.* **43**, 407 (1990).

[13] E. B. Dussan V., E. Ramé, and S. Garoff. On identifying the appropriate boundary conditions at a moving contact line: An experimental investigation. *J. Fluid Mech.* **230**, 97 (1991).

[14] Y. D. Shikhmurzaev. Moving contact lines in liquid/liquid/solid systems. *J. Fluid Mech.* **334**, 211 (1997).

[15] J. Koplik and J. R. Banavar. Corner flow in the sliding plate problem. *Phys. Fluids* **7**, 3118 (1995).

[16] R. C. T. Smith. The bending of a semi-infinite strip. *Austral. J. Sci. Res.* **5**, 227 (1952).

[17] D. D. Joseph. The convergence of biorthogonal series for biharmonic and Stokes flow edge problems. Part I. SIAM *J. Appl. Math.* **33**, 337 (1977).

[18] D. D. Joseph and L. Sturges. The convergence of biorthogonal series for biharmonic and Stokes flow edge problems: Part II. SIAM *J. Appl. Math.* **34**, 7 (1978).

[19] H. C. Kuhlmann. Small amplitude thermocapillary flow and surface deformations in a liquid bridge. *Phys. Fluids* A **1**, 672 (1989).

[20] A. P. Hillman and H. E. Salzer. The roots of $\sin z = z$. *Phil. Mag.* **34**, 575 (1943).

[21] C. I. Robbins and R. C. T. Smith. A table of roots of $\sin z = -z$. *Phil. Mag.* **39**, 1004 (1948).

[22] P. F. Papkovich. Über eine Form der Lösung des biharmonischen Problems für das Rechteck. *C. r. (Dokl.) Acad. Sci. USSR* **27**, 334 (1940).

[23] J. Fadle. Die Selbstspannungs-Eigenwertfunktionen der quadratischen Scheibe. *Ing. Archiv* **11**, 125 (1941).

[24] A. M. J. Davis. Thermocapillary convection in liquid bridges: Solution structure and eddy motions. *Phys. Fluids* A **1**, 475 (1989).

[25] A. Rybicki and J. M. Floryan. Thermocapillary effects in liquid bridges. I. Thermocapillary convection. *Phys. Fluids* **30**, 1956 (1987).

[26] H. K. Moffatt. Viscous and resistive eddies near a sharp corner. *J. Fluid Mech.* **18**, 1 (1964).

[27] F. Pan and A. Acrivos. Steady flows in rectangular cavities. *J. Fluid Mech.* **28**, 643 (1967).

[28] P. G. Drazin and W. H. Reid. *Hydrodynamic Stability.* Cambridge University Press 1981.

[29] Rayleigh, Lord. On the instability of jets. *Proc. London Math. Soc.* **10**, 4 (1879).

[30] M. Abramowitz and I. A. Stegun. *Handbook of Mathematical Functions.* Dover 1972.

[31] S. Ostrach. Motion induced by capillarity (V. G. Levich Festschrift). In Spalding, editor, *Physico-chemical Hydrodynamics* volume 2. Advance Publications 1977.

[32] L. G. Napolitano. Marangoni boundary layers. In *Proceedings of the IIIrd European Symposium on Materials Science in Space* page 249. ESA SP-142 1979.

[33] M. B. Glauert. The wall jet. *J. Fluid Mech.* **1**, 625 (1956).

[34] S. J. Cowley and S. H. Davis. Viscous thermocapillary convection at high Marangoni numbers. *J. Fluid Mech.* **135**, 175 (1983).

[35] J. F. Mercier and C. Normand. Buoyant-thermocapillary instabilities of differentially heated liquid layers. *Phys. Fluids* **8**, 1433 (1996).

[36] P. M. Parmentier, V. C. Regnier, and G. Lebon. Buoyant-thermocapillary instabilities in medium-Prandtl-number fluid layers subject to a horizontal temperature gradient. *Int. J. Heat Mass Transfer* **36**, 2417 (1993).

[37] J. Priede and G. Gerbeth. Influence of thermal boundary conditions on the stability of thermocapillary-driven convection at low Prandtl numbers. *Phys. Fluids* **9**, 1621 (1997).

[38] M. K. Smith and S. H. Davis. Instabilities of dynamic thermocapillary liquid layers. Part 1. Convective instabilities. *J. Fluid Mech.* **132**, 119 (1983).

[39] M. K. Smith and S. H. Davis. Instabilities of dynamic thermocapillary liquid layers. Part 2. Surface-wave instabilities. *J. Fluid Mech.* **132**, 145 (1983).

[40] M. K. Smith. The nonlinear stability of dynamic thermocapillary liquid layers. *J. Fluid Mech.* **194**, 391 (1988).

[41] S. H. Davis. Thermocapillary instabilities. *Annu. Rev. Fluid Mech.* **19**, 403 (1987).

[42] M. K. Smith. Instability mechanisms in dynamic thermocapillary liquid layers. *Phys. Fluids* **29**, 3182 (1986).

[43] C. Canuto, M. Y. Hussaini, A. Quarteroni, and T. A. Zhang. *Spectral Methods in Fluid Dynamics.* Springer 1988.

[44] W. H. Press, B. P. Flannery, S. A. Teukolsky, and W. T. Vetterling. *Numerical Recipes* (FORTRAN). Cambridge University Press 1989.

[45] M. K. Smith. *The instabilities of thermocapillary shear layers.* PhD thesis Northwestern University 1982.

[46] J. R. A. Pearson. On convection cells induced by surface tension. *J. Fluid Mech.* **4**, 489 (1958).

[47] A. K. Sen and S. H. Davis. Steady thermocapillary flows in two-dimensional slots. *J. Fluid Mech.* **121**, 163 (1982).

[48] C. De Saedeleer, A. Garcimartin, G. Chavepeyer, J. K. Platten, and G. Lebon. The instability of a liquid layer heated from the side when the upper surface is open to the air. *Phys. Fluids* **8**, 670 (1996).

[49] P. Laure, B. Roux, and H. Ben Hadid. Nonlinear study of the flow in a long rectangular cavity subjected to thermocapillary effect. *Phys. Fluids* A **2**, 516 (1990).

[50] D. Villers and J. K. Platten. Coupled buoyancy and Marangoni convection in acetone: Experiments and comparison with numerical simulations. *J. Fluid Mech.* **234**, 487 (1992).

[51] Kelvin, Lord. On a disturbing infinity in Lord Rayleigh's solution for waves in a plane vortex stream. *Nature* **23**, 45 (1880).

[52] D. Schwabe, U. Möller, J. Schneider, and A. Scharmann. Instabilities of shallow dynamic thermocapillary liquid layers. *Phys. Fluids* A **4**, 2368 (1992).

[53] J. Schneider. *Strukturen thermokapillarer Konvektion in einem Ringspalt.* PhD thesis Universität Giessen 1995.

[54] M. Wanschura. *Lineare Instabilitäten kapillarer und natürlicher Konvektion in zylindrischen Flüssigkeitsbrücken.* PhD thesis ZARM – Universität Bremen 1996.

[55] C. A. Jones. Numerical methods for the transition to wavy Taylor vortices. *J. Comput. Phys.* **61**, 321 (1985).

[56] D. J. Acheson. *Elementary Fluid Dynamics.* Oxford University Press 1990.

[57] B. M. Carpenter and G. M. Homsy. High Marangoni number convection in a square cavity: Part II. *Phys. Fluids* A **2**, 137 (1990).

[58] G. K. Batchelor. On steady laminar flow with closed streamlines at large Reynolds numbers. *J. Fluid Mech.* **1**, 177 (1956).

[59] R. Velten, D. Schwabe, and A. Scharmann. The periodic instability of thermocapillary convection in cylindrical liquid bridges. *Phys. Fluids* A **3**, 267 (1991).

[60] F. Preisser, D. Schwabe, and A. Scharmann. Steady and oscillatory thermocapillary convection in liquid columns with free cylindrical surface. *J. Fluid Mech.* **126**, 545 (1983).

[61] M. Wanschura, H. C. Kuhlmann, and H. J. Rath. Linear stability of two-dimensional combined buoyant-thermocapillary flow in cylindrical liquid bridges. *Phys. Rev.* E **55**, 7036 (1997).

[62] M. Levenstam and G. Amberg. Hydrodynamic instabilities of thermocapillary flow in a half-zone. *J. Fluid Mech.* **297**, 357 (1995).

[63] G. P. Neitzel, K.-T. Chang, D. F. Jankowski, and H. D. Mittelmann. Linear-stability theory of thermocapillary convection in a model of the float-zone crystal-growth process. *Phys. Fluids* A **5**, 108 (1993).

[64] J.-J. Xu and S. H. Davis. Convective thermocapillary instabilities in liquid bridges. *Phys. Fluids* **27**, 1102 (1984).

[65] S. E. Widnall and Ch.-Y. Tsai. The instability of the thin vortex ring of constant vorticity. *Proc. Roy. Soc. London* **287**, 273 (1977).

[66] M. Prange. Stabilisierung thermokapillarer Konvektion in zylindrischen Flüssigkeitsbrücken mittels axialer Magnetfelder. Master's thesis ZARM – Universität Bremen 1997.

[67] E. L. Koschmieder. *Bénard Cells and Taylor Vortices.* Cambridge University Press 1993.

[68] M. C. Cross and P. G. Hohenberg. Pattern formation outside of equilibrium. *Rev. Mod. Phys.* **65**, 851 (1993).

[69] J. Guckenheimer and P. Holmes. *Nonlinear Oscillations, Dynamical Systems, and Bifurcations of Vector Fields* volume 42 of *Applied Mathematical Sciences.* Springer 1983.

[70] R. Rupp, G. Müller, and G. Neumann. Three-dimensional time dependent modelling of the Marangoni convection in a zone melting configuration for GaAs. *J. Crystal Growth* **97**, 34 (1989).

[71] G. Chen, A. Lizée, and B-Roux. Bifurcation analysis of the thermocapillary convection in cylindrical liquid bridges. *J. Crystal Growth* **180**, 638 (1997).

[72] K. A. Muehlner, M. F. Schatz, V. Petrov, W. D. McCormick, J. B. Swift, and H. L. Swinney. Observation of helical traveling-wave convection in a liquid bridge. *Phys. Fluids* **9**, 1850 (1997).

[73] J. D. Crawford and E. Knobloch. Symmetry and symmetry-breaking bifurcations in fluid dynamics. *Annu. Rev. Fluid Mech.* **23**, 341 (1991).

[74] J. Leypoldt, H. C. Kuhlmann, and H. J. Rath. Three-dimensional numerical simulation of thermocapillary flows in cylindrical liquid bridges. Submitted to J. Fluid Mech. 1998.

[75] R. Savino and R. Monti. Oscillatory Marangoni convection in cylindrical liquid bridges. *Phys. Fluids* **8**, 2906 (1996).

[76] J. Xu and A. Zebib. Oscillatory two- and three-dimensional thermocapillary convection. Submitted to J. Fluid Mech. 1997.

[77] H.-Ch. Hsieh. A numerical study of three-dimensional surface tension driven convection with free surface deformations. In *Fluid Mechanics Phenomena in Microgravity* volume AMD 154 / FED 142 page 85. ASME 1992.

[78] D. D. Joseph and L. Sturges. The free surface on a liquid filled trench heated from its side. *J. Fluid Mech.* **69**, 565 (1975).

[79] A. Zebib, G. M. Homsy, and E. Meiburg. High Marangoni number convection in a square cavity. *Phys. Fluids* **28**, 3467 (1985).

[80] R. V. Birikh. Thermocapillary convection in a horizontal layer of liquid. *J. Appl. Mech. Tech. Phys.* **7**, 43 (1966).

[81] M. Strani, R. Piva, and G. Graziani. Thermocapillary convection in a rectangular cavity: Asymptotic theory and numerical simulation. *J. Fluid Mech.* **130**, 347 (1983).

[82] H. Ben Hadid and B. Roux. Thermocapillary convection in long horizontal layers of low-Prandtl-number melts subject to a horizontal temperature gradient. *J. Fluid Mech.* **221**, 77 (1990).

[83] H. Schlichting. *Grenzschichttheorie.* Springer 1968.

[84] M. Ohnishi, H. Azuma, and T. Doi. Computer simulation of oscillatory Marangoni flow. *Acta Astronautica* **26**, 685 (1992).

[85] H. Ben Hadid and B. Roux. Buoyancy- and thermocapillary-driven flows in differentially heated cavities for low-Prandtl-number fluids. *J. Fluid Mech.* **235**, 1 (1992).

[86] M. Mundrane and A. Zebib. Oscillatory buoyant thermocapillary flow. *Phys. Fluids* **6**, 3294 (1994).

[87] L. J. Peltier and S. Biringen. Time-dependent thermocapillary convection in a rectangular cavity: Numerical results for a moderate Prandtl number fluid. *J. Fluid Mech.* **257**, 339 (1993).

[88] V. Saß, H. C. Kuhlmann, and H. J. Rath. Investigation of three-dimensional thermocapillary convection in a cubic container by a multi-grid method. *Int. J. Heat Mass Transfer* **39**, 603 (1996).

[89] U. T. Bödewadt. Die Drehströmung über festem Grunde. *Z. Angew. Math. Mech.* **20**, 241 (1940).

[90] P. Gillon and G. M. Homsy. Combined thermocapillary-buoyancy convection in a cavity: An experimental study. *Phys. Fluids* **8**, 2953 (1996).

[91] D. Schwabe and J. Metzger. Coupling and separation of buoyant thermocapillary convection. *J. Crystal Growth* **97**, 23 (1989).

[92] F. Daviaud and J. M. Vince. Travelling waves in a fluid layer subjected to a horizontal temperature gradient. *Phys. Rev.* E **48**, 4432 (1993).

[93] M. Mundrane and A. Zebib. Two- and three-dimensional buoyant thermocapillary convection. *Phys. Fluids* A **5**, 810 (1993).

[94] H. C. Kuhlmann, M. Wanschura, and H. J. Rath. Flow in two-sided lid-driven cavities: Non-uniqueness, instabilities, and cellular structures. *J. Fluid Mech.* **336**, 267 (1997).

[95] H. C. Kuhlmann, M. Wanschura, and H. J. Rath. Elliptic instability in two-sided lid-driven cavity flow. *Eur. J. Mech., B/Fluids* (to appear) (1998).

[96] B. J. Bayly. Three-dimensional instability of elliptical flow. *Phys. Rev. Lett.* **57**, 2160 (1986).

[97] R. T. Pierrehumbert. Universal short-wave instability of two-dimensional eddies in an inviscid fluid. *Phys. Rev. Lett.* **57**, 2157 (1986).

[98] F. Waleffe. On the three-dimensional instability of strained vortices. *Phys. Fluids* A **2**, 76 (1990).

[99] M. G. Braunsfurth and G. M. Homsy. Combined thermocapillary-buoyancy convection in a cavity. Part II. An experimental study. *Phys. Fluids* **9**, 1277 (1997).

# DROPS, JETS AND BUBBLES


**J.I.D. Alexander**
**Case Western Reserve University, Cleveland, OH, USA**



ABSTRACT

Drops, jets and bubbles occur in many different technological and scientifically relevant situations and, thus, have been the subject of a great deal of theoretical and experimental work.   In the six lectures presented here, I have focused on the equilibria and dynamics of captive drops (or liquid bridges), the thermocapillary migration of drops. In the last lecture I give a brief introduction to the related problem of contact line motion and dynamic contact angles.   (Contact line and angle behavior is an important aspect of captive, sessile and pendant drop dynamics.)   In Chapter 1, the basic equations governing bridge dynamics and equilibria are introduced and a method for determining the stability of axisymmetric equilibrium shapes is outlined. Selected results for the stability of bridges subject to gravity and isorotation are discussed.   In Chapter 2 we consider 1D models of liquid bridge oscillation. These models are based on models used to describe axisymmetric jets and have been adapted for modeling liquid bridge dynamics.   In Chapter 3, 2D and 3D liquid bridge oscillations are examined and the nonlinear behavior of bridges undergoing large amplitude oscillations is discussed. It is shown that, for nonlinear oscillations, liquid bridges behave like a soft spring.  In Chapter 4 we consider different models of the breaking of jets, pendant drops and liquid bridges.  In Chapter 5 we briefly analyze the Young-Goldstein-Block model of the thermocapillary migration of bubbles. In Chapter 6, we introduce and discuss recent experimental and theoretical work concerning dynamic contact angles and highlight the complexity of the behavior of moving contact lines.

# 1.  LIQUID BRIDGE EQUILIBRIA AND DYNAMICS

## 1.1  Introduction

In this chapter we outline the equations governing the equilibria and equilibria of captive drops or liquid bridges.  In most cases, the presence of a second phase (a fluid or gas surrounding the bridge) is ignored for practical computational reasons.  However, since it may not always be permissible to assume that the second phase is unimportant, two fluids are accounted for in the following description.  In this way, students will be able to assess for themselves the extent of the approximations made in the various models that will be presented and discussed in subsequent chapters.

## 1.2  Governing Equations and Boundary Conditions

To undertake a general treatment of isothermal liquid bridge dynamics we need the equations governing the flow of an isothermal, constant density, viscous fluid.  In addition, we require boundary conditions that enforce conservation of mass and momentum at the free surface of the bridge and properly describe the behavior of the bridge at its solid supports.  These equations, for a constant density Newtonian fluid, are introduced below.

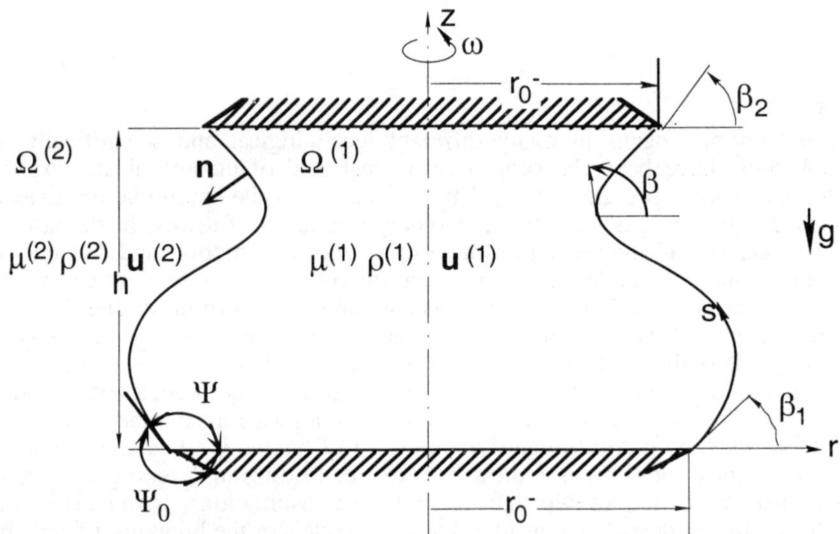


Fig. 1 Schematic of the liquid bridge. The solid line is the projection of the surface $\Sigma$ which separates the liquid bridge $D_1$ and the surrounding medium $D_2$.

Consider a region $\Omega = \Omega^{(1)} \cup \Omega^{(2)}$ that is occupied by two fluids (see Fig. 1.1).  The region $\Omega^{(1)}$ is a liquid bridge of volume $V_O$ that is held between rigid circular disks of radius $r_o^+$ and $r_o^-$.  The liquid bridge is surrounded by a second fluid which occupies the region $\Omega^{(2)}$.  It is assumed that the fluids are immiscible and are Newtonian, with constant dynamic viscosity $\mu^{(i)}$, and density $\rho^{(i)}$, $i = 1,2$.  The boundary, $\Sigma = \Omega^{(1)} \cap \Omega^{(2)}$, between the two immiscible fluids is characterized by a constant surface tension $\gamma$.

## 1.2.1 Continuity and momentum equations

The density of each fluid is assumed to be constant. This leads to the following familiar expression for incompressible flow

$$\nabla \cdot u^{(i)} = 0 \ \ in \ \Omega_i, i = 1, 2 \,, \tag{1.1}$$

where $u^{(i)}$ is the velocity of the fluid "$i$", i = 1, 2.

For incompressible, constant viscosity Newtonian fluids, conservation of linear momentum yields the following equation of motion in each fluid

$$\rho^{(i)}\left(\frac{\partial u^{(i)}}{\partial t} + (u^{(i)}\cdot\nabla)u^{(i)}\right) = -\nabla p^{(i)} + \mu^{(i)} \nabla^2 u^{(i)} + \rho^{(i)} g \,, \ in \ \Omega_i \,, i = 1, 2 \tag{1.2}$$

When $p^{(i)}$, $i = 1,2$ is the pressure in each fluid, $t$ denotes time and $g$ is the gravity vector. It is often convenient to define a "reduced" pressure.

$$p*^{(i)} = p^{(i)} - \rho^{(i)} g \cdot x \tag{1.3}$$

which eliminates the third term in (1.2) when $p^{(i)}$ is replaced with $p*^{(i)}$.

## 1.2.2 Interface and surface boundary conditions

Let $F(x, t) = 0$ represent the free surface, $\Sigma = \Sigma = \Omega^{(1)} \cap \Omega^{(2)}$, of the liquid bridge. At any point $x_\Sigma \in \Sigma$ defined by $F(x, t) = 0$ we shall require that the following conditions be satisfied:
(a) no mass from fluid 1 or 2 will pass through the surface; (b) the two fluids will remain in contact (continuity); (c) if both fluids have non-zero viscosity there will be no slip along $\Sigma$ of one fluid relative to the other; (d) forces at the interface must balance (conservation of linear momentum)

Conditions (a) and (b) require that no mass shall pass from $\Omega^{(1)}$ to $\Omega^{(2)}$ and that the two fluids shall always remain in contact. Thus, at any point $x_\Sigma$ on the surface the component of fluid velocity $u(x_\Sigma, t)$ that is instantaneously normal to $F(x, t) = 0$ must be equal to the speed, $V_n(x_\Sigma, t)$, of the surface along the direction of $n(x_\Sigma, t)$, the unit normal vector to $F=0$ at $x_\Sigma$. This can be written as

$$\lim_{\substack{x \to x_\Sigma \\ x \in \Omega^i}} u^{(i)}(x,t) \cdot n (x_\Sigma, t) = V_n (x_\Sigma, t) \tag{1.4}$$

for all $x_\Sigma \in \Sigma$. Now consider a point $P$ which lies on the surface $\Sigma$ and use $x = \chi_p(t)$ to describe its trajectory as the surface moves. We then have

$$F(\chi_p(t), t) = 0 \tag{1.5}$$

Upon differentiating (1.5) with respect to time we obtain

$$\nabla F \cdot \frac{d\chi_p}{dt} + \frac{\partial F}{\partial t} = 0 \,, \tag{1.6}$$

where $d\chi_P/dt$ simply represents the velocity of the surface at point $P$.  The normal $\boldsymbol{n}$ is taken to point out of fluid 1 and is defined by

$$n = \frac{\nabla F(x_\Sigma, t)}{|\nabla F(x_\Sigma, t)|} \; . \tag{1.7}$$

Thus, from (1.7) and

$$u(x_\Sigma, t) \cdot n = \frac{d\chi_P}{dt} \cdot n(x_\Sigma, t) = \frac{d\chi_P}{dt} \cdot \frac{\nabla F}{|\nabla F|} \; , \tag{1.8}$$

we can then use (1.6) to equate the component of fluid velocity at the surface in the direction of the surface normal, $\boldsymbol{n}$, to the speed $V_n$ of the surface, or

$$u_n(x_\Sigma, t) = \frac{-1}{|\nabla F|} \frac{\partial F(x_\Sigma, t)}{\partial t} \; , \tag{1.9}$$

and, from (1.4) , we obtain the limit

$$\lim_{\substack{x \to x_\Sigma \\ x \in \Omega^i}} u^{(i)} \cdot \nabla F \left( x_\Sigma, t \right) = -\frac{\partial F}{\partial t} \left( x_\Sigma, t \right), \tag{1.10}$$

Equation (1.10) is a kinematic boundary condition for the free surface [1.1] and takes the form

$$\frac{DF}{Dt} = \frac{\partial F}{\partial t} + u^{(i)} \cdot \nabla F = 0 \; \cdot \tag{1.11}$$

It is used together with

$$u^{(1)}(x_\Sigma, t) \cdot n = u^{(2)}(x_\Sigma, t) \cdot n \; , \qquad\qquad x_\Sigma \in \Sigma \; , \tag{1.12}$$

For viscous fluids, the additional "no-slip" condition (condition (c)) generalizes (1.12) to

$$u^{(1)}(x_\Sigma, t) = u^{(2)}(x_\Sigma, t) \; , \; x_\Sigma \in \Sigma \; . \tag{1.13}$$

We assume that the surface, $\Sigma$, is characterized by a constant surface tension, $\gamma$ , such that a force $(= -2\gamma\kappa \, (x_\Sigma, t))$ proportional to its local mean curvature, $\kappa \, (x_\Sigma, t)$ is exerted along the normal vector at a point $x_\Sigma$ on the surface.  Balance of forces at the liquid bridge surface (condition (d)) thus requires that the difference, $F_{2,1}$ between the bulk forces acting across the surface

$$F_{2,1} = T^{(2)}n - T^{(1)}n \; , \tag{1.14}$$

$$T^{(i)} = -p^{(i)}I + \mu^{(i)} \left[ \nabla u^{(i)} + \nabla u^{(i)\,T} \right]$$

which arises from the pressures and viscous stresses in the bulk fluids, should be equal to the surface tension force.  That is,

$$F_{2,1} = -2\gamma\kappa n .\qquad(1.15)$$

where, the mean curvature, $\kappa(x_\Sigma,t)$, is [1.2]

$$2\kappa\left(x_\Sigma, t\right) = -\nabla_\Sigma \cdot n\left(x_\Sigma, t\right),\qquad(1.16)$$

and $\nabla_\Sigma$ is the surface gradient operator [1.2].  Equations (1.14),  (1.15) and (1.3) yield

$$-p^{**(2)} + p^{**(1)} - \Delta\rho[g + a\omega^2 \sin \omega t]\,z + \left[2\mu^{(2)} D^{(2)} n - 2\mu^{(1)} D^{(1)} n\right] \cdot n = -2\gamma\kappa\qquad(1.17)$$

$$\mu^{(1)} D^{(1)} n \cdot \tau = \mu^{(2)} D^{(2)} n \cdot \tau,\qquad(1.18)$$

where $\tau$ is a unit tangent vector to the surface and $D = 1/2\left[\nabla u + \nabla u^T\right]$ is the rate of deformation tensor

### 1.2.3 Initial conditions, boundary conditions and special transformations

Boundary conditions at the support disks will involve the fluid velocities only.  Since we consider the surface to be pinned to the sharp edges of the supporting disks, no constraints will be placed on the contact angle $\varphi$ (see Fig. (1.1)) except that it will remain pinned to the disk edge.  This means that the slope of the surface at the disk edge is free to take on any angle (see sections 1.3.1 and 6.1 for further discussion of this constraint and for thermodynamic and geometrical constraints on the range of admissible angles).

Generally we will have velocity components that satisfy

$$u\left(x^\pm, t\right) \cdot e_r = g^\pm(t), \; u\left(x^\pm, t\right) \cdot e_z = f^\pm(t),\qquad(1.19)$$

for all points $x^\pm$ at the upper $^{(+)}$ or lower $^{(-)}$ disks. where the functions $f^\pm$ and $g^\pm$ are used to describe the time-dependent motion of the support disks if necessary.  Thus,  for a problem which considers free oscillations with fixed supports we would have $f^\pm = g^\pm = 0$,  and

$$F(x, 0) = r - R\left(\theta, z, 0\right) = 0 \quad, \quad u^{(i)}(x, 0) = 0 , \, i = 1,2),\qquad(1.20)$$

where $R\left(\theta, z, 0\right)$ specifies the initial shape of the liquid bridge surface. For problems, such as disk vibration, for example, we might have as conditions at the disk and initial conditions

$$f^+(t) = \omega^+ a^+ \cos\left(\omega^+ t\right), \; f^- = \omega^- a^- \cos \omega^- t,\qquad(1.21)$$
$$g^+ = g^- = 0, \quad r - R\left(\theta, z, 0\right) = 0, \; u^{(i)}(x, 0) = 0 .$$

Here $\omega^+$, and $\omega^-$ are the angular frequencies of the disk oscillations, and $f^\pm$ correspond to the velocities arising from oscillating displacements of amplitude $a^\pm$.  When $a^+ = a^-$ and $\omega^+ = \omega^-$ in (1.21) we can use the transformation $x'^{(i)} = x^{(i)} + a \sin \omega t$ to obtain

$$\frac{Du'}{Dt} = \frac{Du}{Dt} - \omega^2 \, a \sin \omega t \, e_z, \qquad (1.23)$$

This changes (1.2) to

$$\rho^{(i)}\left(\frac{\partial u^{(i)}}{\partial t} + u^{(i)}\nabla u^{(i)}\right) = -\nabla p^{*(i)} + \mu^{(i)} \nabla^2 u^{(i)} - \rho\omega^2 \, a \sin\omega t \, e_z, \qquad (1.24)$$

where, for convenience, we have now dropped the " ′ " superscript.  If we now redefine the reduced pressure to be

$$p^{**(i)} = p^{*(i)} - \rho^{(i)}za\omega^2 \, a \sin \omega t, \qquad (1.25)$$

and substitute (1.25) into (1.24) we can eliminate the body force.  The body force term now only appears in the force balance boundary condition. The boundary conditions are then $u^\pm = 0$ *at* $x = x^\pm$, and, if we take $g = ge_z$ and $\Delta\rho = \rho^{(1)} - \rho^{(2)}$ we have

$$-p^{**(2)} + p^{**(1)} - \Delta\rho[g + a\omega^2 \sin \omega t] \, z + \left[2\mu^{(2)} D^{(2)} n - 2\mu^{(1)} D^{(1)} n\right] \cdot n = -2\gamma\kappa \qquad (1.26)$$

which together with (1.18) and (1.6) define the problem.  In addition to these equations it is usually necessary to enforce a constant volume constraint [1.3].

## 1.3    Liquid bridge stability and equilibria
### 1.3.1 Stability of axisymmetric liquid bridges
Consider a liquid bridge contained between two coaxial disks of equal-radius $r_0$ separated by a distance $h$ . The liquid-solid contact lines coincide with the disk edges and are fixed to them even when the bridge is subjected to perturbations. The liquid is subject to a gravity vector that is parallel to the axis of the disk and has a magnitude $|\,g\,|$. ($g > 0$ when gravity points along the negative z axis of the cylindrical coordinate system $(r,\ \theta,\ z)$.   In addition,the bridge is subject to a centrifugal force due to an isorotation of the system about the disk axis with angular velocity $\omega$. Generally four parameters describe the equilibrium [1.3]: the slenderness, $\Lambda$, the relative volume, $V$, the Bond number, $B$, and the Weber number, $W$.  They are, respectively,

$$\Lambda = h/(2r_0), \quad V = v/(\pi r_0^2 h), \quad B = \Delta\rho g r_0^2/\gamma, \quad W = \Delta\rho\omega^2 r_0^3/(2\gamma) . \qquad (1.27)$$

Here, $v$, $\Delta\rho$ and $\gamma$ are the liquid volume, density difference between the bridge and the surrounding medium and surface tension, respectively.
In equilibrium, (1.26) reduces to

$$\Delta p - \Delta\rho g \, z + \Delta\rho r^2\omega^2 = -2\kappa\gamma \ , \quad \Delta p = p^{*(1)} - p^{*(2)} \qquad (1.28)$$

where $\Delta p$ is the difference between the reference pressure inside the liquid bridge at z= r=0 a and that of the surrounding gas, $2\kappa = (GL + EN - 2FM)/EG$  is twice the mean curvature, and $E$, $F$, $G$, and $L$, $M$, $N$, are the surface's fundamental magnitudes of first and second order [1.4]. For a surface parameterized by $u$ and $v$, these magnitudes are given by

$$E = \left|\frac{\partial r}{\partial u}\right|^2, \quad G = \left|\frac{\partial r}{\partial v}\right|^2, \quad F = \frac{\partial r}{\partial u}\cdot\frac{\partial r}{\partial v},$$

$$L = n\cdot\frac{\partial^2 r}{\partial u^2}, \quad M = n\cdot\frac{\partial^2 r}{\partial u\partial v}, \quad N = n\cdot\frac{\partial^2 r}{\partial v^2}, \quad n = \frac{1}{\sqrt{EG}}\frac{\partial r}{\partial u}\times\frac{\partial r}{\partial v} \tag{1.29}$$

The shape of an axisymmetric equilibrium free surface can be described parametrically with the coordinates $r = r(s)$ and $z = z(s)$ ($s$ is the arc length of axial section $\theta$ = *constant*). For the fundamental magnitudes (1.29) with $u = \theta$ and $v = s$ this yields

$$G = r'^2 + z'^2, \quad E = r^2, \quad F = 0, \quad N = (r''z' - r'z'')(r'^2 + z'^2)^{-1/2},$$

$$L = -rz'(r'^2 + z'^2)^{-1/2}, \quad M = 0, \quad n = (z'e_r - r'e_\theta)(r'^2 + z'^2)^{1/2} \tag{1.30}$$

where a " $'$ " superscript denotes d/ds, $e_r$ and $e_z$ are unit vectors in the cylindrical coordinate system and the mean curvature, $\kappa$, is

$$2\kappa = \frac{r(r''z' - r'z'') - z'(r'^2 + z'^2)}{r(r'^2 + z'^2)^{3/2}} = \frac{-1}{rr'}\left(\frac{rz'}{(r'^2 + z'^2)^{1/2}}\right)' = -\nabla_\Sigma\cdot n. \tag{1.31}$$

Equations (1.30) and (1.27) lead to [1.3]

$$\frac{\gamma}{rr'}\left(\frac{rz'}{(r'^2 + z'^2)^{1/2}}\right)' = \Delta p - \Delta\rho g z + \Delta\rho r^2\omega^2. \tag{1.32}$$

Given the identities, $r' = \cos\beta(s)$ and $z' = \sin\beta(s)$, where $\beta = \beta(s)$ is the angle measured from the $r$-axis to the tangent to the equilibrium profile (the surface axial section) directed in the sense of increasing $s$, (1.31) can be rearranged to give

$$r'' = -z'\beta', \quad z'' = r'\beta', \quad \beta' = -bz + p r^2 + q - z'/r \quad (' = \tfrac{d}{ds}) \tag{1.33}$$

with

$$r(0) = r_0, \quad r'(0) = \cos\beta_1, \quad z(0) = 0, \quad z'(0) = \sin\beta_1, \quad \beta(0) = \beta_1 \tag{1.34}$$

at the point $s = s_1 = 0$ on the lower disk. Here, $b = \Delta\rho g/\gamma$ and $p = \Delta\rho\omega^2/2\gamma$. The constant $q = \Delta p/\gamma$ is proportional to the pressure difference between the reference pressure, $p^{*(1)}$ inside the liquid bridge at the point $r = z = 0$ and that of the surrounding gas, $p^{*(1)}$. The quantity q should be considered as a reference for the mean curvature.

The unknown quantities $q$, $\beta_1$ and the value $s = s_2$ at the end of the arc on the upper disk are determined by the following conditions

$$r\left(s_2\right) = r_0, \quad z\left(s_2\right) = h, \quad \int_0^{s_2} r^2 z'\, ds = v / \pi. \tag{1.35}$$

The integration of the second equation in (1.33) from $s = 0$ to $s = s_2$ and using (1.34) and (1.35) yields the following useful identity valid for any stable axisymmetric liquid bridge [1.3]

$$\sin \beta_2 = \sin \beta_1 - AB\left(1 - V\right), \tag{1.36}$$

where $\beta_2 = \beta(s_2)$ is the slope angle at the top disk.

An axisymmetric equilibrium configuration of a liquid bridge is stable only if it is stable with respect to both axisymmetric and nonaxisymmetric perturbations. The axisymmetric surface perturbation normal to the equilibrium surface is $\varphi_0\,(s)$. The most dangerous nonaxisymmetric perturbations correspond to the first harmonic in the polar angle $\theta$ and have the form $N = \varphi_1(s)\cos\theta$. The perturbations must satisfy the condition of liquid volume conservation which takes the form

$$\int_0^{s_2} r\,\varphi_0(s)\, ds = 0. \tag{1.37}$$

We will also assume that the inequalities, $\psi > \alpha$ and $\psi_0 > \pi - \alpha$, are satisfied on both of the contact lines that coincide with the disk edges. Here, $\psi$ and $\psi_0$ are the angles formed by the liquid and the gas at the contact line (see Fig. 1), and $\alpha$ is the wetting angle formed by the liquid at the contact with the smooth disk surface. Then the most dangerous perturbations are restricted to those which keep the contact lines fixed [1.3], that is

$$\varphi_0\,(0) = \varphi_0(s_2) = \varphi_1\,(0) = \varphi_1\,(s_2) = 0. \tag{1.38}$$

If one of inequalities for $\psi$ is not satisfied on one of the disk edges, no stable equilibrium state for a bridge surface pinned to the edges of disks exists [1.3].

We use the method described in [1.3] to study the stability to perturbations satisfying conditions (1.37) and (1.38). An equilibrium bridge is stable to axisymmetric (to nonaxisymmetric) perturbations provided the function $D(s)$ (function $\varphi_1\,(s)$), does not vanish on the interval $0 < s \le s_2$. Here,

$$D(s) = \varphi_{01}(s) \int_0^s r\varphi_{02}(s)\, ds - \varphi_{02}(s) \int_0^s r\varphi_{01}(s)\, ds, \tag{1.39}$$

and the functions $\varphi_{01}\,(s)$, $\varphi_{02}\,(s)$ and $\varphi_1\,(s)$ are the solutions of the following initial-value problems:

$$L\,\varphi_{01} = 0, \quad \varphi_{01}\,(0) = 0, \quad \varphi_{01}'\,(0) = 1;$$

$$L\,\varphi_{02} = 1, \quad \varphi_{02}\,(0) = 0, \quad \varphi_{02}'\,(0) = 1;$$

$$L\,\varphi_1 + \frac{1}{r^2}\,\varphi_1 = 0, \qquad \varphi_1(0) = 0, \qquad \varphi_1'(0) = 1; \tag{1.40}$$

$$L\,\varphi \equiv \varphi'' + \frac{r'}{r}\,\varphi' + \left( br' + 2prz' + \frac{z'^2}{r^2} + \beta'^2 \right)\varphi.$$

If there is at least one point $s$, $0 < s < s_2$, at which $D\,(s)$ or $\varphi_1\,(s)$ vanishes (in practice, changes sign), then the bridge is unstable.

We define the point $s = s*$ as the first point on the solution of problem (2) and (3) at which one of the functions $D\,(s)$ or $\varphi_1\,(s)$ vanishes. To find the profile of a neutrally stable surface with a prescribed value of $\beta_1$ and a chosen $r_0$, we vary the parameter $q$ until the condition $r\,(s*) = r_0$ is satisfied within the required accuracy. For the construction of a general boundary of the stability region, this procedure is performed for a sequence of possible values of the angle $\beta_1$. For a bridge of cylindrical volume, a neutrally stable surface can be found by variation of the parameters $q$ and $\beta_1$ until both conditions

$$r\left(s*\right) = r_0 \quad \text{and} \quad \int_0^{s*} r^2 z'\, ds = r_0^2\, z\left(s*\right)$$

are satisfied. For a neutrally stable surface, axisymmetric (nonaxisymmetric) perturbations are critical when $D\,(s)\,(\varphi_1\,(s))$ first vanishes at $s = s*$.


## 1.3.2 Axisymmetric liquid bridge stability: selected results

The above approach for analyzing liquid bridge stability has been used extensively by Slobozhanin and co-workers [1.3, 1.5-1.8]. Slobozhanin and Perales [1.5] recently extended these analyses to include the full range of possible contact angles and a wide range of $B$, $V$, $\Lambda$. They showed that for a finite Bond number, the stability boundary can be represented in the $\Lambda - V$ parameter space as a curve defined by three segments (see Fig. 1.2). The upper segment of the curve corresponds to the maximum volume stability limit. Between points O and C stability is lost to non-axisymmetric perturbations. At present no theory exists for the stability of these non-axisymmetric configurations although recent experimental work [1.9] indicates that some of these configurations are stable. For $B < 3.6$ the segment from F to C is characterized by loss of stability to axisymmetric perturbations. This leads to breaking of the bridge into pendant and sessile drops. A small satellite drop is also created. The third segment, OF, of the stability boundary corresponds to small volumes and slenderness ($V < 1$, $\Lambda < 1$). When gravity is directed downward and the liquid is denser than the surrounding medium, loss of stability occurs through detachment from the upper disk. The experimental results of Bezdenejnykh et al. [1.9] are in good qualitative agreement with the theoretical results.

An interesting result recently obtained by Slobozhanin and Alexander [1.8] concerned the combined effects of isorotation and gravity. The stability region in the $(\Lambda, V)$-plane can be constructed for any fixed pair of values of $B$ and $W$. The curves in Fig. 2.3 illustrate the independent effects of gravity and centrifugal force. Each of these forces narrows the stability region in comparison to the case $B = W = 0$. The stability region for their combined action belongs to the intersection of the stability regions for their independent action.

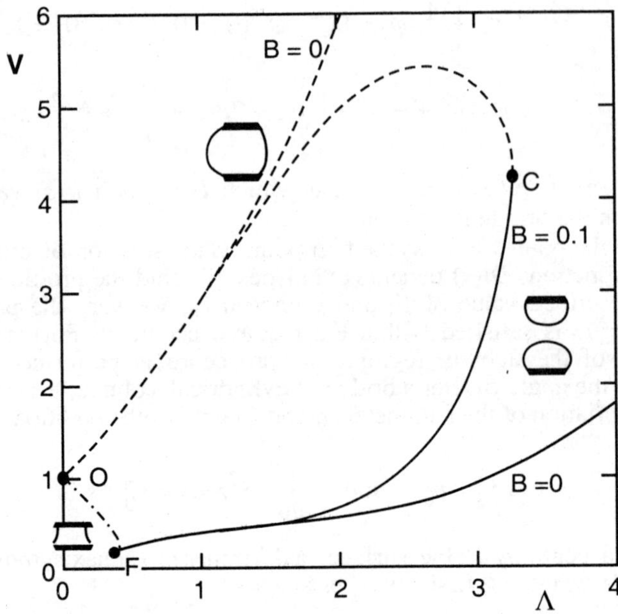

Fig. 1.2 Stability boundary for $B = 0$ and $B = 0.1$ bridges between equal radii coaxial disks. (After [1.5])

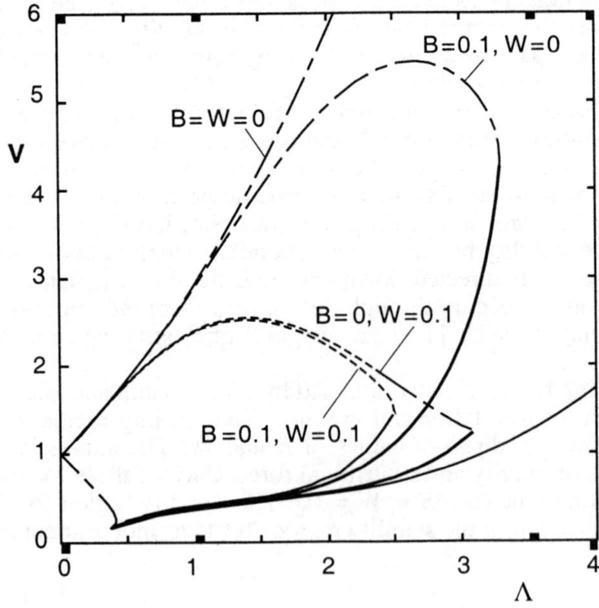

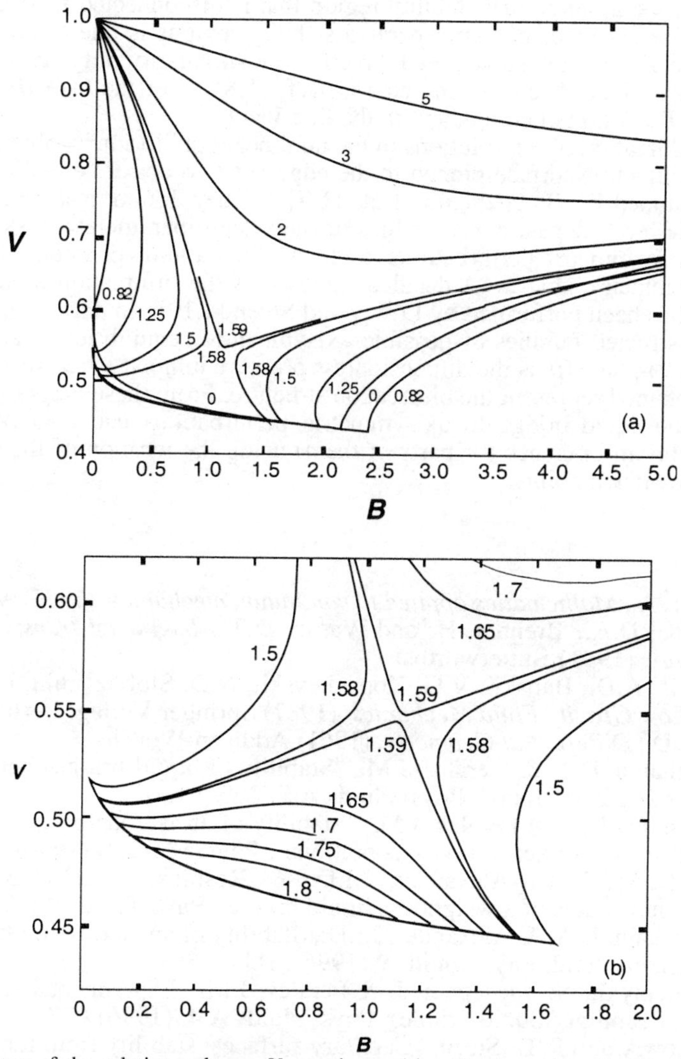

Fig. 1.3 Separate effects of isorotation and gravity on the stability boundary.

Fig. 1.4.   Dependence of the relative volume, $V$, on the positive Bond number, $B$, for critical bridges with $\beta_l = 90°$: (a) general diagram; (b) detail showing the existence of disconnected stability regions. Numbers on the curves denote values of the Weber number, $W$. (After [1.8])

For one of the isorotation cases examined, the slope at the lower disk, $\beta_l$ is fixed at $\beta_l = 90°$. (This is a common growth angle value for materials used in floating zone crystal growth.) Both positive and negative values of $B$ were considered. Examples show that, for $B > 0$, the stability of a bridge cannot be determined generally by the value of $\Lambda$ for critical states [1.1]. For $\beta_l = 90°$, selected results are shown in Fig. 4. For most values of $B$ and $W$, the stability region is connected and is enclosed between the stability boundary and the line $V = 1$. However, for $B > 0$ and the Weber number between $W_0$ $(1.58 < W_0 < 1.59)$ and 1.917, there exists an additional stability region that is disconnected from the previous one (see Fig. 3b). The existence of two separated stability regions is due to the centrifugal force [1.4]. The distinctive features associated directly with singularity of the case $\beta_1 = 90°$ for $B = 0$ and $0 \leq W \leq 1.25$ are discussed in detail in Ref. [1.8]. For the case $\beta_1 < 90°$, the upper boundaries for $B > 0$ do not coincide with the line $V = 1$.

The bifurcation of the solutions of the nonlinear equilibrium problem of a weightless liquid bridge with a free surface pinned to the edges of two coaxial equidimensional circular disks was examined by Slobozhanin et al. [1.7]. They found that along the maximum volume stability limit, depending on values of the system parameters, loss of stability with respect to nonaxisymmetric perturbations results in either a jump or a continuous transition to stable nonaxisymmetric shapes. A detailed analysis of the bifurcation among axisymmetric bridge shapes has been performed by Lowry and Steen [1.10]. In particular, for fixed values of $\Lambda$, they constructed families of possible axisymmetric equilibrium states and presented them in the $(p, V)$-plane ($p$ is the dimensionless pressure jump at a free surface). The $(p, V)$-diagrams so obtained represent the bifurcation structure. From these diagrams, the stability or instability of the liquid bridge to axisymmetric perturbations can also be determined for different equilibrium branches (or parts of them) using the location of the turning points in volume and the branch points.

## References

[1.1]   Segel, L.A., *Mathematics applied to continuum mechanics*, (1977) Macmillan.

[1.2]   Edwards, D.A., Brenner, H. and Wasan, D.T., *Interfacial transport processes and Rheology*, (1991) Butterworths.

[1.3]   Myshkis, A.D., Babskii, V.G, Kopachevskii, N.D. Slobozhanin, L.A. & Tyuptsov, A.D., *Low Gravity Fluid Mechanics* (1987) Springer Verlag, Berlin.

[1.4]   Struik, D., *Differential Geometry*, (1961) Addison-Wellsley.

[1.5]   Slobozhanin, L.A. & Perales, J.M., "Stability of liquid bridges between equal disks in an axial gravity field", Phys. Fluids A **5** (1993) 1305.

[1.6]   Slobozhanin,L.A. & Perales, J.M., "Stability of an isorotating liquid bridge between equal disks under zero-gravity conditions", Phys. Fluids **8** (1996) , 2307.

[1.7]   Slobozhanin, L.A., Alexander, J.I.D. & Resnick, A. H., Bifurcation of the equilibrium states of a weightless liquid bridge, Phys. Fluids **9** (1997) 1893-1905.

[1.8]   Slobozhanin, L.A. & Alexander, J.I.D., Stability of an isorotating liquid bridge in an axial gravity field, Phys. Fluids **9** (1996), 1880-1892.

[1.9]   Bezdenejnykh, N., Meseguer, J. & Perales, J.M., Experimental analysis of stability limits of capillary liquid bridges, Phys. Fluids A **4**, (1996) 677.

[1.10]  B. J. Lowry and P. H. Steen, "Capillary surfaces: stability from families of equilibria with application to the liquid bridge", Proc. R. Soc. London A **449** (1995) 411-439.

# 2. LIQUID BRIDGE DYNAMICS: 1D MODELS

## 2.1.  Introduction and problem definition

Liquid bridges occur in a variety of physical and technological situations and a great deal of theoretical and experimental work has been done to determine axisymmetric equilibria and stability for various disk configurations, aspect ratios, gravity levels and rotation rates. There have also been numerous investigations of liquid bridge dynamics.  Such investigations have been motivated both by practical considerations and basic scientific interest.  The behavior of liquid bridges and drops are important considerations in propellant management problems in liquid fuel chambers and related problems with fluids management in space. Pendular liquid bridges occur widely in the powder technology industry and are a major influence on powder flow processes and mechanical properties [2.9, 2.10]. In porous media flow, liquid-liquid displacement can lead to evolution of pendant and sessile lobes or lenticular bridges [2.11]. The formation of liquid bridges from the gel that coats lung micro-airways is a precursor in lung collapse [2.12].

In this lecture we will examine 1D models of the response of a slender liquid bridge to axial forcing resulting from vibration. Viscous and inviscid cases will be considered. Numerical results obtained from the inviscid model will be compared to analytical results.  A slender liquid bridge is held between two equidimensional rigid circular disks (radius $R_0$) separated by a distance $L$. The disks are aligned coaxially and are subject to an axial sinusoidal vibration such that both disks move in phase.  For a frame of reference attached to the moving disks (see Fig. 2.1), the vibration appears as an axial body force term, $g(t)$, in the equations of motion.  The liquid is assumed to be an isothermal incompressible Newtonian fluid with constant physical properties and is held between the disks by surface tension. We also make the following additional assumptions:

(i) Internal fluid motion in the liquid bridge is caused *only* by capillary pressure gradients due to interface deformation in response to $g(t)$.

(ii) The effect of the atmosphere around the bridge is negligible.

(iii) Only the axisymmetric response of the bridge is considered.

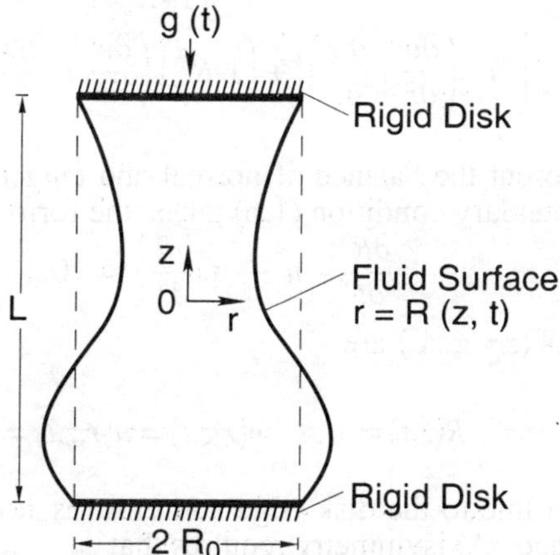


Fig. 2.1 The axisymmetric liquid bridge subject to an axial vibration

### 2.2.1 Governing equations

The dimensionless equations for mass and momentum transfer)
in the liquid $(0 < r < R(z,t), -\Lambda < z < \Lambda)$ are now

$$\frac{1}{r}\frac{\partial(ru)}{\partial r} + \frac{\partial w}{\partial z} = 0 , \tag{2.1}$$

$$\frac{\partial u}{\partial t} + u\frac{\partial u}{\partial r} + w\frac{\partial u}{\partial z} = -\frac{\partial p^*}{\partial r} + Oh\left(\frac{\partial^2 u}{\partial r^2} + \frac{1}{r}\frac{\partial u}{\partial r} + \frac{\partial^2 u}{\partial z^2} - \frac{u}{r^2}\right), \tag{2.2}$$

$$\frac{\partial w}{\partial t} + u\frac{\partial w}{\partial r} + w\frac{\partial w}{\partial z} = -\frac{\partial p^*}{\partial z} + Oh\left(\frac{\partial^2 w}{\partial r^2} + \frac{1}{r}\frac{\partial w}{\partial r} + \frac{\partial^2 w}{\partial z^2}\right) \tag{2.3}$$

where $p^*$ is a reduced pressure given by $p^* = p + B_o g(t)z$, and $\nu$ is the kinematic viscosity. The function $g(t)$ is taken to be $g(t) = 1 + (B_1/B_0) \sin(\omega t)$, where $\omega$ is the dimensionless angular frequency, $Bo = g_0 \rho R_0^2/\gamma$ is the usual static Bond number, and $B_\omega = g_\omega \rho R_0^2/\gamma$ is a dynamic Bond number associated with the time-dependent component arising from the vibration. The steady and time-dependent acceleration magnitudes are $g_0$ and $g_\omega$, respectively and the linear frequency of the disturbance is $f = \omega (2\pi\rho R_o^3 /\gamma)^{1/2}$. In equation (2.3), $Oh = \nu (\rho/\gamma R)^{1/2}$ is the Ohnesorge number and is a measure of the relative strength of viscous and capillary forces.

### 2.1.2 Boundary conditions

The surface is defined by $F(r,z,t) = r - R(z,t) = 0$. The boundary conditions at the surface $r = R(z,t)$ are

$$p^* - \frac{2Oh}{\left(1+R_z^2\right)}[\frac{\partial u}{\partial r} + R_z^2\frac{\partial w}{\partial z} - R_z\left(\frac{\partial w}{\partial r} + \frac{\partial u}{\partial z}\right)] = \frac{1}{\left(1+R_z^2\right)^{3/2}}[\frac{1+R_z^2}{R} - R_{zz}] + B_0\, g(t)z, \tag{2.4}$$

$$\left[2R_z\left(\frac{\partial u}{\partial r} - \frac{\partial w}{\partial z}\right)\right] + \left(1-R_z^2\right)\left(\frac{\partial w}{\partial r} + \frac{\partial w}{\partial z}\right) = 0, \tag{2.5}$$

which, respectively, represent the balance of normal and tangential force components at the surface. The kinematic boundary condition (1.6) takes the form

$$\frac{\partial R}{\partial t} - u + w\frac{\partial R}{\partial z} = 0. \tag{2.6}$$

The conditions at the disks $(z = \pm\Lambda)$ are

$$R(z,t) = 1 , \quad w(r,z,t) = u(r,z,t) = 0. \tag{2.7}$$

which anchors the contact line to the disk edge and ensures no slip of the interior fluid in contact with the disk surface. Axisymmetry requires that

$$u(0,z,t) = 0 , \frac{\partial w}{\partial r}(0,z,t) = 0. \tag{2.8}$$

The initial conditions are

$$R(z,0) \;\; = R_0(z), \;\; w(r,z,0) = u(r,z,0) = 0, \tag{2.9}$$

where $R_0(z)$ represents the equilibrium shape corresponding to the static Bond number $Bo$. The formulation is finally complete upon specifying the following volume constraint

$$\int_{-\Lambda}^{+\Lambda} R_0^2 \, (z) \, dz = \int_{-\Lambda}^{+\Lambda} R^2 \, (z,t) \, dz = 2\Lambda. \tag{2.10}$$

Equations (2.4)-(2.10) define an unsteady free boundary problem since the location (and hence the shape) of the oscillating interface is *a priori* unknown and must be determined along with the velocities and pressure as part of the solution.

## 2.3    The 1D "slice model"

The equations and boundary conditions (2.4-2.10) are approximated by a 1D model due to Meseguer [2.1] that considerably reduces the complexity of the problem and allows an examination of a relatively wide range of parameter without excessive computation times. It has been used in several previous studies of liquid bridge dynamics [2.2,2.3] and holds provided the slenderness $\Lambda$ is large. The model is based on a 1D model for fluid jets [2.4,2.5]. The axial velocity component, $w$, is assumed to depend only on the $z$ coordinate and time. In this case eqs. (2.2) and (2.3) become decoupled and the following system of equations and boundary conditions results:

$$\frac{1}{r}\frac{\partial(ru)}{\partial r} + \frac{\partial w}{\partial z} = 0 \, , \tag{2.11}$$

$$\frac{\partial w}{\partial t} + w\frac{\partial w}{\partial z} = -\frac{\partial p^*}{\partial z} + Oh\frac{\partial^2 w}{\partial z^2}, \tag{2.12}$$

Since $w$ is assumed to depend only on $z$, equation (2.11) is easily integrated and yields

$$u(r,z,t) = -\frac{r}{2}\frac{\partial w}{\partial z}. \tag{2.13}$$

We then take $S=R^2(z,t)$ (which represents a dimensionless measure of surface area), and $Q = S(z,t)w(z,t)$ and substitute (2.13) into the kinematic boundary condition (2.6). This yields

$$\frac{\partial S}{\partial t} + \frac{\partial Q}{\partial z} = 0. \tag{2.14}$$

Similarly, equation (2.12) can be recast in the form

$$\frac{\partial Q}{\partial t} + \frac{\partial}{\partial z}\left(\frac{Q^2}{S}\right) = -S\frac{\partial p^*}{\partial z} + OhS\frac{\partial^2}{\partial z^2}\left(\frac{Q}{S}\right), \tag{2.15}$$

which expresses the conservation of momentum in the axial direction for each cross section of the bridge. The component of force balance normal to the interface, equation (2.4), now reduces to

$$p^* = \frac{Oh}{4S + \left(\frac{\partial S}{\partial z}\right)^2}\left[\left(2\left(\frac{\partial S}{\partial z}\right)^2 - 4S\right)\frac{\partial}{\partial z}\left(\frac{Q}{S}\right) + 2S\frac{\partial S}{\partial z}\frac{\partial^2}{\partial z^2}\left(\frac{Q}{S}\right)\right] +$$

$$4\left[4S + \left(\frac{\partial S}{\partial z}\right)^2\right]^{-\frac{3}{2}}\left[2S + \left(\frac{\partial S}{\partial z}\right)^2 - S\frac{\partial^2 S}{\partial z^2}\right] + Bo\,g(t)z, \tag{2.16}$$

where equation (2.13) and the expressions for $S$ and $Q$ replace the dependent velocity variables $u$ and $w$. Together with the following boundary and initial conditions, equations (2.14) and (2.15) complete our approximate 1D description of the physical system

$$S(\pm\Lambda,t) = 1, \quad Q(\pm\Lambda,t) = 0, \quad S(z,0) = S_0(z), \quad Q(z,0) = 0. \tag{2.17}$$

### 2.3   Linearized model

Meseguer [2.2] employed a linearized version of the above model to study the oscillations of inviscid bridges under microgravity conditions. Equations (2.14), (2.15) and (2.16) are used with $Oh = 0$. For $g(t) = \varepsilon\, b(t)$, linearized solutions are sought of the form

$$\begin{bmatrix} S \\ Q \\ p \end{bmatrix}(z,t) = \begin{bmatrix} S_0(z) \\ 0 \\ p_0 \end{bmatrix} + \varepsilon\begin{bmatrix} s(z,t) \\ q(z,t) \\ p(z,t) \end{bmatrix}. \tag{2.18}$$

Substitution of (2.18) into (2.14)-(2.17) and boundary conditions yields the following system of $O(\varepsilon)$ equations

$$\frac{\partial s}{\partial t} + \frac{\partial q}{\partial z} = 0, \quad \frac{\partial q}{\partial t} + S_0\frac{\partial p}{\partial z} = 0,$$

$$p = 4A^{-3/2}[6\frac{D}{A} - 4 - \frac{d^2 S_0}{dz^2})s + (3\frac{D}{A} - 1)\frac{dS_0}{dz}\frac{\partial s}{\partial z} - S_0\frac{\partial^2 s}{\partial z^2}] + b(t)z, \tag{2.19}$$

$$s(\pm\Lambda,t) = q(\pm\Lambda,t) = 0, \, s(z,0) = q(z,0) = 0.$$

where

$$A = 4S_0 + \left(\frac{dS_0}{dz}\right)^2, \quad D = S_0\left(2 + \frac{d^2 S_0}{dz^2}\right). \tag{2.20}$$

The second equation in (2.19) is differentiated with respect to time and the perturbation variable, $s$, is eliminated from the $\partial^2 p/\partial z\partial t$ term by differentiating (2.14) (four times) with respect to $z$ to obtain

$$\frac{\partial^2 s}{\partial z \partial t} = -\frac{\partial^2 q}{\partial z^2}, \quad \frac{\partial^3 s}{\partial z^2 \partial t} = -\frac{\partial^2 q}{\partial z^3}, \ ..., \ etc., \tag{2.21}$$

and substituting for $s$ and its derivatives. This yields

$$\frac{\partial^2 q}{\partial t^2} = -S_0 \frac{\partial p}{\partial z \partial t} = S\frac{\partial}{\partial z}\left\{ 4A^{-3/2}(6\frac{D}{A} - 4 - \frac{d^2 S_0}{dz^2})\frac{\partial q}{\partial z} + (3\frac{D}{A}-1)\frac{dS_0}{dz}\frac{\partial^2 q}{\partial z^2} - S_0\frac{\partial^3 q}{\partial z^3}\right\} - S_0\frac{db}{dt} \tag{2.22}$$

or, for $b(t) = \beta \sin \omega t$ we look for solutions of the form $q(z,t) = u(z)\sin\omega t$, and, upon expanding derivative on the right hand side, obtain the following fourth-order ordinary differential equation for $u$

$$S_0\frac{d^4 u}{dz^4} + k_3\frac{d^3 u}{dz^3} + k_2\frac{d^2 u}{dz^2} + k_1\frac{du}{dz} - \frac{\beta\omega A^{-3/2}u}{4S_0} = -\frac{\omega^2 A^{-3/2}}{4},$$

$$u(\pm\Lambda) = \frac{du}{dz}(\pm\Lambda) = 0 \tag{2.23}$$

where

$$k_1 = \frac{3}{2A}(\frac{6D}{A} - \frac{d^2 S_0}{dz^2} - 4)\frac{dA}{dz} - \frac{1}{A^2}6(A\frac{dD}{\partial z} - D\frac{\partial A}{dz}) + \frac{d^2 S_0}{dz^2},$$

$$k_2 = \left(\frac{3D}{A}-1\right)\left[\frac{3}{2A}\frac{dS_0}{dz}\frac{dA}{dz} - \frac{d^2 S_0}{dz^2}\right] - \frac{6D}{A} + \frac{d^2 S_0}{dz^2} + 4 + \frac{3}{A^2}(A\frac{dD}{dz} - D\frac{dA}{dz})\frac{dS_0}{dz} \tag{2.24}$$

$$k_3 = (2 - 3\frac{D}{A})\frac{dS_0}{dz} - 3\frac{S_0}{2A}\frac{dA}{dz}.$$

Equation (2.23) is discretized using a fourth-order accurate five-point centered difference scheme [2.2]. This gives rise to an algebraic system of the form

$$\mathbf{M}U(z) = \mathbf{T} = -\frac{\omega^2 A_j^{3/2}}{4}, \quad j = 1,2,3...,m \tag{2.25}$$

where $m$ denotes the number of points used in the discretization, $\mathbf{M}$ is a pentadiagonal matrix and $U(z)$ is the vector representing the unknowns $u(z_i)$, $i=1, ..m$. The problem can now be solved for a given initial shape $S_0(z)$ by inverting $\mathbf{M}$.

## 2.4    Forced oscillation of a liquid bridge

In this section we describe the response of a liquid bridge ($V = 2\pi\Lambda$) to sinusoidal vibration of the supporting disks (g-jitter) using the non-linear 1D "slice" model equations. Single frequency vibration is considered. The sensitivity criterion chosen to characterize the response of the bridge represent two extremes. The first is the deviation of the bridge radius by more than 10% from its equilibrium radius, the second is breakage of the bridge this is taken to be the point at which the bridge radius, $R(z)$, is $O(\Delta z)$, i.e., $R(z) \rightarrow 0$. Note that, since the disks are constrained to vibrate in phase, the eigenfrequencies exhibiting the most sensitive response are those associated with eigenmodes having an odd number of nodes, (i.e., an even number of half waves).

All calculations presented in this section were made for a disk radius of 1.75 cm, and liquids with the following properties, $\rho = 920$ kg m$^{-3}$, $\gamma = 0.02$ N m$^{-1}$, and kinematic viscosities in the range $6.17 \times 10^{-8}$ to $6.17 \times 10^{-5}$ m$^2$ s$^{-1}$ (i.e., $10^{-4} < Oh < 10^{-1}$). These viscosities cover  arrange of typical experimental materials (water, silicone oils, etc.) The effects of viscosity ($Oh$), background acceleration $Bo$, and slenderness, $\Lambda$, on tolerable acceleration are discussed below.  The tolerable acceleration is expressed in units of g=9.8 ms$^{-2}$.

In order to define the sensitivity limits for the liquid zone shape an optimal searching scheme [2.6] was used to delineate the boundary between the regions in parameter space for which solutions either do or do not satisfy our sensitivity criterion.  Six to seven calculations were typically needed to obtain each point on the sensitivity curves.

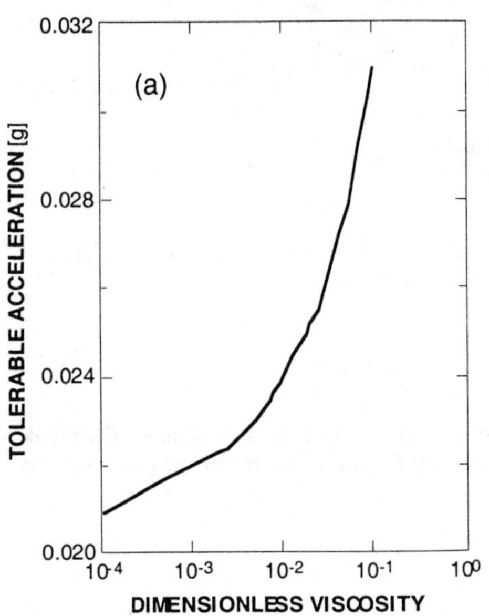

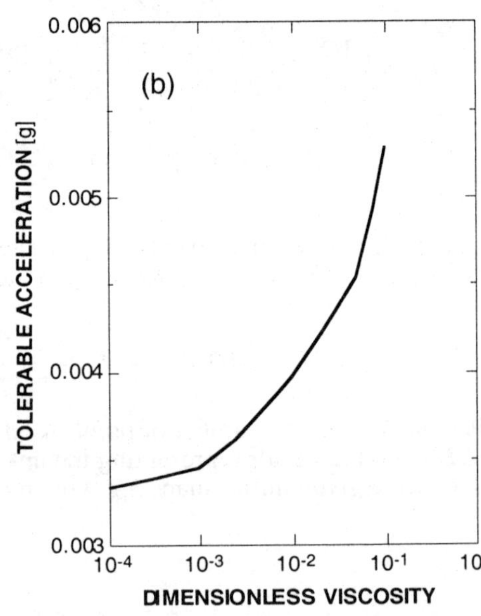


Fig. 2.2  (a) Effect of viscosity ($Oh$) on tolerable acceleration for the breakage criterion at $f$ = 5Hz, $\Lambda$ = 2.6, $Bo$ = 0.002 and $g_0$ = 1.42×10$^{-5}$g. (b) Effect of viscosity on tolerable acceleration for the shape change criterion $f$ = 5Hz, $\Lambda$ = 2.6, $Bo$ = 0.002 and $g_0$ = 1.42×10$^{-5}$g. From [2.6]

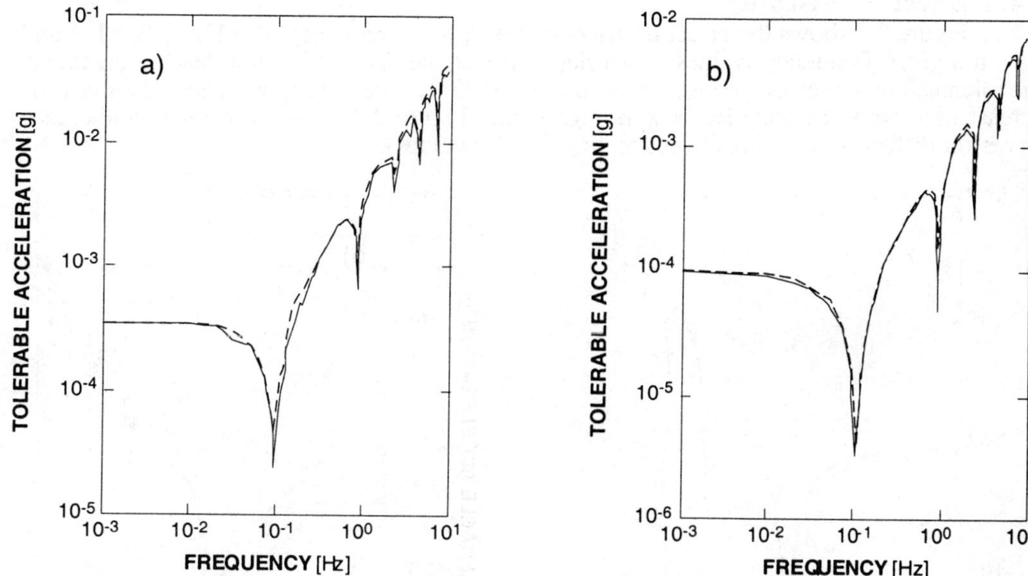



Fig. 2.3 Curves of tolerable acceleration vs. frequency for the breakage criterion (a) and shape change criterion (b) at $\Lambda = 2.6$, $Oh = 0.001$. The solid and the dashed curves are for $Oh = 0.001$ and 0.02 respectively. (From [2.6])

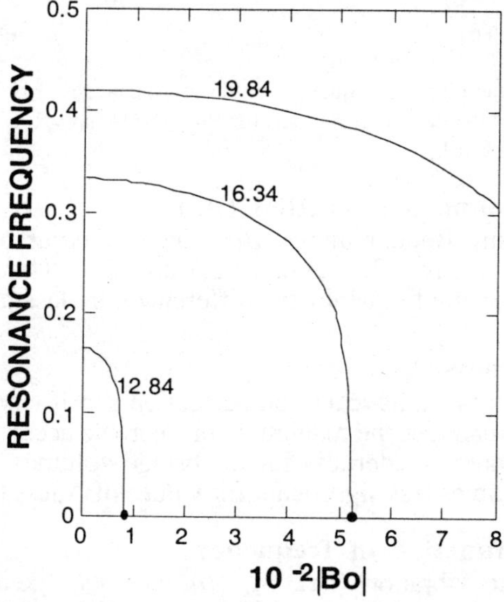



Fig. 2.4 Influence of the steady Bond number on the resonant frequency, $\omega_r$ for $\Lambda = 2.6$. Numbers on the curves denote the bridge volume. (After [2.2])

### 2.4.1 Effect of viscosity

Figure 2.2 shows the effect of viscosity on $g_1$ for a frequency of 5 Hz at fixed $\Lambda$ and $Bo$. at a given frequency, values of acceleration that lie above the curve lead to breakage. The tolerance increases as the viscosity is increased. Only when $Oh$ approaches $10^{-1}$ does the increase in tolerable acceleration become significant. Figure 2.3 shows acceleration tolerance curves for different viscosities (i.e., for different $Oh$ values).

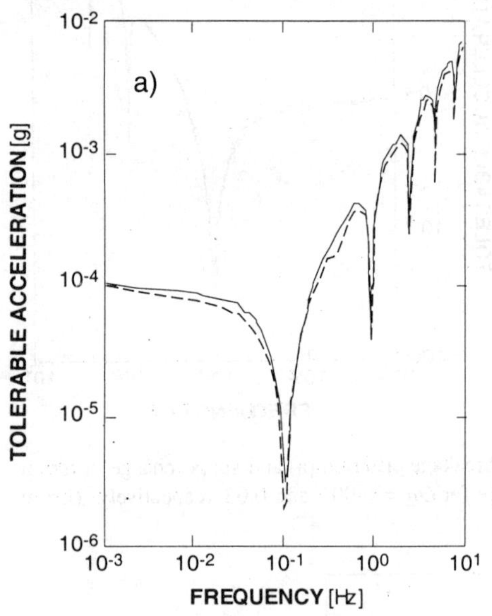
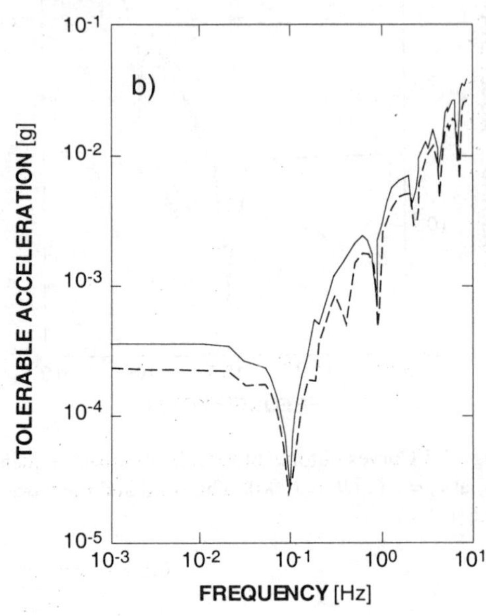

Fig.2.5  Curves of tolerable acceleration vs. frequency for the shape change (a) and breakage (b) criterion at $\Lambda$ = 2.6, $Oh$ = 0.001. The solid and the dashed curves are for $Bo = 0.002$ ($g_0 = 1.42{\times}10^{-5}g$) and $Bo = 0.02$ ($g_0 = 1.42{\times}10^{-4}g$), respectively. (From [2.6].)

### 2.4.2 Effect of background acceleration (Bo)

The value of the static Bond number, $Bo$, has a measurable effect on sensitivity, even at low values as seen in Figs. 2.4 (after [2.2]) and 2.5. This is limited to the lower frequency range. With increasing frequency the differences are less pronounced.

### 2.4.3 Effect of slenderness $\Lambda$

Changes in slenderness, $\Lambda$, have a strong effect on sensitivity  (see Fig. 2.6  (a) and (b)). A small decrease in  $\Lambda$ changes the magnitude of tolerable acceleration markedly when $\Lambda$ is close to $\Lambda_{max}$, the maximum slenderness for the bridge volume.  The change in tolerable acceleration magnitude becomes less significant for values of $\Lambda$ less than 80% of $\Lambda_{max}$.

### 2.4.4 Sensitivity as a function of frequency

The sensitivity of a zone to vibration (i.e., $\Lambda$, $Bo$, and $Oh$ fixed) as a function of the frequency of the disturbance is of most practical interest. This is shown in Figs. 2.3 -2.5. For both sensitivity criteria the general trend shows an increase in tolerable residual gravity with increasing frequency. For each case there are obvious deviations from this trend.  As the frequency of the forcing function approaches an eigenfrequency, the liquid bridge becomes more sensitive to the applied acceleration. Indeed, at the lowest eigenfrequency the deviation from the general trend can be two orders of magnitude lower. For the values of $\Lambda$ and $Bo$ considered the most sensitive frequencies are found in the neighborhood of $10^{-1}$ Hz. Associated maximum tolerable residual gravity levels as low as $10^{-5}$ and $10^{-3}$ g have been calculated. The lowest tolerable acceleration occurs in conjunction with the smallest eigenfrequency. Table 1 gives the maximum tolerable acceleration and the associated eigenfrequencies for the cases examined.

## 2.5 Results from the linearized 1D "slice" model

Figures 2.7 (a) and (b) show results computed from the linearized inviscid slice model described in section **2.3**. As expected, the linearized model shows that the bridge is mostly affected by lower frequencies and that at resonant frequencies (eigenfrequencies) the response is more pronounced than at neighboring frequencies. Figure 2.7 (a) illustrates this trend. Figure 2.7(b) shows the fundamental resonance frequencies, $\omega_r$, for $Bo = 0$ (i.e., no mean gravitational field). The dependence of $\omega_r$ on bridge volume is such that $\omega_r$ decreases as V approaches the minimum volume limit for static bridges with $Bo = 0$. For a fixed volume bridge, the influence of the background acceleration, $Bo$, on $\omega_r$ is depicted. As $Bo$ increases $\omega_r$ becomes smaller until, at the value of $Bo$ where the given value of V represents the minimum volume stability limit, it vanishes ($\omega_r = 0$).

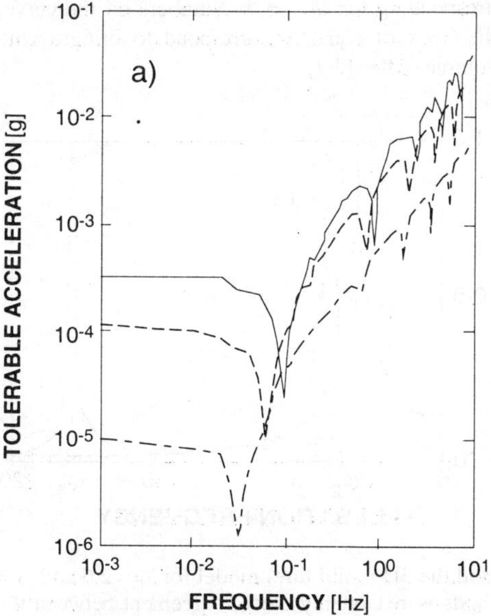
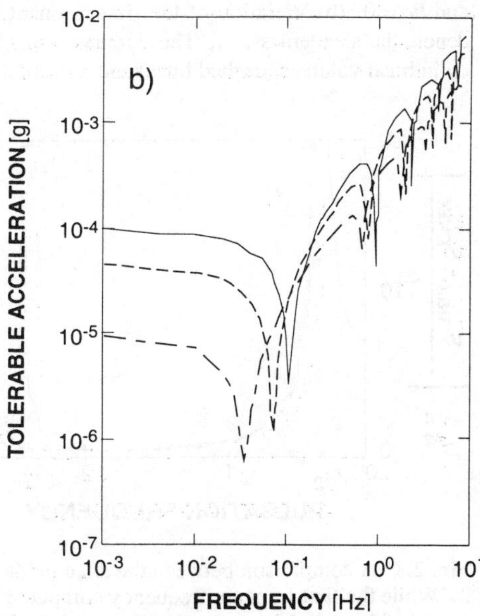

Fig. 2.6 Curves of tolerable acceleration vs. frequency for the breakage criterion (a) and shape change criterion (b) at $Bo = 0.002$ ($g_0 = 1.42 \times 10^{-5}$g) , $Oh = 0.001$. The solid, dotted and the dashed curves are the results for $\Lambda = 2.6$, 2.826 and 3.024, respectively. ( From [2.6].)

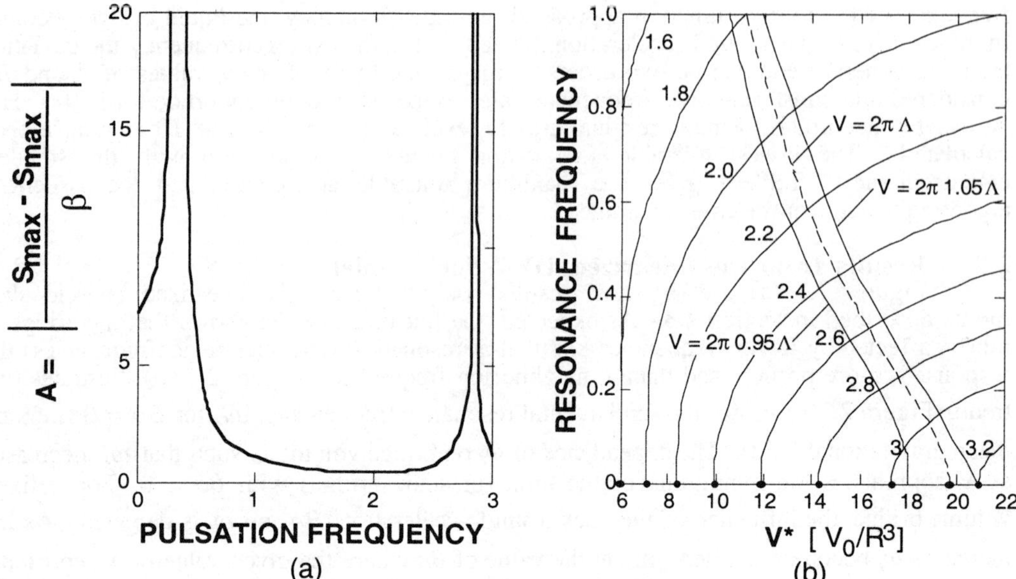



Fig. 2.7        (a) Variation of A = $|s_{max} - s_{min}|/\beta$, the ratio of maximum interface deformation to the perturbation amplitude $\beta$, with the dimensionless angular frequency of vibration, $\omega$, for $\Lambda = 2.6$, $V = 16.34$ and $Bo = 0$. (b) Variation of the first resonant mode $\omega_r$ corresponding for $Bo = 0$. Numbers on the curves denote the slenderness, $\Lambda$, The 3 curves which intersect the lines of equal $\Lambda$ correspond to bridges with cylindrical volumes (dashed lines) and ±5% of a cylindrical volume. After [2.2].

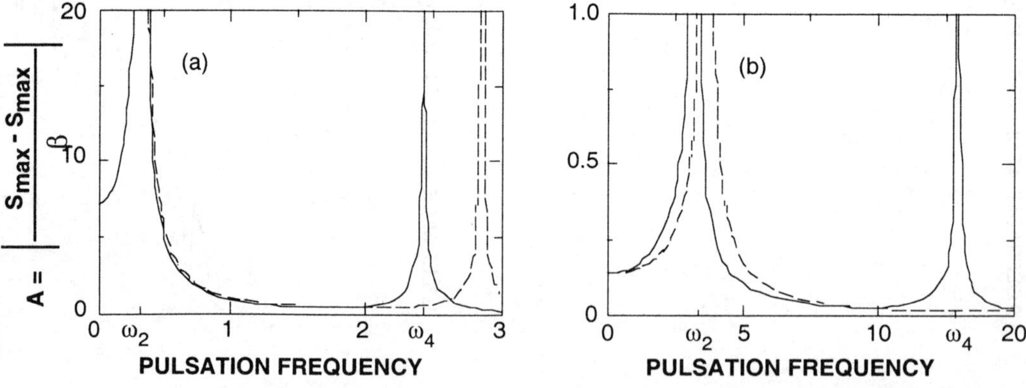



Fig. 2.8   A comparison between the slice model (dash line) and the 3D (solid line) model for $\Lambda = 2.6$ and $\Lambda = 2$. While the first resonant frequency computed by both methods is reasonably close, agreement between the two for the second frequency is poor. This is because at higher frequencies radial momentum effects become more important relative to axial momentum due to the increasing number of nodes associated with the spatial modes excited by the higher frequencies. As the slenderness decreases one also expects to observe increasing disagreement between the two models as radial momentum effects become more important. From [2.8]

## 2.6    Comparison of the linear model with a 3D axisymmetric model

Nicolas [2.8], used a linearized 3D axisymmetric model to approach the problem of response of a bridge to sinusoidal vibration of the supporting disks. A comparison of his results and those of Meseguer's [2.2] are made in Fig. 2.8.

## 2.7    Summary

Different examples of the use of 1D models for g-jitter sensitivity predictions have been examined. The results indicate that the liquid bridge is most sensitive to accelerations with frequencies close or equal to its lowest natural frequency. In terms of predicted and measured levels of residual accelerations experienced on spacecraft, the practical sensitivity range appears to be restricted to disturbances with frequencies in the range $10^{-2}$ - 10 Hz.

## References


[2.1]   Meseguer, J., The breaking of axisymmetric slender liquid bridges, J. Fluid Mech. 130 (1983) 123-151.

[2.2]   Meseguer, J., Axisymmetric long liquid bridges in a time-dependent microgravity field, Appl. Microgravity Tech. 1 (1988) 136-141.

[2.3]   Zhang, Y. & Alexander, J. I. D., Sensitivity of liquid bridges subject to axial residual acceleration, Phys. Fluids A 2 (1990) 1966-1974.

[2.4]   Lee, H.C., Drop formation in a liquid jet, IBM J. Res. Dev., (1974) 18 364-369.

[2.5]   Pimbley, W.T., Drop formation from a liquid jet: A linear one-dimensional analysis considered as a boundary value problem, IBM J. Res. Dev., (1976) 20 148-156.

[2.6]   Zhang, Y. & Alexander, J. I. D., Sensitivity of liquid bridges subject to axial residual acceleration, Phys. Fluids A 2 (1990) 1966-1974.

[2.7]   Meseguer, J., Axisymmetric long liquid bridges in a time-dependent microgravity field, Appl. Microgravity Tech. 1 (1988) 136-141.

[2.8]   Nicolás, J., Frequency response of axisymmetric liquid bridges to an oscillatory microgravity field, Microgr. Sci. Technol. 4 (1991) 188-190.

[2.9]   B. J. Ennis, J. Li, G. Tardos, and R. Pfeiffer, The influence of viscosity on the strength of an axially strained pendular bridge, Chem. Eng. Sci., 45, 3071-3088 (1990).

[2.10] V. P. Mehrota, and K.V.S. Sastry, Pendular bond strength between unequal-sized spherical particles, Powder tech., 25 203-214 (1980).

[2.11] D. N. Mazzone, G.I. Tardos, and R. Pfeffer, The behavior of bridges between two relatively moving particles, Powder Tech., 51 71-83 (1987).

[2.12] L. A. Newhouse, and C. Pozrikidis, The capillary instability of annular layers and thin liquid threads, J. Fluid Mech., 242, 193-209 (1992).

## 3. LINEAR AND NON-LINEAR BEHAVIOR

### 3.1. Linear oscillations
### 3.1.1 Calculation of eigenfrequencies and damping times.

Tsamopolous *et al.* [3.1] examined the linear oscillations of liquid bridges. They were motivated by the desire to evaluate the extent to which forced and free oscillations of liquid bridges can be used to measure the surface tension and viscosity of molten metal materials. They considered a nearly cylindrical bridge with relative volume $V = 1$, and the surrounding environment was not considered. The governing equations (see Ch. **1**) were linearized about a basic state $V_0$ (which corresponds to a static equilibrium shape determined by the aspect ratio and Bond number. These solutions take the form

$$\begin{bmatrix} u(r,\theta,z,t) \\ p(r,\theta,z,t) \\ R(\theta,z) \end{bmatrix} = V_0 + V_p = \begin{bmatrix} 0 \\ p_0(r,z) \\ R(z) \end{bmatrix} + \begin{bmatrix} \hat{u} \\ \hat{p} \\ \hat{f} \end{bmatrix}(r,\theta,z,t), \tag{3.1}$$

with

$$V_p = \begin{bmatrix} \hat{u} \\ \hat{p} \\ \hat{f} \end{bmatrix}(r,\theta,z,t) = \begin{bmatrix} \tilde{u}(r,z) \\ \tilde{p}(r,z) \\ \tilde{f}(r,z) \end{bmatrix} e^{ik\theta} e^{-\sigma t}, \tag{3.2}$$

where Re $\sigma > 0$ gives damped oscillations. The resulting perturbation equations are [3.1]

$$\nabla \cdot \hat{u} = 0 \tag{3.3}$$

$$\frac{\partial \hat{u}}{\partial t} = -\nabla p + Oh \, \nabla^2 \hat{u}, \tag{3.4}$$

$$\hat{u} = 0, \, at \, z = \pm \Lambda \tag{3.5}$$

$$\hat{u} = 0, \quad k = 0,2,3,4..., \quad \hat{u} + \hat{v} = \hat{w}, \, k = 1, \tag{3.6}$$

$$-\hat{p} + 2Oh \, n_0 \cdot \hat{D} n_0 = 2\kappa, \tag{3.7}$$

$$e_\theta \cdot \hat{D} n_0 = 0, \quad t_z \cdot \hat{D} n_0 = 0, \quad r = R(z), \tag{3.8}$$

$$n_0 \cdot e_r \frac{\partial \hat{f}}{\partial t} = n_0 \cdot \hat{u}, \, r = R(z), \tag{3.9}$$

$$\hat{f} = 0 \, at \, z = \pm \Lambda, \quad \int_{-\Lambda}^{+\Lambda} \hat{f} dz = 0, \, \hat{f}(\theta,z,t) = \hat{f}(\theta + 2\pi, z, t). \tag{3.10}$$

Tsamopolous *et al.* solved equations (3.3) - (3.10) using a finite element method. The perturbation quantities $V_p$ were represented in the form

$$\begin{bmatrix} \tilde{\boldsymbol{u}}(r,z) \\ \tilde{p}(r,z) \\ \tilde{f}(r,z) \end{bmatrix} = \begin{bmatrix} \sum\limits_{i=1}^{N} \boldsymbol{u}_i \Phi_i(r,z) \\ \sum\limits_{i=1}^{M} p_i \Xi_i(r,z) \\ \sum\limits_{i=1}^{L} f_i \Omega_i(r,z) \end{bmatrix}, \qquad (3.11)$$

where the $\Phi_i$ and $\Xi_i$, are biquadratic Lagrangian functions and $\Omega_i$ are linear Lagrangian functions. The number of coefficients in each expansion is given by $N$, $M$ and $L$. Following substitution of these approximations into the perturbation equations (3.2) - (3.10) the residual equations are obtained using Galerkin's method [3.2]. That is, each equation is multiplied by the trial functions $\Phi_i$, $\Xi_i$, and $\Omega_i$ and then integrated over the domain [3.0,R]×[3.-$\Lambda/2,\Lambda/2$]. Second derivatives in the integrand are converted to first derivatives by application of the divergence theorem or integration by parts. Boundary conditions are then applied directly onto the boundary integrals thus obtained. The weak form of the *axisymmetric* (i.e. k=0) perturbation equations for continuity, momentum and the kinematic boundary conditions are

$$R_i^C = \int \Xi_i \nabla \cdot \tilde{\boldsymbol{u}} \, rdrdz, \qquad (3.12)$$

$$R_i^M = \int -\left(\sigma\tilde{\boldsymbol{u}} + \frac{1}{r}(\tilde{p} - 2OhD_{\theta\theta})\boldsymbol{e}_r\right)\cdot\Phi_i rdrdz + Oh\int (-\tilde{p}\boldsymbol{I} + 2OhD))\cdot\nabla\Phi_i rdrdz$$

$$-2\int \Phi_i\cdot\kappa_b n_0\tilde{f}\left[1 + \left(\frac{\partial\tilde{f}}{\partial t}\right)^2\right]^{1/2} dz, \quad 0 \le r \le R(z), -\Lambda \le z \le \Lambda, \qquad (3.13)$$

$$R_i^K = \int\left(-\sigma\tilde{f} + \Lambda\frac{\partial\tilde{f}}{\partial t}\tilde{w} - \tilde{u}\right)\Omega_i dz, \qquad (3.14)$$

where

$$0 \le r \le R(z), -\Lambda \le z \le \Lambda.$$

Here $R_i^C$, $R_i^M$ and $R_i^K$ are, respectively, the residuals of the linearized equations for continuity and momentum and the kinematic boundary condition. These equations represented an eigenvalue problem of the form

$$\boldsymbol{Ax} = \sigma\boldsymbol{Bx}, \qquad (3.15)$$

where $\boldsymbol{x}$ is the eigenvector corresponding to the unknown functions $u_i$, $p_i$ and $f_i$. The Peigenvalue problem can be solved using standard routines such as the ISML routine "DGVLRG" ([3.3] and see Tsamopolous et al. [3.1]). They noted that the large memory and storage requirements associated with their approach require an efficient approach such as that developed by Thomas and Brown [3.4] be used to obtain the eigenvectors. This was realized by setting $\tilde{p}(0,\Lambda) = 1.0$ which allows a non-trivial solution to (3.15) to be found. Examples of the results obtained in [3.1] are discussed below.

### 3.1.2 Linear oscillations: results

The normal mode analysis described in the last section yields information concerning the stability of the motion of a given liquid bridge. For unstable motions, $Re\sigma > 0$, while for stable motions $Re\sigma < 0$. Stable motions correspond to either overdamped, $Im\ \sigma \neq 0$, or underdamped, $Im\ \sigma = 0$, oscillations.

Figure 3.1 shows shapes of the liquid bridge subject to an axisymmetric disturbance for B = 0 and B = 1 (a) first, (b) second, (c) third and (d) fourth modes, for $Oh = 0.1$ and $\Lambda = \pi$. In the absence of gravity a cylindrical liquid bridge with $V = 1$ has a circular cylindrical shape provided $\Lambda$ is less than $\pi$. The response of the circular cylinder to small axisymmetric disturbances depends on $Oh$ and $\Lambda$. The damping rates ($Re\sigma$) and frequencies ($Im\sigma$) of the first four modes as a function of the slenderness are shown in Fig. 3.2. For a given ($\Lambda$, $Oh$) both damping rate and frequency increase with increasing mode number, n. At low values of $Oh$ the numerical results agree well with the $\sigma_i$ found from the inviscid theory of Sanz et al. [3.6]. For a given mode, the damping rate, $\sigma_r$, increases as $\Lambda$ decreases. This arises because of the increase in viscous interaction with the solid disks results in a reduction in the intensity of the fluid motion. For the same reasons the eigenfrequencies also increase as $\Lambda$ decreases due to the increase in viscous friction caused by the increase in the ratio of the solid disk area to the liquid surface area for a given slenderness. Figure 3.3 shows the dependence of damping rate and frequency on the Bond number for $Oh = 0.1$ and $\Lambda = \pi/2$ and $\pi/4$. Note that both the damping rate and frequency of a given mode decrease as the stability limit (corresponding to the static stability limit) is approached. (Tsamopoulos et al. note that the accuracy of the results also degrades as the static stability limit is approached.) The decrease in damping and eigenfrequency can be explained by the fact that the shapes corresponding to a static gravity tend to be better described by a combination of the eigenmodes than by a cylindrical shape and are, thus, more readily excited due to the lower energy required to excite the mode (this translates to a lower frequency) and the less energy dissipated by the oscillation (therefore a lower damping rate).

Higueras et al. analyzed linear oscillations of liquid bridges at small values of the Ohnesorge number $Oh$. Two kinds of oscillating modes were identified, inviscid and viscous. Inviscid modes exhibited the following trends for the damping rate $\sigma_r$ and frequency $\sigma_i$

$$\sigma_r = Oh^{1/2}\sigma_{r1} + Oh\sigma_{r2} + O(Oh^{3/2})$$

$$\sigma_i = \sigma_{i0} + Oh^{1/2}\sigma_{i1} + O(Oh^{3/2}), \qquad (3.16)$$

$$\sigma_{i0} > 0,\ \sigma_{r1} = \sigma_{i1} < 0,\ \sigma_{r2} < 0,$$

and the coefficients $\sigma_{r1}, \sigma_{r2}, \sigma_{i0}, \sigma_{i1}$ are dependent on the slenderness, $\Lambda$ and Bond number, $Bo$. Figure 3.4 shows a comparison between the closed form solutions obtained by Higueras et al. [3.7] and the numerical work of Tsamopoulos et al.

The nature of azimuthal variations in the bridge shape are characterized by the wavenumber k. Figure 3.5 shows cross sections of non-axisymmetric modes in the $\theta = 0$, $\theta = \pi$ plane at zero Bond number for $k = 1$. The lowest nonaxisymmetric mode corresponds to $n = 0$ and has a plane of symmetry at $z = 0$. This mode is inadmissible for axisymmetric disturbances since, for a fixed $\Lambda$, volume is not conserved for an axisymmetric deformation with $n = 0$. (The n=0 mode is also known as the "C" mode.)   The effect of increasing $k$ on the damping rates and frequencies is similar to the effect of increasing $n$ (see Fig. 3.6).

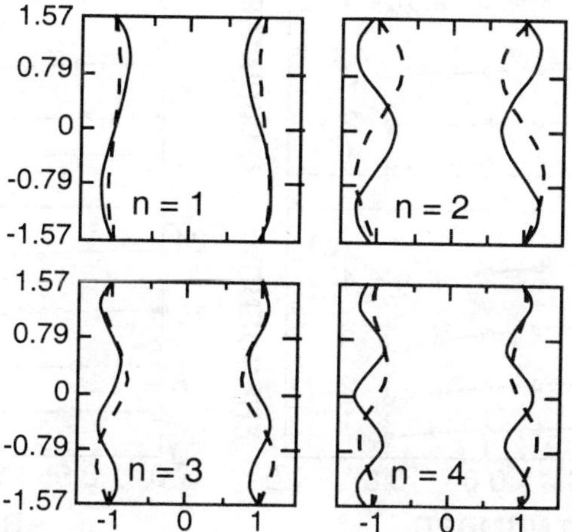


Fig. 3.1 Shapes of an axisymmetric liquid bridge for the first four axisymmetric modes with $B = 0$ (solid line) and $B = 1$ (dash line). From [3.1].

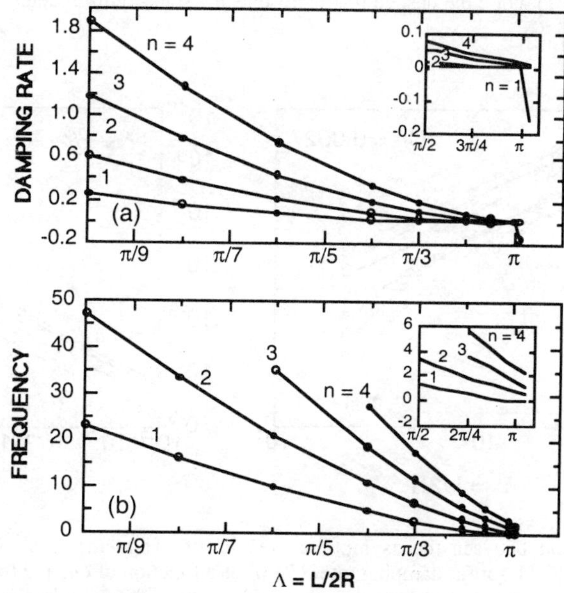


Figure 3.2 Damping rates ($Re\sigma$) and eigenfrequencies ($Im\sigma$) for the first four modes as a function of the slenderness, $\Lambda = L/2R$ for $Oh = 0.002$. The open circles represent the numerical results of Tsamopolous et al. [3.1] while the solid circles represent the results of Sanz and Lopez-Diez [3.6]. In the inset straight lines were used to connect points that were computed numerically. After Tsamopolous et al. [3.1].

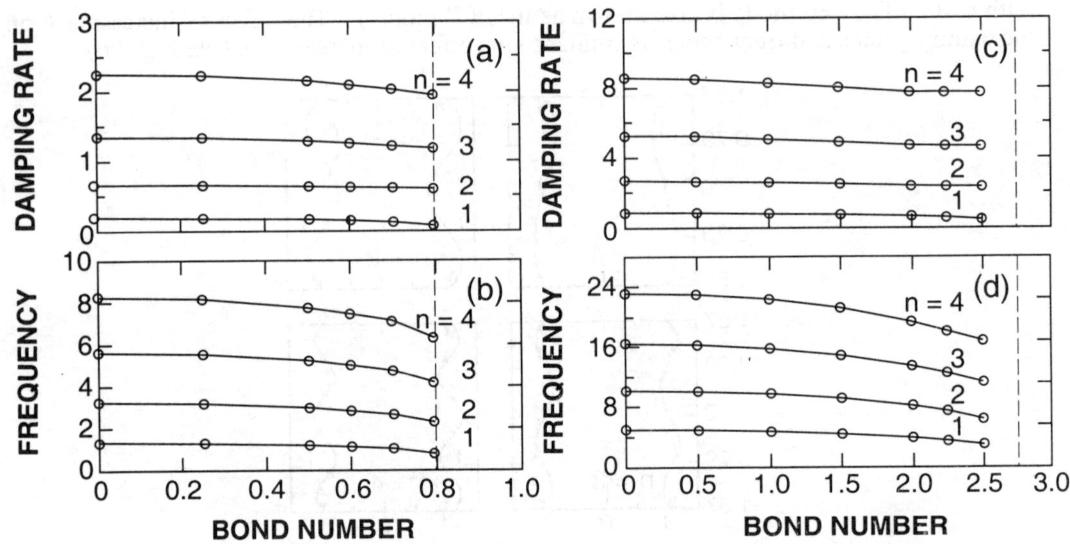

Figure 3.3 Damping rates and frequencies for the first four modes as a function of the Bond number for $Oh = 0.1$ for (a) and (b) $\Lambda = \pi/2$, and (b) and (c) $\Lambda = \pi/4$. The open circles represent the numerical results of Tsamopolous et al. [3.1] while the dashed line represents the stability limit determined from a static analysis. After [3.1].

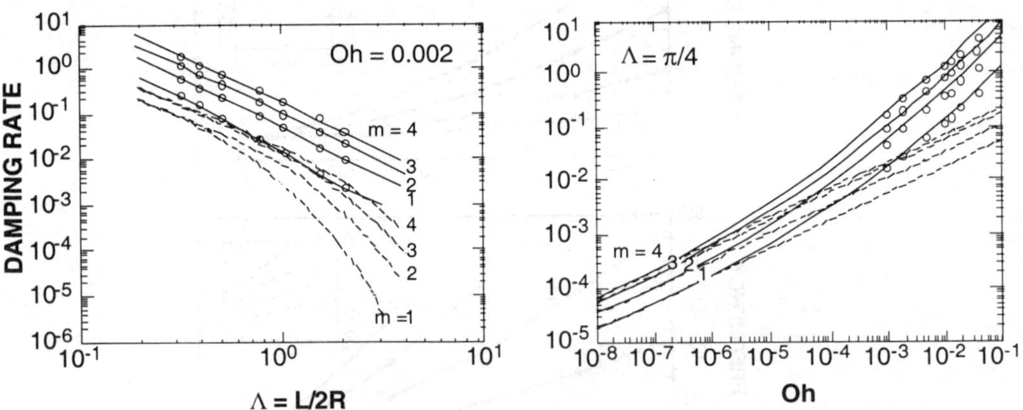

Fig. 3.4  Comparison between the asymptotic results of Higueras et al. [3.5] for small $Oh$, and the numerical results of [3.1] for (a) damping rates ($Re\ \sigma$) as a function of $Oh$ and (b) eigenfrequencies ($Im\ \sigma$) as a function of $\Lambda$.. The open circles denote the numerical results and the dashed curves represent the results for two and the solid line for three terms, of the expression $\sigma = \sigma_0 + Oh^{1/2}\sigma_1 + Oh\sigma_2 + \ldots$

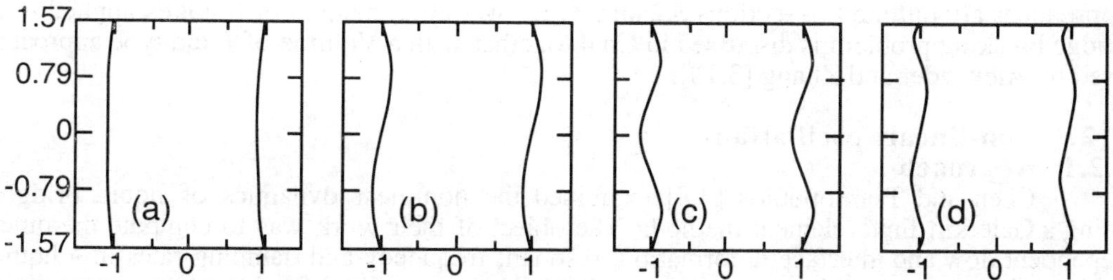

Fig. 3.5  Shapes of a nonaxisymmetric liquid bridge in the absence of gravity for the first second third and fourth modes under a nonaxisymmetric disturbance. From [3.1].

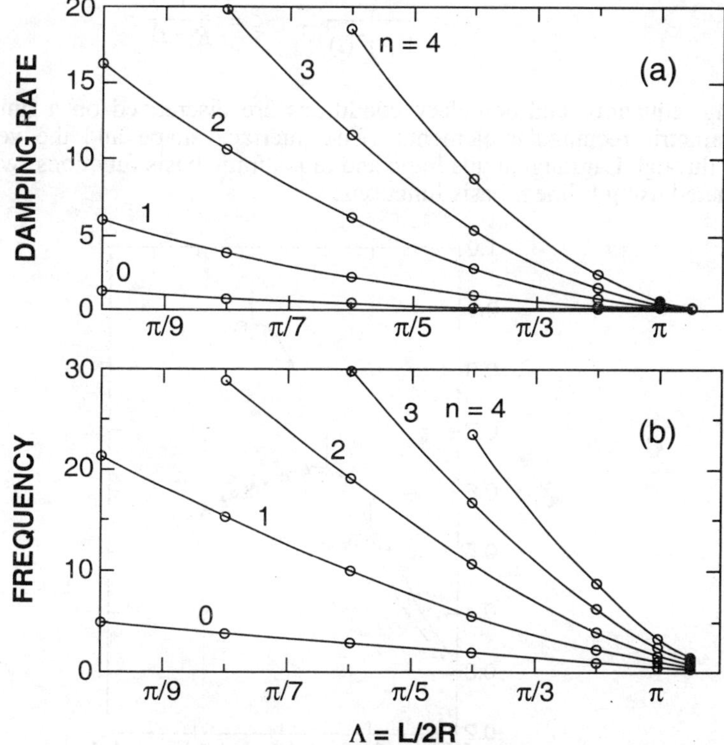

Fig. 3.6  Damping rates ($Re\sigma$) and eigenfrequencies ($Im\sigma$) for the first five nonaxisymmetric modes as a function of slenderness, $\Lambda$, for $Oh = 0.1$. The open circles represent the numerical results of Tsamopolous et al. [3.1].

Nonlinear oscillations of liquid bridges have been examined numerically by Zhang and Alexander [3.7] using finite differences, by Tsamopoulos et al. [3.8] using a finite element method and by Shulkes [3.9] using a boundary element method. The first two approaches are outlined in sections **3.2** and **3.3**,   while the method of Shulkes applied to a bridge breaking problem is discussed in Ch.**4** together with a Volume-of-Fluid type approach used by Alexander and Zhang [3.10].

## 3.2.  Non-linear  oscillations
### 3.2.1  Approach

Chen and Tsamopoulos [3.8] examined the nonlinear dynamics of liquid bridges using a Galerkin finite element method.  The object of their work was to compute the time-dependent flow and interface deformation, resonant frequency and damping rates in a liquid bridge subject contained between rigid coaxial disks to sinusoidal forcing of the upper disk. Similar to Zhang and Alexander they mapped the governing equations onto a fixed cylindrical domain using a non-orthogonal coordinate transformation.  In order to cope with the moving upper disk they used the  mapping

$$\eta = \frac{z}{1 + f^{+}(t)} \quad , \quad \xi = \frac{r}{R(z,t)} \, . \tag{3.17}$$

The resulting  equations and boundary conditions are discretized on a finite element mesh with isoparametric rectangular elements.  The interface shape and the velocity fields are represented through Lagrangian quadratic and biquadratic basis functions, while the pressure is approximated using bilinear basis functions.

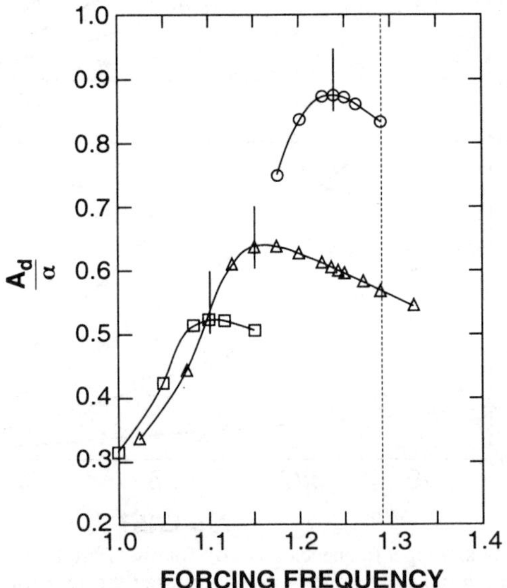


Figure 3.7.  Shift of resonant frequency due to increasing forcing amplitude: $\alpha = 0.1$ (**O**); $\alpha = 0.2$ ($\Delta$); $\alpha =$ 0.3 ($\square$), with $Re = 20$, $\Lambda = \pi/2$. Vertical solid lines indicate the resonant frequencies. The vertical dashed line indicates the linear eigenfrequency calculated by Tsamopolous (1992) et al. [3.2]. From [3.8].

### 3.2.2 Results: the liquid bridge as a soft spring

Selected results from Chen and Tsamopoulos's work are shown in Figs. 3.7-3.9. Only excitations of the first resonant mode were reported in this work. While, linear analyses show that eigenfrequencies of liquid bridges are independent of the disturbance amplitude, weakly nonlinear analyses (see next section) show that the forcing amplitude can modify the resonant frequency. Figure 3.7 through 3.9 show $A_d$, (the ratio of the maximum amplitude of the surface deformation during an oscillation period, to the magnitude, $\alpha$, of the sinusoidal forcing disturbance) as a function of frequency. Figure 3.7 shows the computed effect of different forcing amplitudes on the resonant frequency.

The linear eigenfrequency for this case is denoted by a vertical dashed line. As the amplitude of the response increases, the resonant frequency decreases. The resonant frequency also decreases with increasing slenderness for the same $A_d/\alpha$ as is shown in Fig. 3.8. The effect of viscosity as reflected through the Ohnesorge number, $Oh$, is shown in Fig. 3.9. In contrast to linear theory, which predicts a slight decrease in resonant frequency with increasing $Oh$, the results shown in Fig. 3.9 show that the effect of increasing $Oh$ results in an increase in the resonant frequency values. For small amplitude oscillations, the retarding effect of viscosity on the fluid motion will result in a decrease in the natural oscillation period as the viscosity (or $Oh$) is reduced. Thus, for small disturbances the resonant frequency will increase with increasing $Oh$. At large amplitude oscillations, the viscous dissipation of energy decreases the oscillation amplitude and tends to offset the decrease in resonant frequency arising from the increased inertia associated with the large amplitude response.

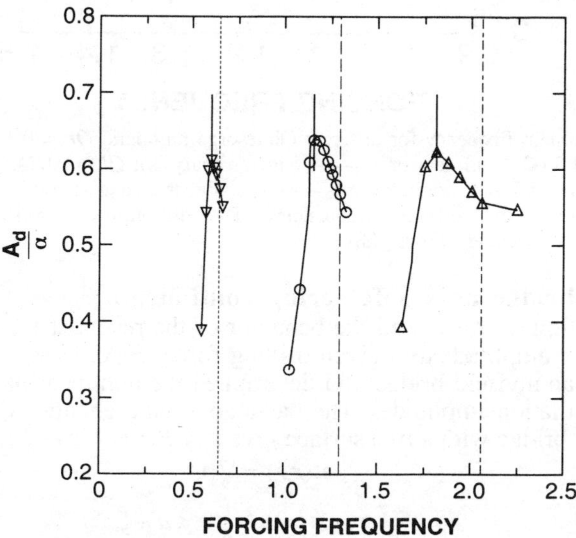

Figure 3.8.    Shift of resonant frequency for different aspect ratios: $\Lambda = 2\pi/3$ ($\nabla$ nonlinear and - - - linear analysis); $\Lambda = \pi/2$ (O, nonlinear and —  — —, linear analysis); and $\Lambda = 2\pi/5$ ($\Delta$, nonlinear and— - —, linear analysis), for $Re = 20$, $\alpha = 0.2$. Vertical solid lines indicate the resonant frequencies. The other vertical lines indicate the corresponding linear eigenfrequencies. From [3.8].

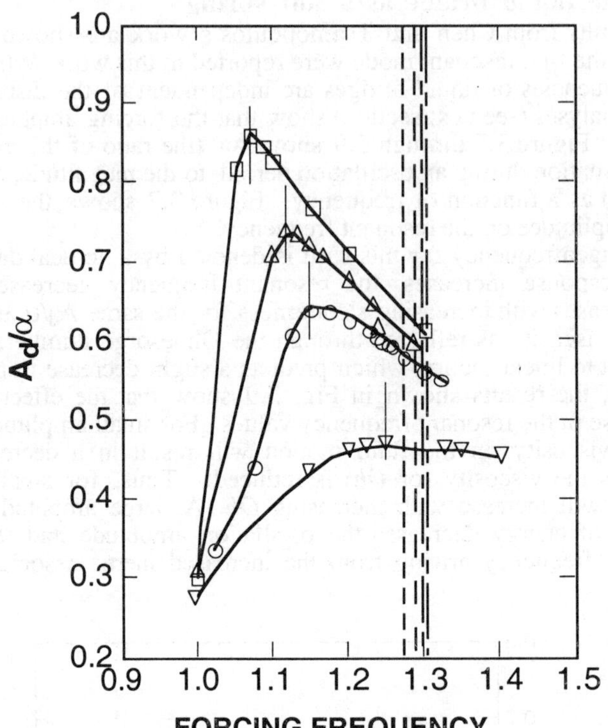



Figure 3.9. Shift of resonant frequency for different Ohnesorge numbers: $Oh = 0.1$ ($\nabla$, nonlinear or - - -, linear analysis); $Oh = 0.05$ (O, nonlinear or ———, linear analysis); $Oh = 0.034$ ($\Delta$, nonlinear or — - —, linear analysis); and $Oh = 0.02$ ($\square$, nonlinear or — - - —, linear analysis); with $\Lambda = \pi/2$ and $\alpha = 0.2$. Vertical solid lines indicate the resonant frequencies. The discontinuous vertical lines indicate the corresponding linear eigenfrequencies. From [3.8].

## 3.3.  The liquid bridge as a soft spring continued...

In the last section we discussed the behavior of the resonant frequency as a function of the surface response amplitude to a given driving force. Eidel [3.9], carried out a weakly nonlinear analysis of an inviscid bridge and determined the nonlinear natural frequencies for finite free surface oscillation amplitudes. The basic governing equations for an axisymmetric motion of an inviscid bridge with a free surface given by $R(z,t) = a + \zeta(z,t)$ are

$$\boldsymbol{u}(r,z) = \nabla \phi,$$

$$\nabla^2 \phi(r,z) = 0, \, 0 < r < a + \zeta,$$

$$\frac{\partial \zeta}{\partial t} = \frac{\partial \phi}{\partial r} - \frac{\partial \phi}{\partial z}\frac{\partial \zeta}{\partial z}, \quad r = a + \zeta, \tag{3.18}$$

$$\rho \frac{\partial \phi}{\partial t} + \frac{\rho}{2}\left[\left(\frac{\partial \phi}{\partial r}\right)^2 + \left(\frac{\partial \phi}{\partial z}\right)^2\right] - 2\gamma\kappa = \frac{\gamma}{a}, \quad r = a + \zeta,$$

$$\frac{\partial \phi}{\partial z} = 0 \, , z = 0, h; \quad \zeta = 0, \ z = 0, h; \quad \int_0^h (2a\zeta + \zeta^2)\, dz = 0.$$

These equations represent continuity $(3.18)_1$ and $(3.18)_2$, the kinematic boundary condition $(5.1)_3$, conservation of momentum and force balance at the interface $(3.18)4$. The boundary conditions at the disk edges are continuity $(3.18)_5$, $(3.18)_6$ represents anchoring the bridge to the disk edges and the constant volume condition is $(5.1)_7$. Eidel evaluates the boundary conditions at $r = a + \zeta$, $a > 0$, assumes that $|\zeta| << a$ and retains terms up to third order in $\zeta$. Then, Laplace's equation for $\phi$ is solved with these weakly nonlinear boundary conditions. The solution for $\phi$ has the form

$$\phi(r,z,t) = \sum_{n=-\infty}^{n=\infty} C_n(t)\, I_0(\frac{n\pi}{h}\, r) exp[\frac{in\pi z}{h}], \quad C_n(t) = C_{-n}(t), \qquad (3.19)$$

where $I_0$ is the zero order modified Bessel's function of the first kind.

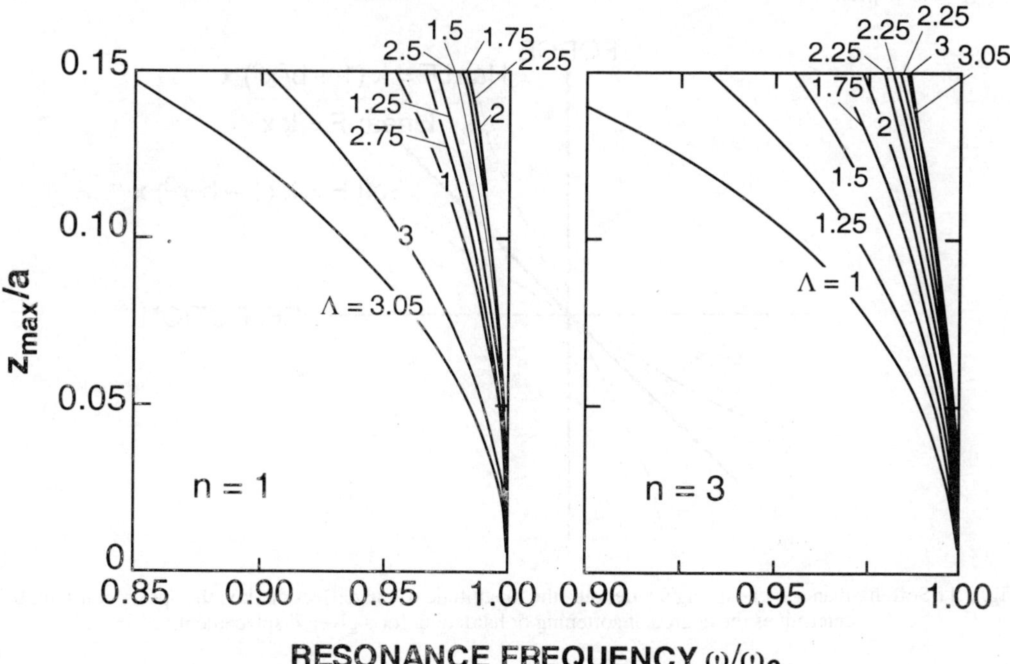


Fig. 3.10 Behavior of n = 1 and n=3 resonant frequencies for different slenderness values. After Eidel [3.15].

The free surface displacement is assumed to have the form

$$\zeta(z,t) = a_0 sin\left(\frac{z}{a}\right) + b_0 cos\left(\frac{z}{a}\right) + \sum_{n=-\infty}^{n=\infty} \overline{c}_n(t) exp\left(\frac{in\pi z}{h}\right), \quad \overline{c}_n(t) = \overline{c}_{-n}(t) \qquad (3.20)$$

Using the anchoring condition $(3.19)_7$ and the fact that

$$sin\left(\frac{z}{a}\right) = \sum_{n=-\infty}^{n=\infty} \alpha_n\, exp[\tfrac{in\pi z}{h}], \quad cos\left(\frac{z}{a}\right) = \sum_{n=-\infty}^{n=\infty} \beta_n\, exp[\tfrac{in\pi z}{h}],$$

$$\alpha_n = \frac{(-1)^n cos(h/a) - 1}{h/a[\gamma_n^2 - 1]}, \quad \beta_n = \frac{(-1)^{n+1} sin(h/a) - 1}{h/a[\gamma_n^2 - 1]}, \quad \gamma_n = n\pi a/h, \tag{3.21}$$

equation (3.20) can be rearranged to give

$$\zeta(z,t) = \sum_{n=-\infty}^{n=\infty} c_n(t)\, exp[\tfrac{in\pi z}{h}], \quad \overline{c}_n(t) = \alpha_n a_0(t) + \beta_n b_0(t) + \overline{c}_n(t). \tag{3.22}$$

Using (3.21) and (3.19) in the weakly nonlinear kinematic and force balance boundary conditions yields a system of coupled nonlinear ordinary differential equations for $C_n(t)$, and $c_n(t)$.

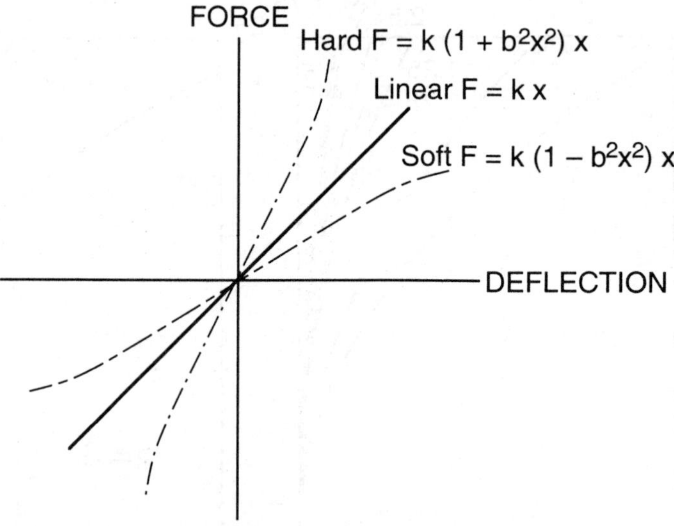


Fig. 3.11 Soft, hard and linear springs, x denotes the magnitude of the deflection, k is the spring constant, b determines the degree of softening or hardening for a given displacement.

Eidel sought coefficients $C_n(t)$, and $\overline{c}_n(t)$ of the form

$$C_n(t) = a\,\overline{X}_n\, sin\,(\omega t), \quad \overline{c}_n(t) = a\,\overline{Y}_n\, cos\,(\omega t),$$

$$a_0(t) = aA_0 cos(\omega t), \quad b_0(t) = aB_0 cos(\omega t). \tag{3.23}$$

and employed the Ritz averaging procedure to obtain a system of non-linear algebraic equations for $X_n$, $Y_n$, $A_0$ and $B_0$, for $n = 1,2,....$ These were solved numerically using a Newton-Raphson method. The non-linear frequencies calculated by Eidel are shown in Fig. 3.10. Consistent with the numerical results shown in the previous section, Eidel's calculations all show a softening behavior. That is, the ratio of the non linear resonant frequency $\omega$ to the linear resonant frequency $\omega_0$ decreases with increasing amplitude, $\zeta_{max}$. In this sense the liquid bridge behaves like a soft spring. To understand how this arises, the restoring force per unit area due to surface tension acting along a stream line, is considered.

Within the weakly non-linear approximation this force can be shown to be given by [3.11]

$$f_c = \frac{2\gamma}{a^2}\left\{ \zeta + a^2 \zeta_{zz} + \frac{1}{a^2}[\zeta^3 - a^2 \zeta_z^2(\zeta + 2a^2 \zeta_{zz})] \right\}. \tag{3.24}$$

For a stable column the linear term, $\zeta + a^2\zeta_{zz}$ is negative for $\zeta > 0$. That is, surface tension acts as a restoring force. Thus, it follows that $\zeta_{zz} < 0$ and the nonlinear term is positive and acts against the first order restoring force. The liquid surface behavior is analogous to that of a soft spring (Fig. 3.11) and it can be considered as a soft "membrane"

## References

[3.1]   Tsamopolous, J., Chen, T. & Borkar, A., Viscous oscillations of capillary bridges, J. Fluid Mechanics 235 (1992) 579-609.

[3.2]   Peyret, R. & Taylor, T.D., Computational Methods for Fluid Flow, (1983) Springer.

[3.3]   ISML Math/Library, *FORTRAN subroutines for mathematical applications,* (1991) IMSL(support@imsl.com).

[3.4]   Thomas, P.D. & Brown, R.A., LU decomposition of matrices with augmented dense constráints, Int. J. Num. Meth. Eng. 24 (1987) 1451.

[3.5]   Higueras, M., Nicolas, J. & Vega J.M., Linear oscillations of weakly dissipative axisymmetric liquid bridges, Phys. Fluids A 6 (1994) 438-450.

[3.6]   Sanz, A. & López-Díez, J., Non-axisymmetric oscillations of liquid bridges, J. Fluid Mech. 205 (1989) 503-521.

[3.7]   Zhang, Y.-Q., and Alexander, J.I.D., The sensitivity of a nonisothermal liquid bridge to residual acceleration, Microgr. Sci. Tech. IV (1991) 128-129; Proceedings of the IUTAM Symposium on Microgravity Fluid Mechanics, Bremen, 1991, ed H. Rath.

[3.8]   Chen, T. & Tsamopolous, J., Nonlinear dynamics of capillary bridges: Theory, J. Fluid Mech. 255 (1993) 373-409.

[3.9]   Shulkes, R.M.S.M., Nonlinear dynamics of liquid columns: A comparative study, Phys. Fluids A 5 (1993) 2121-2130.

[3.10] Zhang, Y., and Alexander, J.I.D., unpublished research (1994).

[3.11] Eidel, W., Weak nonlinear axisymmetric oscillations of an inviscid liquid column with anchored free surface under zero gravity, Microgr. Sci. Technol. 7 (1994) 6.

## 4. BRIDGES AND JETS: NON LINEAR DYNAMICS AND BREAKING

### 4.1    Non-linear dynamics: breaking

Once a liquid bridge has reached its stability limit, it will either change its shape, for example from an axisymmetric to a nonaxisymmetric shape, detach from the support disks, or break up into smaller drops. The process of breaking involves the decay of the bridge radius to zero at some points. When the radius is zero, the bridge separates. This means that the governing equations must be singular at this point. Linear stability theory fails to describe this situation adequately and cannot predict the shape of the surface as breaking is approached, nor does it account for the non-uniform break-up of the bridge, i.e. that separation of the bridge into two or more large drops is accompanied by the formation of much smaller "satellite drops". The breaking of liquid bridges has been studied using 1D models [4.2]. Numerical treatment of the breaking of liquid jets beyond the singularity has been studied recently by Eggers and Dupont [4.3] using a 1D model, and Shulkes [4.4] has compared the predictions of 1D models of inviscid bridges as breaking is approached with the results of a velocity-potential calculation for which no simplifying assumptions were made. Most recently, a modified Volume-of-Fluid method has been applied to the problem of breaking of axisymmetric and non-axisymmetric bridges by Alexander and Zhang [4.5].

First we briefly consider the model of Eggers and Dupont applied to a breaking jet. Then, in the last part of this lecture we examine the work of Shulkes and briefly discuss the advantages and disadvantages of the VOF method.

### 4.2    The breaking of jets

Starting with the equations of motion in cylindrical coordinates, for an axisymmetric column of fluid with shear viscosity $\mu$, density $\rho$ and surface tension $\gamma$ and velocity $\mathbf{u}(r,z) = (u,w)$ (where $u$ and $w$ are the radial and axial velocities, respectively, Eggers and Dupont made the assumption that the thickness of the column was small compared to its length and expanded the dependent variables in powers of $r$   They arrived at the following model equations

$$\frac{\partial w}{\partial t} = -w\frac{\partial w}{\partial z} - \frac{1}{\rho}\frac{\partial p}{\partial z} + \frac{3\mu}{\rho h^2}\frac{\partial}{\partial z}\left(h^2\frac{\partial w}{\partial z}\right) - g, \qquad (4.1)$$

for $0 < r < h$, $-l < z < l$, and at $r = h(z,t)$

$$p = 2\kappa\gamma = \gamma\left[\frac{1}{h(1+h_z^2)^{1/2}} - \frac{h_{zz}}{(1+h_z^2)^{3/2}}\right], \qquad (4.2)$$

where a subscript $z$ denotes a partial derivative with respect to z. Note that the retention of lower order terms in (4.2) yields the exact potential energy for equilibrium surfaces. (These terms do not appear at this order if $p$ is expressed in a consistent expansion in powers of $r$. and a different form of the potentiasl enrgy woul result.) The kinematic boundary condition takes the form

$$\frac{\partial h}{\partial t} = -w\frac{\partial h}{\partial z} - \frac{h}{2}\frac{\partial w}{\partial z}. \qquad (4.3)$$

For a given problem the boundary conditions have the form

$$h(\pm l, t) = h_\pm , \quad w(\pm l, t) = w_\pm , \tag{4.4}$$

where $h$ and $w$ are the prescribed radius and velocities at the ends of the column. It is useful to check the model equations by examining the stability of a fluid column subject to and infinitesimal sinusoidal perturbation with wavelength $\lambda = 2\pi/k$ . That is,

$$h(z, t) = h_0[1 + \varepsilon exp(\omega t)cos(kz)], \quad v(z, t) = \varepsilon v_0 exp(\omega t)sin(kz)], \tag{4.5}$$

This yields the dispersion relation

$$\omega = \omega_0\left\{\left[\tfrac{1}{2}(kh_0)^2(1 - (kh_0)^2) + \tfrac{9l_v}{4h_0}(kh_0)^4\right]^{1/2} - \tfrac{3}{2}(kh_0)^2\left(\tfrac{l_v}{h_0}\right)^{1/2}\right\}, \tag{4.6}$$

with

$$l_v = \tfrac{\mu^2}{\rho\gamma}, \quad \omega_0 = \tfrac{\gamma}{\rho h_0^3} . \tag{4.7}$$

For an expansion to lowest order in $kh_0$, the low viscosity and high viscosity limits of (4.6) coincide with the results obtained by Chandrasekhar [4.6]. Results obtained using (4.1)-(4.4) were compared to experimental studies of the breakup of a liquid jet that emerges from a nozzle [4.7]. In the experiment, water is pumped through a nozzle at high speed so that a liquid column is formed that is practically unaffected by gravity. The jet is subjected to a controlled periodic perturbation. The jet is allowed to reach a steady state so that, at some distance from the nozzle, the jet has broken into droplets. Eggers and Dupont carried out simulations of this process using their 1D model. Due to limitations of the model, the steady state obtained in the experiment cannot be reached. However, the simulation can be carried out up to the point of the first singularity. The result is shown in Fig. 4.1.

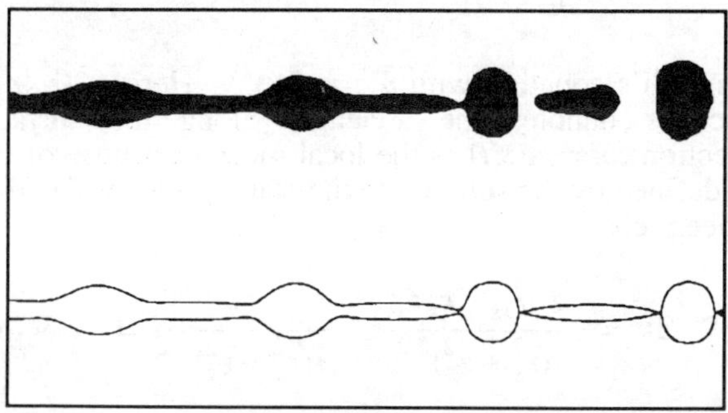

Fig. 4.1 Comparison of a decaying water jet (upper) (from [4.7]) and the simulations of Eggers and Dupont (lower). The nozzle is to the left and the point where the first drop detaches from the jet in the experiment has been aligned with the corresponding point in the simulations. $2\pi/kr_0 = 14.57$, $l_v/r_0 = 6.54(10)^{-4}$, We = $\rho r_0 V^2/\gamma = 239$.

The comparison with experiment up to the first singularity is good. The success of the model equations can be attributed to the fact the expansion method allows for the dissipative effects of viscosity to be accounted for while the consideration of the exact rather than the approximate curvature term. This allows for strong variations in $h$.

## 4.3     Breaking of inviscid bridges
### 4.3.1 Definition of the problem and basic equations
The inviscid slice model  was described chapter 2. To test the range of validity of the slice and Cosserat [4.2, 4.4] models as a bridges proceeds to breaking,  Shulkes [4.4] posed the velocity-potential problem for an axisymmetric bridge in the following form

$$\nabla^2 \phi\,(r,z) = 0,$$

$$\frac{\partial \phi}{\partial r}(1,\pm\frac{\Lambda}{2}) = \frac{\partial \phi}{\partial z}(1,\pm\frac{\Lambda}{2}) = 0, \tag{4.8}$$

while the boundary conditions at the free surface are expressed in terms of its arc length, $s$. The kinematic boundary condition $(4.1)_3$ then takes the form

$$\frac{\partial r(s,t)}{\partial t} + \nabla\phi\cdot\nabla r(s,t) = \frac{\partial \phi}{\partial r} = n_r\frac{\partial \phi}{\partial n}(s,t) - n_z\frac{\partial \phi}{\partial s}(s,t),$$

$$\frac{\partial z(s,t)}{\partial t} + \nabla\phi\cdot\nabla z(s,t) = \frac{\partial \phi}{\partial z} = n_r\frac{\partial \phi}{\partial n}(s,t) + n_z\frac{\partial \phi}{\partial s}(s,t), \tag{4.9}$$

while the dynamic boundary condition

$$\frac{\partial \phi}{\partial t} + \frac{1}{2}|\nabla\phi|^2 = 2\kappa\gamma + p_0 - Bo\,z. \tag{4.10}$$

is obtained from Bernoulli's equation with $p = -2\kappa\gamma$. Here $= (n_r,n_z)$ is the unit normal vector to the surface, $s$ denotes the arclength along the surface (increasing in a counterclockwise direction), $\kappa = \kappa(s,t)$ is the local mean curvature of the surface, and $p_0$ is the pressure constant defined by the solution to the static problem for $Bo$. The curvature and unit normals are, respectively,

$$2\kappa = \frac{z_s\,r_{ss} - r_s\,z_{ss}}{(r_s^2 + z_s^2)^{3/2}} - \frac{z_s}{r(r_s^2 + z_s^2)^{1/2}} = -\nabla_{\Sigma}\cdot n$$

$$n_r = \frac{z_s}{(r_s^2 + z_s^2)^{1/2}}, \quad n_z = \frac{-r_s}{(r_s^2 + z_s^2)^{1/2}}. \tag{4.11}$$

The fluid is assumed to be initially at rest.  The surface shape is initially a right circular cylinder.  The objective of the calculation is to track the evolution of the bridge as it proceeds to break.  Thus for times $t > 0$, the bridge responds to a finite Bond number. The solution to equation (4.8) along with the boundary conditions (4.9)-(4.10) is then found by solving the following integral equation

$$\alpha\phi(x) + \int_{\Sigma} \phi(x')\frac{\partial G}{\partial n'}(x,x')rds(x') = \int_{\Sigma} \frac{\partial \phi}{\partial n}(x')G(x,x')r'ds(x') \qquad (4.12)$$

where $\alpha$ is the solid angle at a point $x$ on the boundary subtended by the fluid domain [4.6] and $G(x,x')$ is Green's function for Laplace's equation in cylindrical coordinates and is given by

$$G(x,x') = \frac{4K[\rho]}{[(r+r')+(z-z')]^{1/2}}, \quad \rho = 1 - \frac{[(r-r')^2 + (z-z')^2]^{1/2}}{[(r+r')^2 + (z-z')^2]^{1/2}}, \qquad (4.13)$$

where $K(\rho)$ is an elliptic integral of the first kind.

### 4.3.2 Solution method

The above boundary integral equation was solved using a finite element method. The surface $\Sigma$ was divided into N elements. A linear interpolating function was used to approximate the velocity potential such that the potential on an element $\Delta\Sigma_i$ with nodes $x_k$ and $x_{k+1}$, is given as

$$\phi_{\Delta\Sigma_k}(\xi) = \phi_k(1-\xi) + \xi\phi_{k+1},$$

$$\phi(x_k) = \phi_k, \quad 0 \le \xi \le 1. \qquad (4.14)$$

Upon substitution of (4.14) into the integral equation (4.12) we obtain the following discrete equation relating $\phi$ and $\partial\phi/\partial n$

$$\sum_{j=1}^{N} H_{jk}\phi_k = \sum_{j=1}^{N} G_{jk}\frac{\partial\phi_k}{\partial n},$$

$$H_{ij} = \int_{\Delta\Sigma_{j-1}} \xi\frac{\partial G}{\partial n'}(x(\xi),x_i)r(\xi)ds(\xi) + \int_{\Delta\Sigma_j} (1-\xi)\frac{\partial G}{\partial n'}(x(\xi),x_i)r(\xi)ds(\xi) \qquad (4.15)$$

$$G_{ij} = \int_{\Delta\Sigma_{j-1}} \xi G(x(\xi),x_i)r(\xi)ds(\xi) + \int_{\Delta\Sigma_j} (1-\xi)G(x(\xi),x_i)r(\xi)ds(\xi).$$

For $i \ne j$ the integrals are regular and can be evaluated using standard numerical integration methods. The elliptic functions are evaluated using polynomial approximations (see Abramowitz and Stegun [4.8]). For $i = j$ there are logarithmic singularities in the integrals $G_{ij}$ which are evaluated using a four point Gauss formula for integrals with singularities. The diagonal elements $H_{ii}$ can be found readily upon noting that for a constant potential the right-hand side of (4.15)$_1$ is zero. If we take $\phi = 1$, then it follows that

$$H_{ii} = \sum_{\substack{j=1 \\ i \neq j}}^{N} H_{jk} \tag{4.16}$$

The dynamic and kinematic boundary conditions are treated as follows. The unit normal and the curvature at a node $x_i$ are found by fitting a quadratic spline through $x_i$ and its neighboring points, that is

$$x(\xi) = \tfrac{1}{2}\xi(\xi - 1)x_{i-1} + (1 - \xi^2)x_i + \tfrac{1}{2}\xi(\xi+1)x_{i+1}, \quad -1 \leq \xi \leq 1. \tag{4.17}$$

Equation (4.17) is used to obtain the derivatives of r and z with respect to s. For example,

$$\left.\frac{\partial r}{\partial s}\right|_{x_i} = \left.\frac{\partial r}{\partial \xi}\frac{d\xi}{ds}\right|_{x_i}, \quad \left.\frac{\partial z}{\partial s}\right|_{x_i} = \left.\frac{\partial z}{\partial \xi}\frac{d\xi}{ds}\right|_{x_i}, \dots \tag{4.18}$$

where the derivative $d\xi/ds$ is assumed to be constant (and thus can be factored out of the equations for $\mathbf{n}$ and $\kappa$). The derivative $\partial\phi/\partial s$ is evaluated in a similar fashion. Using the above procedure to discretize the kinematic and dynamic boundary conditions, the following system of discrete equations is obtained

$$\frac{\partial r}{\partial t} + \nabla\phi\cdot\nabla r = f_1(\phi,\psi,r,z;s), \quad \frac{\partial z}{\partial t} + \nabla\phi\cdot\nabla z = f_2(\phi,\psi,r,z;s),$$

$$\frac{\partial\phi}{\partial t} + \nabla\phi\cdot\nabla\phi = f_3(\phi,\psi,r,z;s), \tag{4.19}$$

$$H\phi = G\psi, \quad \psi = \frac{\partial\phi}{\partial n}$$

where $f_1, f_2, f_3$, are nonlinear functions of their arguments. Since the evolution of the free surface is not uniform, there are large variations in the distribution of the Lagrangian nodes. This can lead to significant discretization errors when areas with depleted nodes have high curvatures. In addition, short wavelength instabilities will tend to occur in regions with a high concentration of Lagrangian nodes. For these reasons, Shulkes used a regridding procedure which is described in detail in [4.9].

### 4.3.3 Comparison of 1D models with the velocity potential model

Figures 4.2 - 4.4 show the results of a comparison between the 1D slice mode, the Cosserat model and the velocity potential calculations. An initially circular cylindrical liquid column with $\Lambda = L/2R = 3$, was subjected to an axial acceleration equal to $Bo = 0.1$. In Fig. 4.2 we see the evolution of the surface toward bifurcation for the three models. The volume of the upper region of the drop is approximately equal for the inviscid slice (IS) and the velocity-potential (VP) calculation. The Cosserat or first order model (FO) predicts a smaller volume. Figure 4.3 shows a comparison between the three cases for $Bo = 0.5$. At $t = 1.0$ the differences between the three are small, however, at $t = 2.75$ the IS model deviates significantly form the trends exhibited by the FO and VP results. In particular, the IS model predicts a large free surface slope at the top disk. Short wavelength instabilities forced

Shulkes to discontinue the IS calculations shortly after $t = 2.8$. Fig. 4.4 shows a comparison of the FO and VP results at later times. Here we see the major shortcomings of 1D models. The velocity potential model shows the downward collapse of the column into the lower volume of the bridge and the bending of the bridge well below the lower disk edge. This cannot be calculated with either the IS or the FO model.

Finally, Figure 4.5 compares the bifurcation times predicted by the three models. The bifurcation time is defined by the time taken for the neck of the column to vanish. The faster bifurcation times predicted by the IS model are due to the neglect of radial acceleration effects.

In summary, at large $Bo$ the combination of the neglect of radial momentum effects and the susceptibility to short wavelength instability limits the applicability of the IS model to breaking problems. The agreement between the VP and the FO model is generally better, but the FO model fails to properly account for situations in which large accelerations cause the bridge surface to move below the lower disk edge.

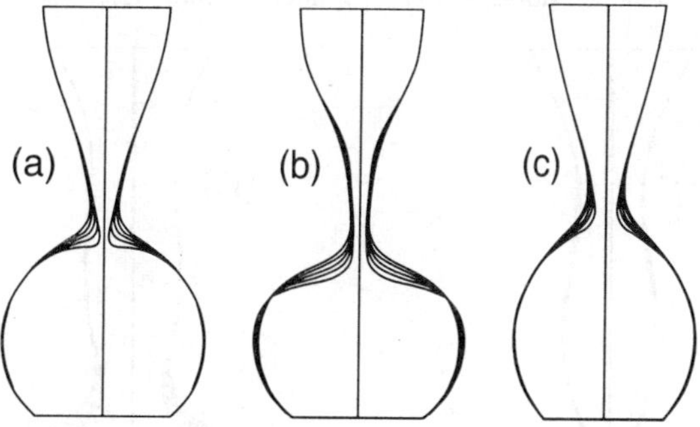


Fig. 4.2 Evolution of a liquid bridge as predicted by (a) the VP model, $6.15 \le t \le 6.53$; (b) the FO model, $6.5 \le t \le 6.7$; (c) the IS model $5.9 \le t \le 6.53$; $\Delta t = 0.05$, $Bo = 0.1$. From [4.4]

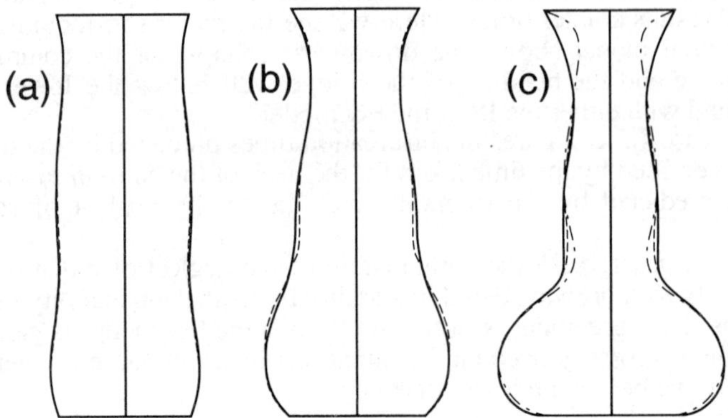

Fig. 4.3 Evolution of a liquid bridge for $Bo = 0.5$, $\Lambda = 3$, (a) $t = 1$ (b) $t = 2.0$, (c) $t = 2.75$. (Solid line - VP model, dash line - FO model, dot-dash line - IS model. From [4.4]

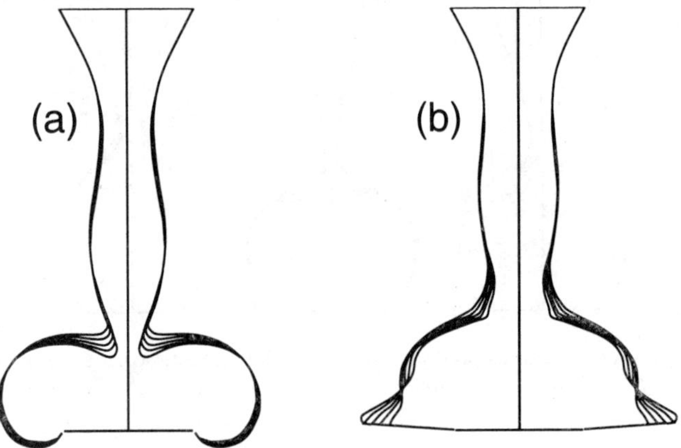

Fig. 4.4 Evolution of a liquid bridge as predicted by (a) the VP model, $3.45, \leq t \leq 3.65$; (b) the FO model, $3.1 \leq t \leq 3.3$. From [4.4]

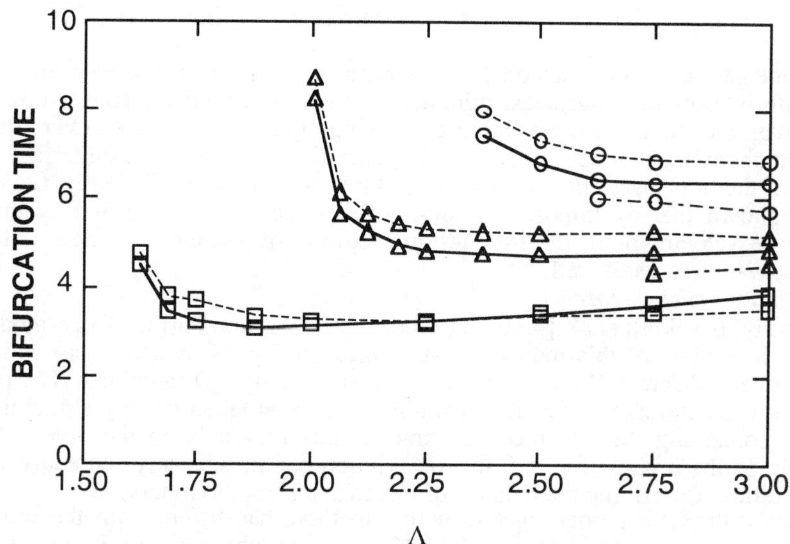

Fig. 4.5 The bifurcation time as a function slenderness, $\Lambda$ for $Bo$ = 0.1 (open circles),
$Bo$=0.2 (open triangles) and $Bo$ = 0.5 (open squares). The solid line corresponds to the VP
model, the dotted line to the FO model and the dot-dash line to the IS model.

## 4.4   A volume of fluid method applied to the problem of viscous breaking

In the Volume-of-Fluid (VOF) method the interface between two fluids is described
by defining a function C which takes on the value 1 in fluid 1 and 0 in the second fluid.  The
average value of C in any computational cell will then represent the fraction of the cell volume
occupied by fluid 1.  Thus, it follows that cells with values of $C$ between 0 and 1 must
contain the interface between the fluids.  For a given cell once $C$ and its normal vector are
calculated a line can be drawn through the cell which approximates the interface location
within that cell.  Once the location is known the appropriate boundary conditions can be set.
The key to the success of the VOF method is in the method used to describe the evolution of
the $C$ field.  The time dependence of F is governed by the kinematic equation

$$\frac{\partial C}{\partial t} + \nabla \cdot (\boldsymbol{u} C) = 0 \tag{4.20}$$

This says that the interface moves with the fluids and represents continuity of the two fluids,
i.e. a particle of fluid on the interface always remains on the interface. In a Lagrangian mesh,
the value of $C$ is conserved, in an Eulerian mesh, the flux of F through the cells must be
computed.  If standard finite difference techniques are used to compute this flux, the $C$
function is smeared over the cell and the interface loses its definition.  In other words, the
discontinuity of $C$ associated with the interface is not preserved. However, since $C$ is a step
function with values of 0 and 1 there is a flux approximation, the donor-acceptor method
[4.10], which retains the discontinuity.  In the vicinity of a flux boundary, the donor-
acceptor method uses information about the upstream and downstream values of $C$ and
establishes a crude approximation to the interface to compute the flux.  The slope of the
surface is used to improve the fluxing algorithm and the $C$ function is used to define a surface

location for the application of boundary conditions (including those involving surface sources).

Advantages: the VOF method follows regions rather than surfaces and so avoids problems with intersecting interfaces, minimum storage requirements (one word per mesh cell). The main advantage is the volume preserving flux scheme. However, despite this useful feature, there is a problem caused by the generation of "flotsam" as interfaces advance. Elimination of the flotsam results destroys the volume conservation. Recently, Lafaurie et al. [4.11] have shown that by imposing symmetries between the phases and by enforcing a divergence free velocity field, the troublesome regions of flotsam can be eliminated and volume conservation is recovered.

*Continuous surface force*

Recently, Brackbill et al. [4.10] introduced a continuum surface force model (CSF). The essential ingredient of this model is that surface tension is treated as a continuous 3D effect rather than a discrete 2D quantity evaluated on a sharp 2D interface. The position of the interface is characterized by a color function $C$ which is taken to vary smoothly in some small region containing the interface. In practice this region is on the order of the grid resolution $\delta h$. In the limit as $\delta h \to 0$ this description can be shown to be equivalent to the classical conditions describing the balance of forces at a sharp boundary.

We used the CSF model together with a method that differs from the original VOF method and subsequent modification [4.11,12] for two-phase flows in that rather than employing a fluxing scheme, equation (4.20) is solved directly. The averaging associated with the direct solution of (4.20) using standard differencing techniques can be a problem [4.13] due to numerical diffusion which results in "smearing". This refers to gradients in the color function which may be steep in reality but become shallower in the numerical representation. This is suppressed in our approach by employing a high accuracy third order TVD upwind scheme [4.14]. This scheme has been shown to be suitable for problems with large gradients and discontinuities (shock capturing, for example).

The surface tension force associated with a portion of the surface is calculated from (see reference [4.12])

$$F_{SV}(x) = 2\gamma\kappa(x)\frac{\nabla C}{[C]},$$

where the VOF function C has the values

$$C = \begin{cases} 1, & \text{in fluid 1} \\ 0 < C < 1, & \text{in the interface region} \\ 0, & \text{in fluid 2} \end{cases}$$

and $[C] = C_2 - C_1$. The curvature, $\kappa$ is calculated from

$$2\kappa = -\nabla \cdot n = -\frac{1}{|N|}\left(\frac{N}{|N|} \cdot \nabla|N| - \nabla \cdot N\right),$$

where $N = \nabla C$ is the gradient of the color function and is normal to the surface.

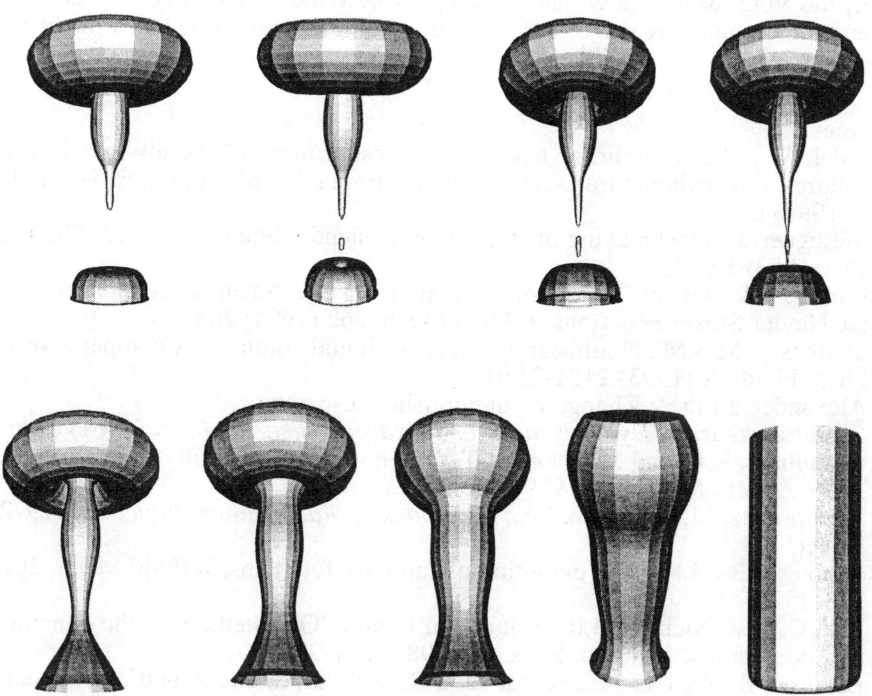

Fig. 4.6 Sequence of bridge configurations calculated using a modified VOF method. From right to left dimensionless time = 0, 2, 2.75, 3.45, 4.08, 4.546, 4.55, 4.567,4.607. $\Lambda = 3$, $Bo = 0.5$.

*Governing equations and discretization*

The governing equations of motion have the form

$$\nabla \cdot \boldsymbol{u} = 0, \quad \text{in } D$$

$$\rho \frac{D\boldsymbol{u}}{Dt} = -\nabla p + \mu \Delta \boldsymbol{u} + \rho \boldsymbol{g} + \boldsymbol{F}_{SV}, \quad \text{in } D$$

with $\boldsymbol{u} = \boldsymbol{0}$ on $\partial\Omega$ . These are solved together with the VOF equation (4.20) and the appropriate boundary conditions. The governing equations are discretized in space using an implicit finite volume scheme [4.14]. The advantage of VOF type schemes is that it can cope with highly distorted interfaces. Surfaces that are multivalued functions of the bridge height are not a problem, and fragmentation of breakage of the bridge can be modeled. However, as the point of fragmentation or breakage is approached the physics of the process will be lost. It may be possible to incorporate microscopic models that allow the proper continuation of the calculation through and beyond the breaking process. This could be a fruitful area for future research. Figure 4.6 shows a result computed for the breakage of the bridge with the same initial conditions as for Fig. 4.3 and 4.4. Obviously, in this case the two liquids must have a finite viscosity and the motion of the outer liquid is taken into account. However, we used a viscosity of water for the bridge and air for the outer fluid. As shown in Figs. 4.3 and 4.4 two areas of necking evolve, and at the initial stages the lower neck is thinner.

However, the VOF results show that at the upper neck the radius begins to decrease more rapidly and the eventual breakage, into two large volumes and a satellite drop forms at the upper neck.

**References**

[4.1] Eidel, W., Weak nonlinear axisymmetric oscillations of an inviscid liquid column with anchored free surface under zero gravity, Microgr. Sci. Technol. 7 (1994) 6.

[4.2] Meseguer, J., The breaking of axisymmetric slender liquid bridges, J. Fluid Mech. 130 (1983) 123-151.

[4.3] Eggers, J. & Dupont, T.F., Drop formation in a one dimensional approximation of the Navier-Stokes equations, J. Fluid Mech. 262 (1994) 205.

[4.4] Shulkes, R.M.S.M., Nonlinear dynamics of liquid columns: A comparative study, Phys. Fluids A (1993) 2121-2130.

[4.5] Alexander, J.I.D. & Zhang, Y., unpublished research

[4.6] Chandrasekhar, S., *Hydrodynamic and Hydromagnetic Stability* (1961) Clarendon.

[4.7] Chaudhary, K.C. and Maxworthy, T., The nonlinear instability of a liquid jet. Part 2, A. Rev. Fluid Mech. **11** (1979) 207-228.

[4.8] Abramowitz, M. & Stegun, I.A., *Handbook of Mathematical Functions* (1972) Dover.

[4.9] Shulkes, R.M.S.M., The evolution of capillary fountains, J. Fluid Mech. 261 (1994) 223-252.

[4.10] Hirt, C.W. & Nichols, B.R., Volume of Fluid (VOF) method for the dynamics of free boundaries, J. Comp. Phys. 39 (1981) 201-225.

[4.11] Lafaurie, B, Nardone, C., Scardovelli, R., Zaleski, S. and Zanetti, G, Modelling merging and fragmenting in multi-phase flows with SURFER, J. Comp. Phys. 113 (1994) 134-147.

[4.12] Brackbill, J.U., Kothe, D.B. & Zemach, C., A continuum method for modelling surface tension, J. Comp. Phys. 113 (1994) 134-147.

[4.11] Jun, L. & Spalding, D.B., Numerical simulation of flows with moving interfaces, PCH PhysicoChemical Hydrodynamics, 10 (1988) 625-637.

[4.13] Chakravarty, S.R., A new class of high accuracy TVD schemes for hyperbolic conservation laws, Proc. AIAA 23rd Aerospace Sciences Meeting, (1985).

[4.14] Patankar, S.V. *Numerical Heat Transfer and Fluid Flow*, (1980) Hemisphere.

# 5. THERMOCAPILLARY MIGRATION OF DROPS AND BUBBLES

## 5.1    Introduction

When a fluid surface or interface is subjected to a temperature gradient, the temperature dependence of surface tension will generally lead to a force that results in motion of the adjacent bulk fluids [5.1].    The first experimental work that dealt with the thermocapillary motions caused by non-isothermal bubbles and drops was carried out by Young, Goldstein and Block [5.1]. They were able to use thermocapillary flow in a bubble in a heated liquid to counter the tendency of the bubble move in response gravity by balancing the gravitational and thermocapillary forces.

Thermocapillary migration of drops and bubbles occurs in a variety of natural and technological processes ranging from materials processing (e.g., alloys from melts with two immiscible phases) to fluid storage and transfer operations in spacecraft. The thermocapillary effect can become very important under microgravity conditions when the temperature dependence of surface tension may become a dominant driving force. There have been a number of theoretical and experimental studies of the thermocapillary migration of single drops and bubbles (see [5.2] and [5.3] for a review).  In addition, thermocapillary transport in droplet and bubble suspensions has also been examined [5.4-5.7].    In this lecture we derive the Young-Goldstein-Block equation for a single droplet or bubble and discuss a result obtained from a mean field approximation for a suspension of droplets [5.7].

## 5.2    Approximate solution for the steady drop velocity

Consider an incompressible single drop of radius $R$, shear viscosity $\mu^{(1)}$ and density $\rho^{(1)}$ moving at a constant speed $U_0$ in a viscous incompressible fluid with viscosity $\mu^{(2)}$ and density $\rho^{(2)}$.  The interfacial tension between the droplet and the surrounding fluid is assumed to be large enough that deformation of the droplet can be neglected. We choose a frame of reference that moves with the drop such that the fluid velocity $u$ at the (spherical) interface between the drop and the surrounding medium is zero. Under these conditions and assumptions, the steady equations of motion take the form

$$\rho^{(i)}u^{(i)}.\nabla u^{(i)} = -\nabla p^{*(i)} + \mu^{(i)}\nabla^2 u^{(i)}, \quad \mathrm{div}\, u^{(i)} = 0, \; i = 1,2, \qquad (5.1)$$

where, $p^{*(i)} = (p^{(i)} + \rho^{(i)}gz)$ is the modified or reduced pressure, $\mu$ is the shear viscosity, $\rho$ is the density and $i = 1$ inside the drop and $i = 2$ outside the drop.  Here it we have assumed that the cylindrical coordinate system is oriented such that such that gravity is directed along the negative z-direction.  The temperature field satisfies

$$u^{(i)}.\nabla T^{(i)} = \kappa\nabla^2 T^{(i)}, \quad i = 1,2, \qquad (5.2)$$

where $\kappa$ is the thermal diffusivity. For a drop moving freely in an isothermal fluid under the creeping flow conditions for which Stokes' law holds, the nonlinear terms in (5.1) can be neglected.  Furthermore for Pe $= U_0R/\kappa \ll 1$ where $U_0$ is the speed of a spherical drop with radius $R$ moving at a constant rate, the left hand side of (5.2) can also be neglected. This leads to the following equations for the velocity, pressure and temperature fields in and around the bubble

$$\frac{\partial}{\partial r}p^{*(i)} = \mu^{(i)}\nabla^2 u_r^{(i)} - \frac{2u_r^{(i)}}{r^2} - \frac{2}{r^2 sin\,\theta}\frac{\partial(u_\theta^{(i)} sin\,\theta)}{\partial\theta},$$

$$\frac{1}{r}\frac{\partial}{\partial\theta}p^{*(i)} = \mu^{(i)}\nabla^2 u_\theta^{(i)} + \frac{2}{r^2}\frac{\partial u_r^{(i)}}{\partial\theta} - \frac{u_\theta^{(i)}}{r^2 sin^2\,\theta},$$

$$\frac{\partial u_r^{(i)}}{\partial r} + \frac{1}{r}\frac{\partial u_\theta^{(i)}}{\partial\theta} + \frac{2u_r^{(i)}}{r} + \frac{u_\theta^{(i)}cot\,\theta}{r} = 0, \tag{5.3}$$

$$\nabla^2 = \frac{1}{r^2}\frac{\partial}{\partial r}\left(r^2\frac{\partial}{\partial r}\right) + \frac{1}{r^2 sin\,\theta}\frac{\partial}{\partial\theta}\left(sin\,\theta\frac{\partial}{\partial\theta}\right),$$

where $z = rcos\,\theta$.   The $u_r$ velocity components vanish at $r = R$ and far from the drop we have $u \to v = U_0 k$ , $p^* \to 0$, and $T \to T_0 + Gz$.

Taking the curl of the right hand side of (5.1)   (i.e. the linearized momentum equations) and using the fact that $div\,u = 0$ yields $\nabla^2\,curl\,u = 0$.  Since $u = U$ at infinity, and since we can write $u = u^* + U$ , where $u^* = 0$ at infinity we have $div\,u = div\,u^* = 0$. This means that $u = curl\,A + U$ where $A$ is an axial vector (the curl of any polar vector is an axial vector).  The velocity $u$ (and, thus, $A$) depends on $r$ (the radius vector from the center for the drop) and the parameter $U$.  Both $u$ and $U$ are polar vectors.  It follows that [5.8],

$$A = f'(r)g\times U, \text{ or } u = curl\,(gradf\times U) + U, \tag{5.4}$$

$$A = f'(r)g\times U, \text{ or } u = curl(gradf\times U) + U = curlcurl\,(fU) + U. \tag{5.5}$$

It can be shown that [5.8], $f$ must satisfy
$$\nabla^2(\nabla^2 gradf) = 0, \text{where } f^{(2)} = A^{(2)}r + \frac{B^{(2)}}{r} \text{ for } r > R, \tag{5.6}$$

$$\text{and } f^{(1)} = \frac{A^{(1)}}{4}r^2 + \frac{A^{(1)}}{8}r^4, \text{ for } r < R.$$

This yields general solutions for $u^{(1)}$ and $u^{(2)}$ of the form [5.8, 5.9]

$$u^{(1)} = -A^{(1)}U + B^{(1)}r^2[e_r(U\cdot e_r)-2U],$$

$$u^{(2)} = U - A^{(2)}\frac{U + e_r(U\cdot e_r)}{r} + B^{(2)}\frac{3\,e_r(U\cdot e_r)-2U}{r^3}. \tag{5.7}$$

and the modified pressure satisfies

$$p^{(i)} = -\mu^{(i)} \, U \cdot grad\nabla^2 f^{(i)} + p_0^{(i)}, \quad i = 1,2.$$

The temperature field inside and outside the drop is (neglecting the terms on the left-hand side of (5.2))

$$T^{(2)}(r,\theta) = T_0 + G\left(r + \frac{C}{r^2}\right)\cos\,\theta, \quad T^{(1)}(r,\theta) = T_0 + GDr\cos\,\theta .$$

$$(5.8)$$

$$D = \frac{3k^{(2)}}{(k^{(1)} + 2k^{(2)})}, \quad C = R^3(D-1).$$

The constants $A^{(i)}$ and $B^{(i)}$ are determined from the condition that $u_r^{(i)}(R,\theta) = 0$, $i = 1,2$, and the conditions

$$u_\theta^{(1)} = u_\theta^{(2)}, \quad \left[\mu^{(1)}\left(\frac{\partial u_\theta^{(1)}}{\partial r} - \frac{u_\theta^{(1)}}{r}\right) - \mu^{(2)}\left(\frac{\partial u_\theta^{(2)}}{\partial r} - \frac{u_\theta^{(2)}}{r}\right) - \frac{\gamma_T}{R}\frac{\partial T}{\partial \theta}\right] = 0, \quad (5.9)$$

The constants are

$$A^{(2)} = -R^2\frac{3Gk^{(2)}\gamma_T}{6(\mu^{(1)}+\mu^{(2)})\,(k^{(1)}+2k^{(2)})U_0} + \frac{R}{4}\frac{(2\mu^{(2)}+3\mu^{(1)})}{(\mu^{(1)}+\mu^{(2)})}$$

$$B^{(2)} = R^2 A^{(2)} - R^3, \quad B^{(1)} = 2\frac{A^{(2)}}{R^3} - \frac{3}{2R^2}, \quad A^{(1)} = -B^{(1)}R^2.$$

$$(5.10)$$

Here $\gamma_T = d\gamma/dT$. The projection in the direction of $U = U_0 k$ of the force exerted by the fluid on the surface of the drop is given by

$$F = \oiint\left[-p^{*(2)}\cos\,\theta + 2\mu^{(2)}\frac{\partial u_r^{(2)}}{\partial r} - \mu^{(2)}\left(\frac{1}{r}\frac{\partial u_\theta^{(2)}}{\partial \theta} + \frac{\partial u_\theta^{(2)}}{\partial r} - \frac{u_\theta^{(2)}}{r}\right)\right]ds ,$$

with $p^{*(2)} = p^*_0 - \frac{2A^{(2)}}{r^2}\mu^{(2)}U_0\cos\theta$. The integral is simply [5.8]

$$F = 8\pi A^{(2)}\mu U_0.$$

$$(5.10)$$

(Equation (5.10) is true for any velocity with the form of $u^{(2)}$ in (5.7). This force must be equal to the buoyancy force $(4/3)\pi R^3(\rho^{(2)} - \rho^{(1)})g$ or

$$8\pi\mu^{(2)}\left[-R\frac{3Gk^{(2)}\gamma_T}{6(\mu^{(1)}+\mu^{(2)})\,(k^{(1)}+2k^{(2)})U_0} + \frac{R}{4}\frac{(2\mu^{(2)}+3\mu^{(1)})}{(\mu^{(1)}+\mu^{(2)})}\right]U_0 = \frac{4}{3}\pi R^3(\rho^{(2)} - \rho^{(1)}),$$

which leads to the following expression for $U_0$

$$U_0 = \frac{2Gk^{(2)}\gamma_T R}{(2\mu^{(2)} + 3\mu^{(1)})(k^{(1)} + 2k^{(2)})} + (\rho^{(2)} - \rho^{(1)})g\frac{2R^2(\mu^{(1)} + \mu^{(2)})}{3\mu^{(2)}(2\mu^{(2)} + 3\mu^{(1)})}, \qquad (5.11)$$

Equation (5.11) was first derived by Young-Goldstein and Block [5.1] and has subsequently led to a number of experimental and theoretical studies [5.2-5.4].  For a bubble, in the limit $\mu^{(1)} \to 0$, $k^{(1)} \to 0$. this expression reduces to

$$U_0 = \frac{1}{3\mu^{(2)}}\left(\frac{3}{2}G\gamma_T R + \Delta\rho g R^2\right), \qquad (5.12)$$

The validity and applicability of the quasistatic theory for the migration of a bubble under the combined action of gravity and thermocapillarity was shown experimentally by Merrit and Subramanian [5.9].  Later work by Dill [5.10], showed that the quasistatic theory should apply provided that the ratio $\iota$ is sufficiently small, where

$$\iota = \frac{\Delta\rho \, |\dot{R}(t)| \mu^{(2)}}{\rho^{(2)}[-Rg_T|G|]}. \qquad (5.13)$$

### 5.3    Discussion

If a temperature gradient is applied to the bubble such that it increases monotonically from one pole of the bubble to the other, the consequent gradient in interfacial tension will produce a flow.  The thermocapillary flow will move from the warmer pole to the cooler pole within the drop or bubble.  This will result in an overall motion of the drop or bubble in the direction of the temperature gradient.  That is, the bubble will move toward the hot region.

In practical situations, drops and bubbles often occur as somewhat dense suspensions, rather than the case of an isolated bubble or drop discussed in the previous section.  Recent work has dealt with interactions between droplet pairs and also on drop suspensions.  Balasubramanian et al. [5.11] carried out experiments on pairs of drops and showed that a small drop that moves ahead of a large drop in a temperature gradient significantly retards the motion of the larger trailing drop while it appears to be unaffected by the larger drop.

While it is beyond the scope of this lecture to give a detailed account of how one approaches the theoretical descriptions it is worth noting a recent theoretical result due to Felderhof [5.7].  In a mean field approximation of the thermocapillary mobility of a suspension of droplets he was able to find a "Clausius-Mossotti" type relation for the bubble mobilities.  For a single drop the thermocapillary mobility is defined as $\mu^{tc}$ where

$$U = \mu^{tc} \nabla T, \quad \mu^{tc} = -\frac{2\gamma_T R}{\mu^{(2)}(2 + 3\alpha)(2 + \beta)}$$

For a suspension of droplets, Felderhof employed a multiple scattering expansion method. This involves describing the average dynamics of  a suspension using an a set of macroscopic equations and by assuming that probability distribution of the droplets is known. In a mean field approximation to his general results, he found that

$$\mu_{mf}^{tc} = \mu^{tc} \frac{1 - \phi}{1 - \phi\lambda}, \quad \lambda = \frac{k^{(1)} - k^{(2)}}{k^{(1)} + 2k^{(2)}}$$

where $\phi$ is the volume fraction occupied by the droplets.    For bubbles, we neglect the thermal conductivity of the bubbles in comparison to the surrounding liquid.  The mean field approximation to the thermocapillary mobility for bubbles is then

$$\mu_{mf}^{tc} = \mu^{tc} \frac{1 - \phi}{1 - 0.5\phi},$$

which is analagous to the Clausius-Mossotti relation between the dielectric constant and the electronic polarizability in solid-state physics.


**References**


[5.1]   Young, N.O., Goldstein, J.S. & Block, M.J., "The motion of bubbles in a vertical temperature gradient", J. F.M. **6** (1959) 350-356.

[5.2]   Wozniak, G., Siekmann, J. & Srulijes, J., "Thermocapillary bubble and drop dynamics under reduced gravity-survey and prospects", Zeit. Flugwissen. Weltraumforsch. **12** (1988) 137.

[5.3]   Subramanian, R.S., The motion of bubbles and drops in a reduced gravity environment." in *Transport Processes  in Bubbles Drops and Particles*, Chabra, R. and De Kee, D., eds., (Hemisphere, New York, 1922) pp. 1-41.

[5.4]   Anderson, J.L., "Droplet interactions in thermocapillary motion", Int. J. Multiphase Flow, **11** (1985) 813.

[5.5]   Meyyapan, M., Wilcox, W.R. & Subramanian, R.S., "The slow axisymmetric motion of two bubbles in a thermal gradient" J. Colloid Interface Sci. **94** (1984) 243.

[5.6]   Wang, Y. Mauri, R., & Acrivos, A., J.F.M. **261** (1994) 47.

[5.7]   Felderhof, B.U., "Thermocapillary mobility of a suspension of droplets in a fluid" Phys. Fluids **8** (1996) 1705-1714.

[5.8]    Landau, L.D., & Lifshitz, *Fluid Mechanics*, Course of Theoretical Physics **6** (Pergamon, Oxford, 1959).

[5.9]   Merrit, R.M. & Subramanian, R.S., "The migration of isolated gas bubbles in a vertical temperature gradient", J. Colloid and Interface Sci. **125** (1988) 333.

[5.10] Dill, L.H., "on the thermocapillary migration of a growing or shrinking drop", J. Colloid and Interface Sci. 146 (1991) 533.

[5.11] Balasubramanian, R., Lacy, C.E., Wozniak, G., Subramanian, R.S., "Thermocapillary migration of bubbles and drops at moderate values of the Marangoni number in reduced gravity", Phys. Fluids **8** (1996) 872-880.

# 6. SPREADING AND WETTING: DYNAMIC CONTACT ANGLES

## 6.1    Introduction
The liquid bridges considered in the previous lectures have the special property that they have contact lines that are pinned to the sharp edges of the supporting disks (see section 1.2.3).  For the contact lines to remain anchored to the disk edges the angles $\Psi$ and $\Psi_0$ formed by the bridge liquid and the surrounding fluid at the disk edges must satisfy the inequalities [6.1]

$$\Psi > \alpha, \ \Psi_0 > \pi - \alpha, \tag{6.1}$$

where $\alpha$ is the wetting angle formed by the liquid with the smooth surface of the disk.   If either of these inequalities is violated then the contact line will no longer remain pinned to the disk edges.  It will either spread onto the underside of the support disk or recede onto the top surface of the disk depending on the relative volume of the bridge. .
The motion of a contact line is not a simple matter.  In this lecture, as an introduction to the complexity of contact line behavior during spreading and wetting, we will consider some recent experimental and theoretical results on dynamic contact angles.  Because the spreading of liquids on solid surfaces occurs in a variety of natural and technological processes, there has been considerable interest in furthering the understanding of the behavior of liquids in the vicinity of moving contact lines.  In particular, there has been interest in the development of theories that can provide useful macroscopic boundary conditions.  Classical hydrodynamic models usually involve the assumption that the velocity is continuous at surfaces bounding the fluid. For solid bounding surfaces this results in a no-slip condition   at   the   fluid-solid   interfaces.   Due   to   the   latter   condition,   theoretical investigations of spreading and wetting encounter the problem of a non-integrable shear stress singularity [6.2,6.3] associated with the three-phase contact line.  The question of the dependence of contact angle on contact line speed and material properties also arises [6.3-6.6].
The stress-field singularity causes a problem in that it prevents the use of the contact angle as a boundary condition for the interface shape. Thus,  while we use the equilibrium contact angle as a boundary condition for calculating the shape of an equilibrium liquid bridge contained between two flat plates or two solid spheres, its use in a classical hydrodynamic model for same bridge in a dynamical situation is prohibited if the contact line moves.  The difficulty posed by moving contact lines is of concern whenever surface tension forces are large than viscous or gravitational forces.  Such situations will occur whenever the Bond number, $Bo = \rho g L^2/\gamma$ and $Ca = U\mu/\gamma$ are small (here $U$ is the speed of the contact line, $\mu$ is the fluid viscosity and $\gamma$ is the surface tension of the fluid.  The problem of determining the appropriate modifications of the classical model in the vicinity of a moving contact line has been discussed by several authors [6.2-6.6] and has yet to be fully resolved. Some authors have replaced the no-slip condition in the vicinity of the contact line with slip-type boundary conditions. This eliminates the non-integrable singularity.  The most common type of slip condition is a Navier type condition [6.5] of the form

$$\boldsymbol{t\cdot Tn} = \beta \boldsymbol{t\cdot u} \tag{6.2}$$

where $u$ and $T$ are the velocity field and the stress tensor at the solid surface, $n$ is the outward pointing normal to the solid and $t = |u|/u$. Several types of slip conditions have been proposed [6.2-6.6] and it has been shown that different local models yielded similar macroscopic behavior [6.3, 6.7]. Molecular dynamics (MD) simulations of the behavior of immiscible fluids with an interface that meets a solid surface [6.8,6.9] show that there is general agreement between the nature of the velocity fields in the MD simulations and that predicted by classical hydrodynamics. However, there is a region within a few molecular spacings of the contact line where the classical hydrodynamic theory breaks down [6.9]. In particular, Thomson and Robbins find that their MD simulations agree well with solutions to the Navier-Stokes equations with a no-slip boundary condition in the neighborhood of the contact line. The slip length was a few molecular spacings.

Dussan, Ramé and Garoff (DRG) [6.5] developed a model that describes interface shapes near moving contact lines in the limit of low capillary number, $Ca$. Here $Ca = U\mu/\gamma$ where $U$ is a characteristic velocity, $\mu$ is the fluid's shear viscosity and $\gamma$ is the surface tension. We describe the DRG model in more detail below and discuss its range of applicability.

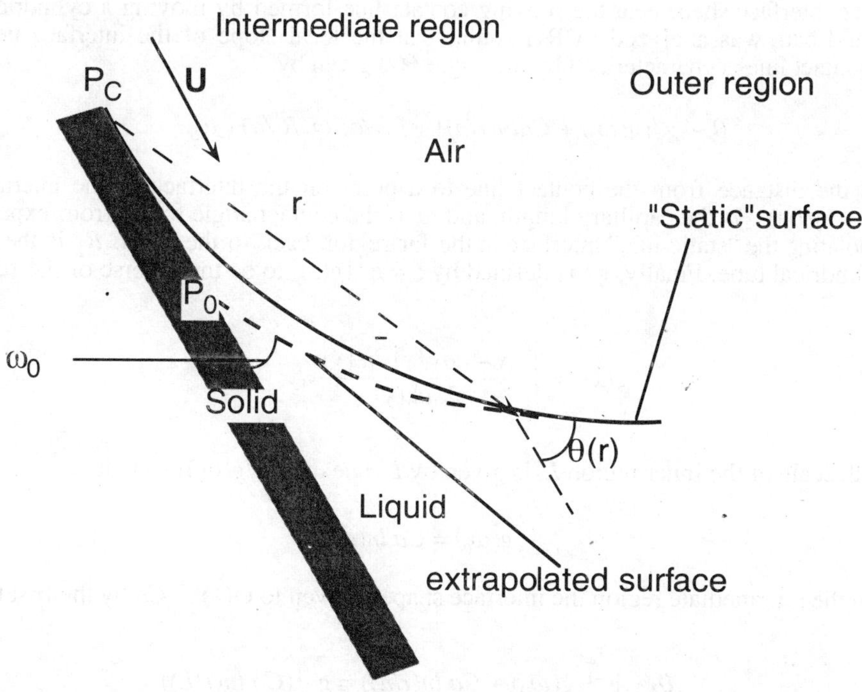


Fig. 5.1 Interfacial regions: In the outer region the air-liquid interface (thick solid line) coincides with a static interface shape. In the intermediate region, viscous effects cause the interface to deviate from the static profile. The thick dashed line corresponds to the extrapolation of the static shape back to the solid where it meets the solid at an angle $\omega_0$ at the point $P_0$. $P_C$ marks the position of the observed contact line.

## 6.2    Interfacial regions: The DRG model

Consider a system in which a fluid solid and gas meet at a common line on the surface of a solid rod (see Fig. 6.1).  The rod is displaced into the fluid at a velocity $U$ that causes the contact line to move relative to the rod.   The basis for the model is that the terms dominating the local balance of forces at the interface changes along the interface as the contact line is approached. Three physical regions are recognized. For low values of $Ca$ and at locations far from the moving contact line, the interface shape is determined by the balance between gravitational and capillary forces.  This is the "outer region" where viscous forces are unimportant.  For the system considered in the development of the DRG model the characteristic length scale of this outer region is $(\gamma/\rho g)^{1/2}$ (i.e. the capillary length).  Near the contact line there is an intermediate region where classical hydrodynamics still holds but the interface shape is now determined by a balance between viscous and surface tension forces. In this region the interface shape is strongly affected by viscous effects even as $Ca \rightarrow 0$ [7-9].  In the third region, the inner region, the classical hydrodynamic assumptions must be replaced with new conditions.  The characteristic lengthscale for the inner region is denoted by $L$ and is a function only of the physics of the inner region [6.5].

The interface shape near the moving contact line formed by moving a cylindrical rod into a liquid bath was analyzed.  DRG found that the local slope of the interface near the moving contact line is characterized by the angle $\theta(r)$ given by

$$\theta \sim g^{-1}(g(\omega_0) + Ca\ln(r/a)) + f_0(r/a; \omega_0, R_T/a) - \omega_0. \tag{6.3}$$

Here $r$ is the distance from the contact line to a point on the interface in the intermediate region, $a = (\gamma/\rho g)^{1/2}$ is the capillary length, and $\omega_0$ is the contact angle found from experiment by extrapolating the "static-like" interface in the far region back to the solid. $R_T$ is the radius of the cylindrical tube. Finally, $g^{-1}$ is defined by $z = g^{-1}(g(z))$ to be the inverse of the function $g$ where

$$g(x) = \int_0^x \frac{y - \cos{(y)}\sin{(y)}}{2\sin{(y)}}\, dy \tag{6.4}$$

The length scale of the inner region $L$, is given by $L = a\exp[-Ca/g(\omega_0)]$ so that

$$g(\omega_0) = Ca\, \ln(a/L) \tag{6.5}$$

In the intermediate region the interface shape is given to O(1) in $Ca$ by the first term in (6.3)

$$\theta_I \sim g^{-1}(g(\omega_0) + Ca\, \ln(r/a)) = g^{-1}(Ca\, \ln(r/L)), \tag{6.6}$$

The second term in (6.3) gives the shape of the interface in the outer region and is the same as for a static interface moving up a rod inserted into the liquid.  In the outer region the slope of the interface is given by

$$\theta_{outer} \sim f_0(r/a; \omega_0, R_T/a) \tag{6.7}$$

## 6.3 Experimental results

Ramé and Garoff [6.10] undertook an experimental investigation of interface shapes within distances of 1700 µm of a contact line. They examined shapes with small capillary numbers in the range $10^3 < Ca < 10^{-1}$. Their experiment consisted of immersing a vertical Pyrex tube at a constant speed $U$ into a container of polydimethylsiloxane (PDMS). (The PDMS wetted the glass completely and the static meniscus rose up and attained a zero static contact angle). When the tube was moved into the fluid at a velocity $U$, the meniscus was flattened due to viscous effects. Below some critical value of $U$ the meniscus remained above the level of the fluid bath. At higher $U$, the meniscus was bent below the bath fluid level. Images of the meniscus were obtained using a microscope attached to a CCD camera. Images were stored digitally. Image analysis involved the identification of the location and slope of the meniscus and fitting the pieces of the slope data to equations (6.3) or (6.7).

It was expected that since the extent of viscous deformation of the interface increases with increasing $Ca$, the distance from the contact line to the limits of the inner boundary and the outer boundary should increase as $Ca$ increased. To identify these regions from the experimental data, Ramé and Garoff used the following procedure:

First they fit slope data in the region $50\mu m < r < 300$ µm to equation (6.3). Then by examining the deviations of data in the region beyond 300 µm from the best fit they were able to test the full extent of the region described by equation (1). For the case of $Ca = 0$, the deviations of the measured slopes of the static meniscus from the best theoretical fit form a cloud about 1 to 2 degrees wide (see Fig. 6.2). These deviations are due to pixel noise in the camera and the cloud thickness determines the standard deviation, $\sigma$, of the slope data. For $Ca \neq 0$, the extent of the region defined by (6.3) is found by locating the point where the deviation cloud differs from zero by $\sigma/3$. This was done by fitting the deviation data to a third order polynomial (see Fig. 6.3). Figure (6.4) shows how (6.3) describes a section of the interface shape for $Ca = 0.005$ and $Ca = 0.1$. Having determined the extent of the region described by (6.3) they then fit different pieces of the data in a range $x^* < x < 1500\mu m$ for $100 m < x^* < 1200$ µm to equation (6.7). These fits are used to determine the extent of the static interface.

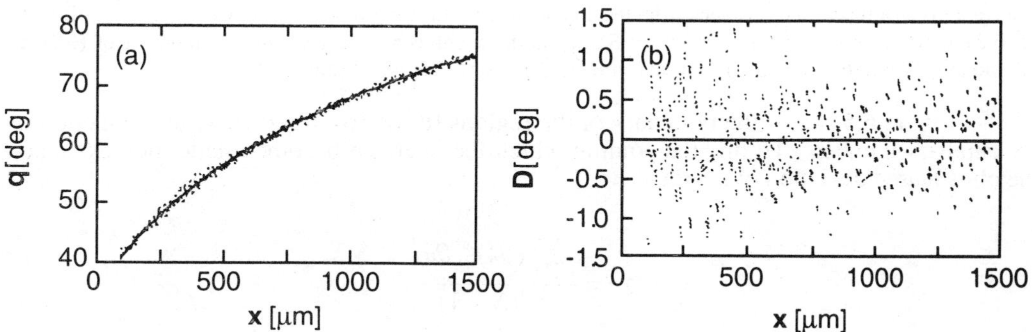

6.2. (a) Data and best fit for the static interface, (b) difference between data and best fit. From [6.10]

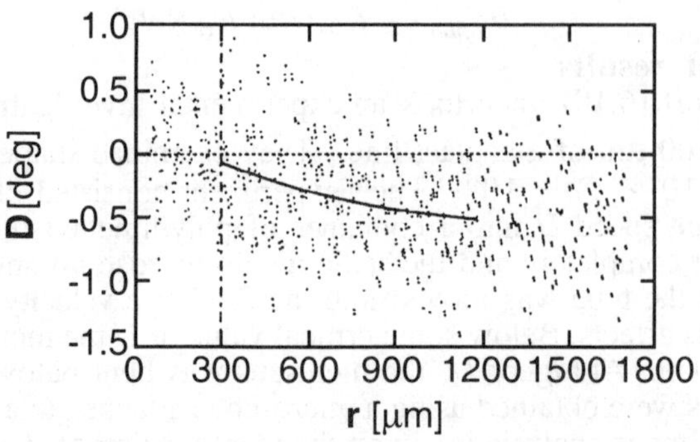

Fig. 6.3.

Difference between data and best fit of (6.3) for 50 μm < $x$ < 300 μm with $Ca = 0.005$. The dot-dash line shows the boundary of the fitted region. The solid line is the 3rd order polynomial fit of the data cloud in the region where the deviation from zero crosses $\sigma/3$. From [6.10]

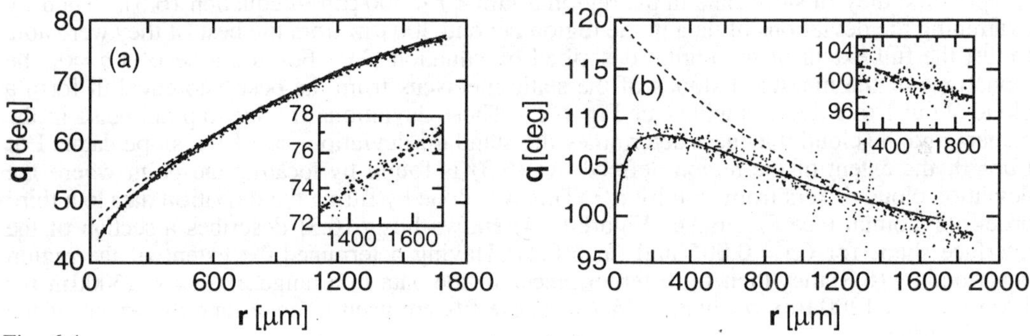

Fig. 6.4

Data and fits as a function of distance from the contact line for (a) $Ca = 0.005$ and (b) $Ca = 0.1$.  ---- Best fit of (6.3) in 50 μm < $x$ < 350 μm, — — — Static interface with contact angle = $\omega_0'$ from the fit of (6.7) for (a) and to $\omega_0$ from (6.3) for (b). The insets show fit for the contact line. From [6.10]

Figure 6.5 shows the boundary of the regions fit by (6.3) and (6.4) as a function of the capillary number. To find the point at which the interface becomes static they examined the chi-squared deviation of the data

$$\chi^2 = \frac{\sum_{n=1}^{N} (\theta_n - \theta_n)}{\sigma^2(N-1)}$$

as a function of $x^*$. Fig. 6.5 also shows that considerable viscous deformation was found at distances less than 1400 μm from the contact line even with $Ca = 0.01$. Since the capillary length, $a$, for these experiments is 1.5 mm we can see that at these distances the interface should be in the outer region. Notice also that, for these experimental conditions, there is no overlap of the regions well described by (6.3) and (6.7). This region is neither static-like nor well described by (6.3). The lack of overlap is explained by the fact that there is a region where gravitational and viscous forces are of the same order and act together to balance the surface tension forces. For any $Ca$, there is a distance $r'$ such that for $r > r'$ the effect of gravity is comparable to viscous forces before the viscous forces become negligible. In this region the interface appears to affected by the geometry of the outer region. The extent of the geometry free region ($r < r^*$) increases with increasing $Ca$.

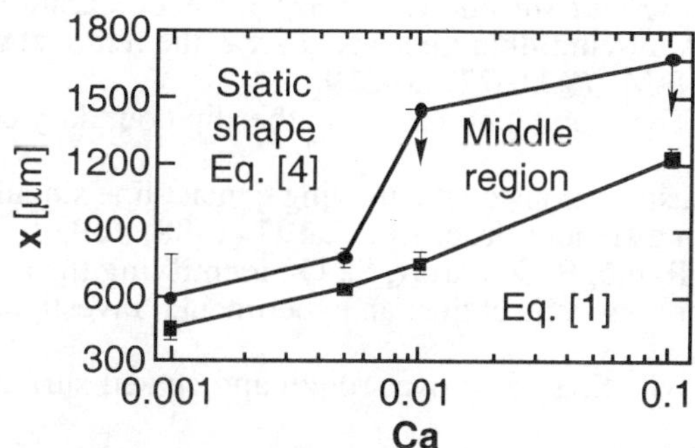


Fig. 6.5. Boundary of regions fit by (6.3) and (6.4) as a function of $Ca$. From [6.10]

In addition to determining the extent of the regions described by (6.3) and (6.7), the interpretation of $\omega_0$ as the contact angle of the static limit of the interface shape far from the contact line was tested. This was done by comparing the values of $\omega_0$ obtained through fits of (6.3) to slope data near the contact line, to $\omega_0'$ obtained from fits of data far from the contact line to equation (6.7). For $Ca = 0.001$ the difference was $-0.26° \pm 0.2°$ (inside the calibration limits for this experiment). For $Ca = 0.005$ the deviation was $-1.07 \pm 0.03°$. These results are consistent with equating $\omega_0$ and $\omega_0'$. This appears to be good evidence that $\omega_0$ is the correct boundary condition for the macroscopic shape.

## 6.4    Summary
The above work showed that
(1) The region of the fluid interface described by the static shape with a static contact angle $\omega_0$ moves out as $Ca$ is increased. Even at relatively small $Ca$, it was shown that significant viscous deformation could occur out to around 1400 μm. This is more than the capillary length that defines the edge of the outer region of the experimental geometry.
(2) The region described by (6.3) also increases in extent with increasing $Ca$ but does not overlap with the region that is described well by static theory. This lack of overlap is due to the presence of a third geometry-dependent region in which both gravity and viscous forces compete to balance surface tension.
(3) The model parameter $\omega_0$ obtained form fitting data in the region where viscous deformation is significant is the boundary condition of the "static-like" shape far from the contact line.

Experiments involving the spreading of carried out by Chen, Ramé and Garoff at higher $Ca$ values [6.11] have shown that for $Ca > 0.1$ the model represented by equation

(6.3) breaks down. They examined the fundamental assumptions of the model and concluded that the O(1) model breaks down at higher Ca. The pioneering work discussed above shows that there is much to be done to further the understanding of dynamic contact angles and contact line motion.

## References


[6.1] Myshkis, A.D., Babskii, V.G, Kopachevskii, N.D. Slobozhanin, L.A. & Tyuptsov, A.D., *Low Gravity Fluid Mechanics* (1987) Springer Verlag, Berlin.

[6.2] Hocking, L.M., A moving fluid interface. Part 2, the removal of the force singularity by a slip flow, J.F.M. **79** (1977) 20-229.

[6.3] Dussan V, E.B., The moving contact line: the slip boundary condition, J.F.M **77** (1976) 665.

[6.4] Durbin, P.A., Considerations on the moving contact line singularity with application to frictial drag on a slender drop, J.F.M. **197** (1988) 169.

[6.5] Dussan V, E.B., Ramé, E. & Garoff, S., On identifying the appropriate boundary conditions at a moving contact line: an experimental investigation, J.F.M **230** (1991) 97-116.

[6.6] Miksis, M. & Davis, S.H., Slip over rough and coated surfaces, J.F.M. **273** (1994) 125-139.

[6.7] Haley, P.J. and Miksis, M.J., The effect of the contact line on droplet spreading, J.F.M. **223** (1991) 57-81.

[6.8] Koplick, J., Banavar, J.R. & Willemsen, J.F., Molecular dynamics of Poiseuille flow and moving contact lines, Phys. Rev. Lett. **60** (1988) 1282.

[6.9] Thompson, P. & Robbins, M.O., Simulations of contact-line motion: slip and dynamic contact angle, Phys. Rev. Lett. **63** (1989) 766

[6.10] Ramé, E. & Garoff, S., Microscopic and macroscopic dynamic interface shapes and the interpretation of dynamic contact angles, J. Colloid Interface Sci. **177** (1996) 234-244.

[6.11] Chen, Q., Ramé, E. & Garoff, S., The breakdown of asymptotic models of liquid spreading at increasing capillary number, Phys. Fluids **7** (1995) 2631-2639.

# PERSPECTIVE ON EULERIAN
# FINITE VOLUME METHODS FOR
# INCOMPRESSIBLE INTERFACIAL FLOWS


**D.B. Kothe**
**Los Alamos National Laboratory, Los Alamos, NM, USA**


## Abstract


*Incompressible interfacial flows* here refer to those incompressible flows possessing multiple distinct, immiscible fluids separated by interfaces of arbitrarily complex topology. A prototypical example is free surface flows, where fluid properties across the interface vary by orders of magnitude. Interfaces present in these flows possess topologies that are not only irregular but also *dynamic*, undergoing gross changes such as merging, tearing, and filamenting as a result of the flow and interface physics such as surface tension and phase change. The interface topology requirements facing an algorithm tasked to model these flows inevitably leads to an underlying Eulerian methodology. The discussion herein is confined therefore to Eulerian schemes, with further emphasis on finite volume methods of discretization for the partial differential equations manifesting the physical model.

Numerous algorithm choices confront users and developers of simulation tools designed to model the time-unsteady incompressible Navier-Stokes (NS) equations in the presence of interfaces. It remains difficult to select or devise algorithms whose shortcomings are not manifested while modeling the problem at hand. In the following, many algorithms are reviewed briefly and commented on, but special attention is paid to projection methods for the incompressible NS equations, volume tracking methods for interface kinematics, and immersed interface methods for interface dynamics such as surface tension. At present, the quest for improved interfacial flow algorithms continues and the future looks very promising. This perspective will hopefully provide "field guidance" useful in devising algorithms whose weaknesses are not magnified when applied to your problem.

# 1   PERSONAL PERSPECTIVE

Flows possessing multiple distinct, immiscible fluids bounded by topologically complex interfaces are ubiquitous in natural and industrial processes [44]. Simulation of interfacial flow problems, via numerical (discrete) solution of appropriate partial differential equations (PDEs), is arguably the principal path to a fundamental understanding of these flows. Motivating numerical simulation of interfacial flows is a common "Achilles heel", or analytical intractability: the presence of interfaces that both move with *and* act back on the local flow field. Interface position is known only at initial time; thereafter it must be determined as part of the overall flow solution. It is not surprising, then, that the quest for accurate numerical simulation of these processes has attracted countless talented researchers over the past half century. At times, for many researchers (including myself), this quest seems futile.

From my perspective, schemes that statically partition the computational domain (via a stationary grid) *must* be the basis for a robust method touted to simulate flows possessing interfaces that are allowed to have complex topology. When moving-mesh (Lagrangian) methods are tasked to model such flows, solution accuracy and robustness deteriorates in a manner proportional to the complexity of the interface topology. Eulerian methods must therefore be embraced. It is interesting, however, that only ten years ago it was not clear that Eulerian methods could adequately meet the interfacial flow challenge. This last decade, however, has witnessed remarkable and exciting progress: many talented researchers have flocked to this field, where they have collectively advanced Eulerian interfacial flow methodology to the point where these methods can often be expected to yield quantitatively accurate flow solutions. The wide applicability of interfacial flow simulations has finally enticed enough people and resources to pay noticeable dividends.

Can we now "close the book" on Eulerian interfacial flow methodology? Hardly. Research in this field is akin to joining in on an arduous journey that is far from complete. New, unbiased minds with fresh ideas must continue to contribute innovative algorithms for this journey to make forward progress. The principal challenges ahead are motivated by the need for consistently robust *and* accurate (two competing requirements) *three-dimensional* Eulerian interfacial flow methods. I am confident, however, that these challenges will be met and overcome in the future.

I entered this field quite by accident back in 1985 when I realized during the course of my Ph.D. research on implosion symmetry of multi-layered ion-driven inertial confinement fusion targets [92] that I needed an interfacial flow simulation capability to help me understand the relevant physical phenomena. To my surprise at that time, however, I found that I needed to develop my own tool, and I chose to base it upon the wonderful and modernized particle-in-cell (PIC) FLIP method developed by J. U. Brackbill [21,23]. I have since been hopelessly addicted to this vibrant field of interfacial flow modeling, having written simulations tools known as Ripple [94,95], Pagosa [89], and, most recently, Telluride [90, 93, 147]. While each of these tools[1] simulate very different physical processes, all are based upon algorithms seeking solutions to the compressible or incompressible NS equations in the presence of interfaces. Despite my experiences, I am by no means an expert in this challenging field, having instead thought of myself as one who enjoys "playing around" with interfacial flow algorithms. Being an engineer at heart, my efforts, although manifested principally by these simulation tools, have always been driven by the need to understand these complex interfacial flows so that answers to difficult engineering questions can be found.

In the following, discussion of algorithms I have personally devised or used will be undoubtedly pedagogical because of my undeniable emotional bonds to these methods. Many algorithms discussed herein were devised and perfected by other (smarter) people who have as a result unknowingly become my benefactor during my endeavor to model these flows. I therefore take care to cite references liberally and aptly acknowledge collaborators and mentors. Honest experiences and frank opinions are offered when I am suitably impassioned. Any inconsistencies and inaccuracies undoubtedly present are due to my own ignorance and incompetence.

## 2  INTRODUCTION

We seek to devise numerical methods capable of generating reliable models for the multi-dimensional flow of incompressible fluids possessing densities that vary both discontinuously and smoothly. For these incompressible interfacial flows, smooth density variations might arise because the fluids are non-isothermal, whereas abrupt variations are manifested at immiscible fluid interfaces. Interfaces are present as a specified initial condition, as a result of phase change processes (evaporation, condensation, solidification, etc.), or because the fluid is undergoing gross topological change such as merging or breakup.

We desire an incompressible flow algorithm that maintains solution accuracy *and* simulation robustness in the presence of these density variations that are in general discontinuous (such as across interfaces). We further burden our method with *the* principal design constraint: the method must model the gross topological change of any interface created within or carried along with the flow. Finally, in addition to robustness and accuracy, our method must be high fidelity, i.e., make effective use of the discrete data and be sensitive to (or *resolve*) its spatial variation. High fidelity methods, while desirable in two-dimensional (2-D) methods, are especially important in three-dimensional (3-D) methods since current computational resources dictate inadequately-resolved partitioning of computational domains.

So, just what is an *accurate, robust, high fidelity* method? These terms are certainly commonly used (or abused) in this field. *Accuracy* is defined as the quality of

---

[1]For further information on Ripple, consult `http://gnarly.lanl.gov/Ripple/Ripple.html`, for further information on Pagosa, consult `http://gnarly.lanl.gov/Pagosa/Pagosa.html`, and for further information on Telluride, consult `http://www.lanl.gov/home/Telluride`.

conforming to fact. For our purposes, this definition is refined as the measured error for a given solution. A distinction must be made between the types of accuracy, for example, between the actual accuracy observed on a given computational grid and the rate at which accuracy increases with increased grid resolution (order of accuracy). For reasonable grid resolution, methods discretized with a higher order of accuracy can often be accompanied by significantly larger numerical error than a lower order method. *Fidelity* is defined as exact correspondence with fact. A solution that possesses fidelity is one that is validated as physically meaningful, i.e., one that is faithful to observances (experiment). We consider a method high fidelity when it produces solutions that are accurate relative to the computational resources (the grid size) applied to them. For example, interface tracking algorithms can increase solution fidelity by maintaining interface discontinuities as the interface advects and undergoes topological change. *Robustness* is the property of being powerfully built or sturdy. A robust method will not fail in a catastrophic manner, instead "degrading gracefully". A robust algorithm can be used with confidence on a difficult problem, where it should generate plausible, physically-reasonable solutions beyond the point where accuracy is expected or achieved.

In Section 3 a solution algorithm for the incompressible NS equations is presented that I believe possesses many important aspects of accuracy, fidelity, and robustness properties we desire in a solution method. Here I focus on the projection methods fathered by Harlow [67], Chorin [34], and Temam [186]. This primitive-variable algorithm belongs to a popular class of modern fractional-step [86, 208] projection methods that have been particularly successful in generating accurate solutions for *transient* flows possessing localized (interfacial) physics. Although the algorithm can be used to time-march solutions to steady state, it is not designed nor is it optimal for finding steady state flow solutions. Other viable NS solution algorithms such as pressure correction schemes (e.g., from the SIMPLE family) [30, 76, 135, 145, 193], streamline-vorticity methods, and artificial compressibility approaches [33, 187] are not discussed. Within the context of projection methods, we review algorithms for the projection operator, multi-dimensional advection, Newtonian stress effects, linear algebra, and velocity filters. Most of the algorithms presented are relatively independent of the choice of fluid variable arrangement on the grid (e.g., staggered or colocated), but colocated (cell-centered) schemes are emphasized when necessary. The current preference for cell-centered schemes is driven more by "conservation of implementation difficulty" considerations on 3-D unstructured grids rather than favorable accuracy comparisons with staggered schemes. Turbulence models [107], their formulation [104] and implementation [174, 203], are beyond the scope of this paper.

Section 4 is devoted to algorithms for interface kinematics, i.e., those designed to track topologically complex interfaces on fixed grids. I focus on recent developments in volume tracking methods [157], but offer ample information about many other viable methods for tracking interfaces. Here a method for interface kinematics must model the advection (movement) of interfaces given some prescribed velocity. The method should keep the interface width compact (of order $h$, the grid spacing) as the interface undergoes arbitrary topological change. The current list of methods from which to choose is long and impressive: front tracking, boundary integral, moving-mesh, level sets, phase field, continuum advection, volume tracking, and particle-based methods. Not surprisingly, more detail is provided for the method (volume tracking) that has occupied a significant portion of my research efforts these past few years.

The accurate simulation of free surface flows also requires modeling physical processes specific to and localized at fluid interfaces. Obvious examples are surface tension and phase change. Section 5 details the immersed interface approach [110] for modeling interface dynamics (e.g., surface tension), which is a methodology dating back to the pioneering work of Peskin [139]. These ideas inspired the popular continuum surface force (CSF) model for surface tension [22] and its improved and enhanced variants. In immersed interface numerical methods for modeling surface tension, surface tension is modeled by applying its effects to fluid elements everywhere within numerically resolvable interface transition regions. Surface processes are replaced with volume processes whose integral effect properly reproduces the desired interface physics. This methodology is the underlying feature in the CSF method for surface tension. The CSF method, while having proven successful in a variety of studies [59, 113, 121, 149, 150, 165, 171], is still in need of improvements yielding increased accuracy (both absolute and convergence) and fidelity. These are briefly touched upon. Section 6 concludes with parting thoughts.

# 3　INCOMPRESSIBLE FLOW ALGORITHM

## 3.1　Overview

Los Alamos National Laboratory (LANL) has a long history in computational fluid dynamics (CFD). A particularly important contribution came in 1965 with the Marker-and-Cell (MAC) method [4, 62, 67] for incompressible multiphase flows, which is a method along with its successors that remains a popular choice to this day. This method enabled researchers to investigate a number of multiphase phenomena such as drop/splash dynamics and Rayleigh-Taylor instabilities [39, 40, 64–66, 68]. An example of a MAC successor is the SOLA-VOF method, which coupled the SOLA (solution algorithm) methodology with a volume tracking method for fluid interfaces [71]. A more recent example is the Ripple code that utilizes the CSF model discussed later in this paper [94, 95].

In the nearly thirty years since the inception of these algorithms, numerous advances have been made. These advances have in some cases greatly improved our ability to generate accurate solutions of incompressible interfacial flows. Examples are the recent coupling of high-order Godunov advection and interface tracking schemes with modern projection methods [144, 159]. Improvements in finding efficient solutions to linear systems of equation (often the most expensive portion of the solution algorithm) have also greatly impacted the range of flows for which solutions can be obtained.

After introducing a useful and widely applicable one-field model in Section 3.2, a brief "CFD 101" digression follows in Section 3.3. A semi-implicit projection method that employs high-order Godunov advection methods [3,10,13,101,152,158] is then described in Section 3.4. The method is second order accurate in both time and space, and does not suffer from cell Reynolds number stability restrictions. Next, in Section 3.5, the important issue of detecting and "filtering" non-solenoidal velocities is discussed. Filtering is highly recommended (and often required) for maintaining solution quality with colocated schemes, but it also appears to be beneficial for staggered schemes. The construction of higher-order, multi-dimensional monotonic continuum advection schemes is discussed in Section 3.6. This task is especially crucial for incompressible flows, which are multi-dimensional by nature. Next, in Section 3.7, the approximation of Newtonian viscous terms is discussed. This task might seem straightforward at first, but it is tricky when fluid interfaces aligned arbitrarily to the computational mesh are present. Solutions to the linear systems of equations arising from the projection and diffusion steps must be obtained, hence linear solution algorithms are briefly discussed in Section 3.8. Three approaches are highlighted: Krylov subspace methods, multi-grid (MG) methods, and hybrid methods that combine ideas and algorithms from the Krylov *and* MG schemes.

The resulting incompressible flow algorithm, as constructed, is able to model a wide range of flow regimes, including interfacial flows where density ratios across interfaces can be arbitrarily large (e.g., $\sim$800:1 for water:air). This feature is greatly facilitated by the use of a suitable interface tracking algorithm. In conjunction with approximate projection methods [3,101,152], low error formulations [153], and the filtering of non-solenoidal modes [100,101,154], the method remains robust for difficult problems. This method also merges quite naturally with physical models such as the CSF model for surface tension [22]. A very similar flow algorithm has recently been published by E. G. Puckett and coworkers [144].

## 3.2 A One-Field Model

Consider $k$ incompressible, immiscible Navier-Stokes fluids, each bounded from the other $k - 1$ fluids by a discernible interface. We wish to model the dynamics of this system while resolving interface length scales such as local curvature. With interface topologies resolved in this manner, we invoke a *one-field* model assumption, whereby each of the $k$ fluids is assumed to move with the local center-of-mass:

$$\mathbf{u}_k = \mathbf{u}, \tag{1}$$

where $\mathbf{u} = (u, v, w)^{\mathrm{T}}$ is the center-of-mass velocity field. A one-field model assumption is a good one when a representative volume element (RVE) of the system (i.e., a computational cell) does not contain *homogeneous mixtures* of two or more fluids. The RVEs are instead assumed to resolve the interface topology between each of the

$k$ fluids. Interfaces are allowed to snake arbitrarily through our fixed-grid computational domain, yet most of the cells ($> 80\%$) will contain just one fluid, and those cells possessing two or more fluids will in general be smaller than local radius of curvature of the interface separating the fluids. This is in contrast to multi-phase (the more apt description is *multi-field*) models, where the RVEs are of sufficient size to contain homogeneous mixtures, hence the system is best modeled with fluids that possess individual velocity fields [54]. With the one-field assumption in (1), $k$ mass and momentum equations are replaced by *one* mass and momentum conservation equation for the system center-of-mass.

Solutions sought are those to the incompressible NS equations for our one-field model. Conservation of mass is manifested as a solenoidal condition for the center-of-mass velocity field:

$$\nabla \cdot \mathbf{u} = 0, \tag{2}$$

or, equivalently, as a statement of Lagrangian invariance for each of the $k$th fluid densities:

$$\frac{\partial \rho_k}{\partial t} + \mathbf{u} \cdot \nabla \rho_k = 0. \tag{3}$$

Fluid densities $\rho_k$, equal to the mass $M_k$ of material $k$ per unit total volume $V$, can be expressed as the product of volume fractions $f_k$ and constant material densities $\rho_k^0$,

$$\rho_k = \frac{M_k}{V} = \left(\frac{V_k}{V}\right) \cdot \left(\frac{M_k}{V_k}\right) = f_k \rho_k^0. \tag{4}$$

The volume fractions $f_k$ are bounded by $0 \leq f_k \leq 1$, where

$$f_k = \begin{cases} 1, & \text{inside fluid } k\,; \\ > 0, < 1, & \text{at the fluid } k \text{ interface}\,; \\ 0, & \text{outside fluid } k\,. \end{cases} \tag{5}$$

Since fluid volumes are volume-filling, volume fractions must sum to unity, $\sum_k f_k = 1$, throughout the domain. Since $\rho_k^0$ is constant, substitution of (4) into (3) gives

$$\frac{\partial f_k}{\partial t} + \mathbf{u} \cdot \nabla f_k = 0. \tag{6}$$

The volume fractions in (5) delineate the presence (or absence) of each fluid, so $f_k$ serves as a Heaviside (characteristic) function for each material $k$. Equation (6) is therefore a very useful and important relation, as it gives an evolution equation for the location of each fluid. In this regard, any approximation for a material Heaviside function can be used in place of volume fractions $f_k$, e.g., a level set function in level set methods, an indicator function in front tracking methods, a phase field in phase field methods, or particle identities in particle-based methods. This will be discussed in more detail in Section 4.

Average density is recovered from the volume fractions by

$$\rho = \sum_k f_k \rho_k^0 , \tag{7}$$

which is simply an algebraic statement of local mass conservation. An expression for a fluid-averaged viscosity, on the other hand, is not constrained by a conservation law, hence is written as a general function $\Psi_k$ of the material viscosities, volume fractions, interface orientation (having normal $\hat{\mathbf{n}}_k$), and local fluid velocity:

$$\mu = \sum_k \Psi_k(f_k, \mu_k, \hat{\mathbf{n}}_k, \mathbf{u}) . \tag{8}$$

The functional form for $\Psi_k$ is derived from considerations of continuity of Newtonian stresses across the interface. The important issue of deriving a proper average viscosity is briefly touched upon in Section 3.7.

If we assume that all fluid $k$ pressures $p_k$ are equal (to $p$), invoke our one-field assumption in (1), and use averaging expressions (7) and (8), all fluid $k$ conservation of momentum equations can be summed to give one system conservation of momentum equation:

$$\frac{\partial \mathbf{u}}{\partial t} + \nabla \cdot (\mathbf{uu}) + \frac{1}{\rho}\nabla p = \frac{1}{\rho}\nabla \cdot \mu \left(\nabla \mathbf{u} + \nabla \mathbf{u}^T\right) + \mathbf{f} , \tag{9}$$

where $\mathbf{f}$ is an arbitrary body force, e.g., incorporating the effects of gravity or surface tension.

The system of equations for our one-field model consists of equations (2), (6), and (9) for the unknowns $\mathbf{u}$, $f_k$, and $p$. For a 2-D two-fluid model, we have four equations and four unknowns: $(u, v)$, $p$, and $f_1$. Densities $\rho$ and viscosities $\mu$ follow from $f_k$ using subsidiary equations (7) and (8). This system of equations is solved via the projection method described in the following sections.

## 3.3 Digression: CFD 101

Here we briefly digress into a few important CFD-related issues that must be addressed before any numerical solutions can be realized: the choice of grid topology, control volume and fluid variable positioning, and PDE discretization method. This discussion is admittedly too brief, so the reader should consult the numerous books and articles cited in the text for further information.

### 3.3.1 Partitioning the Physical Domain: Choosing a Grid Topology

Of the many different types of grids (or meshes) available and used today [188], most fall into one of three categories: block structured (BS), boundary-fitted (BF), and generalized-connectivity unstructured (GU) meshes [47]. For a given grid, two distinguishing features determine its categorization: the shape of each cell (element) and the cell-to-cell connectivity. A mesh is considered orthogonal if each of its cells has edges (faces) that meet at right angles (either in physical or logical space), and a mesh is considered "structured" if the number of cells surrounding each grid point is constant. Meshes with regular (structured) connectivities and orthogonal cells are considered the "simplest" for a number of reasons: mesh generation is trivial, algorithm implementation is straightforward, and discretization error can be quantified and most easily minimized. From this point of view, a BS, orthogonal mesh is the simplest and a GU mesh is the most complex. Cells in GU meshes do not in general have uniform shapes and they are typically connected together in arbitrary ways.

As the complexity of the mesh topology increases, it becomes more and more difficult to generate high fidelity (e.g., second order and higher) solutions without unreasonable investments in computational work and memory storage. Algorithms targeted for GU meshes often consume three to five times more computer memory and an order of magnitude more computational work than their BS, orthogonal mesh counterparts. So why use a GU mesh? Because solution fidelity is frequently dictated by the ability to resolve and partition the geometric complexities along the boundary of a domain, and these boundaries often can only be partitioned by GU meshes. As a rule of thumb, choose a grid topology that is the simplest possible, yet adequately partitions the computational domain.

### 3.3.2  Control Volume and Fluid Variable Positioning

Discrete PDE solutions are sought at each control volume, which is the smallest resolvable volume on the mesh. For a given discretization method, the position of each control volume (its geometric centroid) does not necessarily have to coincide with those of the cell volumes that are bounded by grid line intersection points (referred to as grid vertices or grid points). Furthermore, the location of discrete fluid variables (e.g., $\rho$, $\mathbf{u}$, and $p$) does not necessarily have to lie at cell volume centroids or grid points. Thus, the notion of "staggering" applies to both control volumes and fluid variables: a method is considered staggered if its control volumes do not coincide with cell volumes. In some cases, a method will have control volumes coincidental with cell volumes (nonstaggered), but the fluid variables might still be staggered, i.e., located at positions other than control volume centroids.

For the incompressible NS equations, four basic forms of control volume and fluid variable positioning are prevalent in the literature:

- *MAC positioning* [68]: mass control volumes (and associated variables) are positioned at cell centroids, whereas momentum control volumes (and associated variables) are positioned at cell face centroids;

- *ALE positioning* [69]: mass control volumes (and associated variables) are positioned at cell centroids, whereas momentum control volumes (and associated variables) are positioned at grid points;

- *colocated positioning* [159]: all control volumes and fluid variables are positioned at either cell centroids or cell grid points;

- *hybrid positioning* [10]: all control volumes are positioned at either cell centroids or cell grid points, but some variables (e.g., $p$) are staggered.

Unfortunately no one choice of control volume and fluid variable positioning is optimal for all meshes, i.e., disadvantages accompany each choice. For example, MAC positioning is most popular and effective on 2-D BS and BF meshes, but the need for multiple control volumes leads to increasingly complex implementation in three dimensions and on GU meshes. The use of only one control volume in colocated and hybrid positioning is an attractive implementation simplification that is often embraced for 3-D GU meshes. Is there any positioning choice that leads to consistently more accurate solutions? In general no, as the answer to this question is usually problem specific and mesh dependent. Interesting studies centered around this question can be found in references [43, 136, 172].

### 3.3.3  Discretization Method

We seek discrete, numerical solutions to the PDEs comprising our system of conservation equations. Consider the representative conservation equation for an arbitrary quantity $\phi$:

$$\frac{\partial \phi}{\partial t} + \boldsymbol{\nabla} \cdot (\mathbf{u}\phi) = S(\phi) , \tag{10}$$

where $S$ is a source term that is in general dependent upon $\phi$. Given a physical domain partitioned into a number of discrete volume elements ("cells") denoted by subscript $i$, we seek solutions to (10) for a discrete $\phi_i$ in each cell. Most discretization methods for PDEs in conservation law form fall loosely into one of the following categories: finite difference methods [151, 179], finite volume methods [47, 198], finite element methods [211], spectral methods [58], and particle-based methods [118].

Spectral methods typically express $\phi$ as a Fourier series, which easily enables high order approximations to spatial derivatives of $\phi$. Spectral methods, however, are most suited for uniformly-partitioned, periodic domains having structured meshes. Spectral methods can also encounter difficulties in resolving discontinuous data, where Gibbs phenomena can be problematic. Particle methods, e.g., smoothed particle hydrodynamics [120, 122, 123] and particle-in-cell [91] techniques, are quite powerful by virtue of their Lagrangian description of the fluid. They typically model the nonlinear advection terms with discrete particle motion, then rely upon finite difference, finite volume, or finite element discretization techniques for the remaining portions of the PDEs. In finite difference methods, discrete numerical derivatives for $\phi$ in space and time are formed with local differences of $\phi_i$. Finite difference methods are the most popular and perhaps most effective choice on orthogonal, structured meshes. Great care must be taken in forming the derivatives on generally nonorthogonal, complex topology (unstructured) meshes [168]. Finite element methods, arguably the most popular choice today for engineering analysis, are quite effective in forming discrete parabolic and elliptic PDEs on complex unstructured meshes. Only within the last decade, however, have accurate solutions to hyperbolic PDEs been attainable with finite element methods.

Finite volume methods are often assumed synonymous with finite difference methods, when in fact they are a "closer cousin" to finite element methods because discretizations are derived from integral (weak) forms of the PDEs. Finite volume methods are attractive because discretely conservative discretizations are inherent by design, unstructured meshes are a simple extension, hyperbolic systems are as easily discretized as other systems, and the basic form of the PDEs is not "lost" in a maze of interpolation coefficients and shape functions. Many researchers unfamiliar with finite volume methods assume such methods are amenable only for structured, orthogonal meshes, which is simply untrue.

The basic approach in finite volume methods is to begin with a volume integral form of the PDE, e.g., for (10) we have,

$$\frac{1}{\int dV} \int \left[\frac{\partial \phi}{\partial t} + \boldsymbol{\nabla} \cdot (\mathbf{u}\phi) = S\left(\phi\right)\right] dV\,, \tag{11}$$

which, using the Gauss divergence theorem, can be expressed as

$$\int \frac{\partial \phi}{\partial t} dV + \oint d\mathbf{A} \cdot (\mathbf{u}\phi) = \int S\left(\phi\right) dV\,, \tag{12}$$

where the volume integral is over some representative control volume (the cell) and the surface integral is over the surface bounding the volume. Finite element methods also begin with (11), except that the equation is additionally multiplied by some weight function. For finite volume methods, the integrals in (12) are approximated with discrete numerical quadratures, giving a discrete evolution equation for $\phi_i$:

$$\phi_i^{n+1} - \phi_i^n + \frac{1}{V_i} \sum_f \delta t \langle \mathbf{A} \cdot \mathbf{u}^{n+1/2}\rangle_f \langle \phi \rangle_f^{n+\frac{1}{2}} = \frac{\delta t}{V_i} S_i^{n+\frac{1}{2}}\,, \tag{13}$$

where the surface integral has been approximated as a sum over discrete control volume faces $f$ having an area vector $\mathbf{A}_f$, $V_i$ is the control volume, and $\phi_i$ is a local integrally-averaged value,

$$\phi_i = \frac{\int \phi dV}{\int dV} = \frac{1}{V_i} \int \phi dV\,, \tag{14}$$

and similarly for $S_i$. Note also in (13) that we have assumed a second order mid-point time integration scheme, but this choice is arbitrary. For spatially second order schemes, $\phi$ is assumed to vary linearly in space, hence $\phi_i$ is simply expressed as

$$\phi_i = \phi(\mathbf{x}_c)\,, \tag{15}$$

where $\mathbf{x}_c$ is the geometric centroid of the control volume. This is simply a one-point quadrature. For higher order (third order and above) schemes, the integral in (14) must be evaluated numerically with at least two quadrature points [198]. In a similar manner, second order schemes allow one-point quadratures for the face integrals, e.g., face quantities $\langle\phi\rangle_f$ are given at the geometric centroid of each face. For higher order schemes, face integrals of $\phi$ over the flux volume $\delta t\langle\mathbf{A}\cdot\mathbf{u}\rangle_f$ must be estimated with higher order quadrature rules. As pointed out clearly in [80], the truncation errors associated with performing these integrals have two sources, namely from discretely approximating the volume integral in (11) as a surface integral having discrete faces, and from the approximation of $\phi$ on each face.

Finite volume methods often reduce to finite difference methods on BS, orthogonal meshes. For example, substitution of suitable expressions in (13) for $\mathbf{A}_f$, $\mathbf{u}_f$, and $\langle\phi\rangle_f$ for a mesh that is uniform and orthogonal leads to a discrete equation that is often identical to that derived by finite differencing (10).

## 3.4   Projection Methods

In the projection method, solutions to the unsteady NS equations are obtained by first time-advancing the velocity field without regard for its solenoidality constraint in (2), then recovering the proper solenoidal velocity field, $\mathbf{u}^d$ ($\nabla\cdot\mathbf{u}^d = 0$). The means to this end is a projection operator, $\mathcal{P}$, which projects $\mathbf{u}^d$ out of $\mathbf{u}$:

$$\mathbf{u}^d = \mathcal{P}(\mathbf{u}) . \tag{16}$$

This projection is derived by drawing upon the Helmholtz-Hodge vector decomposition theorem [35], whereby the velocity $\mathbf{u}$ can be decomposed into a solenoidal vector, $\mathbf{u}^d$, and a curl-free vector, expressed as the gradient of a potential, $\nabla\varphi$. This decomposition is written

$$\rho\mathbf{u} = \rho\mathbf{u}^d + \nabla\varphi , \tag{17}$$

or

$$\mathbf{u} = \mathbf{u}^d + \sigma\nabla\varphi , \tag{18}$$

where $\sigma = 1/\rho$. Taking the divergence of (18) yields an elliptic equation for $\varphi$:

$$\nabla\cdot\mathbf{u} = \nabla\cdot\mathbf{u}^d + \nabla\cdot\sigma\nabla\varphi \rightarrow \nabla\cdot\mathbf{u} = \nabla\cdot\sigma\nabla\varphi . \tag{19}$$

Once the solution $\varphi$ is obtained, $\mathbf{u}^d$ results from the correction

$$\mathbf{u}^d = \mathbf{u} - \sigma\nabla\varphi . \tag{20}$$

For our purposes, $\varphi$ can be considered to be either the pressure $p$ or pressure increment $\delta p$, depending upon the nature of the projection.

Now let us consider a class of projection algorithms known as *approximate* projection methods. Approximate projection methods are extensions of Chorin's classic projection method [34], and its modernization by Bell and coworkers for solving the constant-density [10] and variable-density [13] incompressible NS equations. Approximate projection methods were first introduced by Almgren, Bell and Szymczak [3] (see also [100, 152, 153]), and were motivated primarily by the desire for robust projection methods on schemes utilizing colocated or hybrid positioning. Classical projection methods can often be termed as *exact* projections because the discrete projection operator behaves similarly to the analytic operator. In approximate projection methods, the discrete projection operator does not necessarily algebraically match the conditions for being a projection. In short, the discrete $\nabla\cdot$ and $\nabla$ operators used on the LHS of (19) and RHS of (20) are not necessarily consistent with the Laplacian $\nabla\cdot\sigma\nabla$ on the RHS of (19). Instead, the most straightforward means to discretize each operator ($\nabla\cdot$, $\nabla$ and $\nabla\cdot\sigma\nabla$) is chosen in approximate projection methods.

For our projection method, we wish to use a fractional-step [208] algorithm for advancing the velocity field $\mathbf{u}$ in (9) at a cell $i$ in a colocated scheme (i.e., $\mathbf{u}$ is positioned at the cell $i$ centroid). The first step in the algorithm consists of a *predictor* step, in which the solenoidal nature of $\mathbf{u}$ is ignored, followed by a *corrector* step, in which a projection recovers the solenoidal part of $\mathbf{u}$. For the predictor step, a known (time level $n$) velocity in cell $i$, $\mathbf{u}_i^n$, is advanced in time to level $*, n+1$ by discretizing (9):

$$\frac{\mathbf{u}_i^{*,n+1}}{\delta t} = \frac{\mathbf{u}_i^n}{\delta t} - (\mathbf{u}\cdot\nabla\mathbf{u})_i^{n+\frac{1}{2}} + \sigma_i^{n+\frac{1}{2}}\mathbf{G}_i\phi^{n-\frac{1}{2}} - \frac{\nu\sigma_i^{n+\frac{1}{2}}}{2}L_i\left(\mathbf{u}^n + \mathbf{u}^{*,n+1}\right) - \mathbf{f}_i^{n+\frac{1}{2}}, \quad (21)$$

where $\mathbf{G}_i$ and $L_i$ are discrete gradients and Laplacians, respectively, and a second-order time discretization has been used. Not that the viscous stresses have been treated above in an implicit Crank-Nicholson fashion, but this choice is arbitrary. Fully implicit or fully explicit treatments are also valid. The $(\mathbf{u}\cdot\nabla\mathbf{u})$ advection term in (21) is discretized with an unsplit high-order Godunov method [10, 36, 152] that is derived from Taylor-Series (TS) expansions in space (to cell face $i+\frac{1}{2}$) and time (to level $n+\frac{1}{2}$). This will be visited briefly in Section 3.6. Further details on the discretization of (21) can be found in [13, 144].

Several variations of the projection implied by (21) are possible. By removing the gradient of pressure from (21), $\varphi$ in (19) is actually a pressure rather than an increment in pressure. The form of $\mathbf{u}$ in the discrete divergence on the LHS of (19) can be chosen several ways, e.g., the advanced-time predictor velocity, $\mathbf{u}^{*,n+1}$, or the predicted change, $(\mathbf{u}^{*,n+1} - \mathbf{u}^n)$. One might assume these differences to be higher order effects, but experience has shown otherwise for both exact and approximate projections [152, 159].

An exact projection results when the Laplacian operator ($L$) on the RHS of (19) is derived from the discrete discrete divergence ($D$) and gradient ($\mathbf{G}$) operators: $L = D\cdot\sigma\mathbf{G}$. The discrete velocity divergence in an exact projection is zero to within the convergence tolerance of the solution to (19). Exact projections, however, have been found to have some practical difficulties [3, 100, 152]. Numerical instabilities sometimes tend to grow, originating from a pressure/velocity decoupling that is seeded by strong localized source terms, such as those brought about by chemical reactions, surface tension, or abrupt density jumps across interfaces. Additionally, this local decoupling renders multigrid techniques cumbersome [75], and hampers the implementation of adaptive grid techniques [2, 75]. Approximate projections, then, were motivated and introduced in [3] to address these problems. Very simply, the operator $L$ in an approximate projection is *not* derived from a $D \cdot G$ pair, but rather from a straightforward discretization of the continuous Laplacian operator. The discrete velocity divergence in an approximate projection is not zero, but is rather a function of the truncation error. The operators $D$ and $G$ have the same form as the exact projection, but the Laplacian is modified.

### 3.4.1 Robust Projection Methods

In approximate projections, the velocity divergence is not constrained to be zero (but rather to some small tolerance), hence one must take care to insure these nonzero velocity divergences remain bounded. The principal problem with approximate projections is the presence and growth of null spaces in the discrete operator $D \cdot \mathbf{u}$. *Null spaces* are those spatial modes in $\mathbf{u}$ that are not solenoidal, $\nabla \cdot \mathbf{u} \neq 0$, yet the discrete operator $D$ does not "see" them, returning $D \cdot \mathbf{u} = 0$. The growth of these $\mathbf{u}$ null spaces is typically manifested as high-frequency noise. This is currently controlled by identifying and filtering unphysical velocity modes [102, 153] or by carefully formulating the form of the approximate projection [154]. The filtering and damping of these unphysical modes can act on the flow either through an explicit addition of a high-order viscosity or through the use of an iterated projection derived from a discrete stencil that differs from the approximate projection stencil. These methods are most effective when used in concert [154]. Without these steps, approximate projection methods are prone to failure on more difficult problems, e.g., flows with interfaces across which density ratios are high.

The formulation of the projection directly affects the time evolution of the discrete divergence. If the divergence on the LHS of (19) is the difference between the predicted and old time velocity,

$$\nabla \cdot \left( \mathbf{u}^{*,n+1} - \mathbf{u}^n \right) , \tag{22}$$

then the discrete divergence errors *accumulate* in time. On the other hand, if a predictor velocity is used,

$$\nabla \cdot \mathbf{u}^{*,n+1} , \tag{23}$$

than the discrete divergence errors *do not accumulate* in time, which is preferable. The formulation of the Poisson pressure equation can also have a profound impact on the solutions' quality. I have found the following formulation to be robust and accurate:

$$\nabla \cdot \frac{1}{\rho^{n+\frac{1}{2}}} \boldsymbol{\nabla} \phi = \frac{1}{\delta t} \nabla \cdot \mathbf{u}^{*,n+1} , \tag{24}$$

where $\phi$ is the *total* pressure and $\mathbf{u}^{*,n+1}$ is the predicted velocity given by an implicit solution of (21):

$$\mathbf{u}^{*,n+1} = \mathbf{A}^{-1} \left[ \mathbf{u}^n - \delta t \nabla \cdot (\mathbf{uu})^{n+\frac{1}{2}} - \frac{\delta t}{\rho^{n+\frac{1}{2}}} \boldsymbol{\nabla} \phi^{n-\frac{1}{2}} + \delta t \mathbf{f}^{n+\frac{1}{2}} \right.$$

$$+ \frac{1}{2} \frac{\delta t}{\rho^{n+\frac{1}{2}}} \nabla \cdot \mu^{n+\frac{1}{2}} \left( \nabla \mathbf{u} + \nabla \mathbf{u}^T \right)^n \Bigg] + \frac{\delta t}{\rho^{n+\frac{1}{2}}} \nabla \phi^{n-\frac{1}{2}} , \qquad (25)$$

where

$$\mathbf{A} = \mathbf{I} - \frac{\delta t}{2} \frac{1}{\rho^{n+\frac{1}{2}}} \nabla \cdot \mu^{n+\frac{1}{2}} \left( \nabla + \nabla^T \right) . \qquad (26)$$

Finding $\mathbf{u}^{*,n+1}$ in (25) requires simultaneous solutions a linear system of equations because of the implicit treatment of the viscous stress. An explicit treatment, however, is also allowed, and this approach yields a simple algebraic expression for $\mathbf{u}^{*,n+1}$. The projection in (24) based on total pressure and absolute predicted velocity controls the presence of the discrete divergence without sacrificing the accuracy of the method. A full exposition on this subject is given in [152, 153, 159].

## 3.5   Non-Solenoidal Velocities: Detection and Filtering

In approximation projection methods the velocity divergence is nonzero, being of the same order of the spatial truncation errors, e.g., $\nabla \cdot \mathbf{u} \sim \mathcal{O}(h^2)$ is expected for a second order discretization. These errors can (and should) be used as a measure of the solution's quality: if $\nabla \cdot \mathbf{u} \sim \mathcal{O}(h^2)$ is the expected error, then the discrete divergence $D \cdot \mathbf{u}$ should exhibit the same errors, and, more importantly, these errors should not grow. To prevent growth, the null space of the discrete divergence $D \cdot \mathbf{u}$ must be quantified and reduced to manageable levels. For example, if $D \cdot \mathbf{u}$ has a central difference form [152], non-solenoidal modes in $\mathbf{u}$ not detected by a central difference operator can persist if they are unaffected by the projection (which is often the case). These modes add to the already nonzero discrete divergence, and can contribute to a net growth in error. For this reason, when using approximate projections, it is important to control the growth of divergence errors. This can be accomplished if non-solenoidal velocity modes can be reliably detected and "filtered" (removed) [154].

Reference [159] accentuates the striking differences that can arise in non-filtered solutions of high density ratio interfacial flows using both standard exact and approximate projection methods. Without filtering, both projection solutions can exhibit spurious velocity field features such as decoupling, noise, or asymmetry. Despite the use of a smaller time step in integrating the flow, approximate projection solutions for such flows often remain compromised. Solutions can be improved if the projection in (19) projects the absolute predictor velocity $\mathbf{u}^{*,n+1}$ (rather than the velocity difference) while obtaining solutions for the total pressure (rather than the pressure increment). This is especially true for exact projection solutions, whereas approximate projection solutions often require additional velocity filters before equivalent solution quality is obtained [159].

A number of filters have been devised to remove or damp spurious non-solenoidal velocity modes from the flow field solution. Two types of filters have appeared in recent publications [152, 154]: a projection filter that is derived from a solution to (19) with a RHS that is typically a vertex- or face-based velocity divergence; and a velocity filter that detects and damps divergent modes based upon physical arguments. The projection filter follows the same prescription as a regular projection, i.e., a solution $\varphi$ to (19) is used to "correct" the velocity field of non-solenoidal modes. Fortunately, however, a projection filter potential $\tilde{\varphi}$ that is only approximate (e.g., resulting from a single iterative Jacobi sweep) is often adequate in damping error modes [154]. Velocity filters, on the other hand, attempt to recognize and heuristically remove erroneous divergent modes via the application of a fourth-order damping term. Both filters are typically applied every computational cycle.

If divergent velocity modes are ignored, especially for approximate projections, solution accuracy and stability can deteriorate. This is especially true in flows experiencing large density jumps, large local source terms (such as surface tension), and long integrations (e.g., to steady state). Can exact projection methods constructed from the popular MAC positioning of fluid variables and control volumes allow divergent velocity modes to seed, persist, or even grow? In short, yes, if the discrete divergence operator has a null space. It is advisable, therefore, to either minimize the discrete divergence null space or detect and filter divergence modes, regardless of the projection scheme used. The interested reader should consult reference [101] for a general discussion of filters and [100, 102] for additional examples of the application of filters specific to incompressible reactive flows.

## 3.6   Advection

Advection methods here refer to those algorithms that are devised for accurate numerical solutions to the simple hyperbolic equation

$$\frac{\partial \phi}{\partial t} + \mathbf{u} \cdot \boldsymbol{\nabla} \phi = 0 , \tag{27}$$

for some arbitrary quantify $\phi$, which we can rewrite in conservative form as

$$\frac{\partial \phi}{\partial t} + \nabla \cdot (\mathbf{u}\phi) = \phi \nabla \cdot \mathbf{u} . \tag{28}$$

Following (13), a finite volume discretization of this equation can be expressed as

$$\phi_i^{n+1} - \phi_i^n + \frac{1}{V_i} \sum_f \delta t \langle \mathbf{A} \cdot \mathbf{u}^{n+1/2} \rangle_f \langle \phi \rangle_f^{n+\frac{1}{2}} = \frac{\phi_i^{n+\frac{1}{2}}}{V_i} \sum_f \delta t \langle \mathbf{A} \cdot \mathbf{u}^{n+1/2} \rangle_f , \tag{29}$$

where a second order (midpoint) time integration has been used. Spatial accuracy of the discretization in (29) is dictated by the estimation of $\langle \phi \rangle_f$. For example, a second order Taylor series (TS) expansion in space and time gives

$$\phi(i)_{i+\frac{1}{2}}^{n+\frac{1}{2}} = \phi_i^n + \frac{\delta t}{2} \left( \frac{\partial \phi}{\partial t} \right)_i^n + (\mathbf{x}_{i+\frac{1}{2}} - \mathbf{x}_i) \cdot (\boldsymbol{\nabla}_{\mathrm{L}} \phi)_i^n$$

$$= \phi_i^n + \left[ \alpha_{\mathrm{L}}(\mathbf{x}_{i+\frac{1}{2}} - \mathbf{x}_i) - \frac{\delta t}{2} \mathbf{u}^n \right] \cdot (\boldsymbol{\nabla} \phi)_i^n , \tag{30}$$

where $i + \frac{1}{2}$ is assumed to reside at a cell face $f$, $(i)$ indicates the expansion originates from cell $i$, and $\nabla_\mathrm{L}$ denotes a *limited gradient* [108, 194] needed to insure that the spatial portion of the TS expansion does not introduce new extrema for $\phi$ (relative to centroid data). The limited gradient $\nabla_\mathrm{L}$ in (30) has been expressed as the product $\alpha_\mathrm{L} \nabla$, where $\alpha_\mathrm{L}$ is a scalar limiter, and the actual PDE for $\phi$, equation (27), has been substituted for the temporal derivative. If monotonicity considerations are applied as constraints during construction of $\nabla_\mathrm{L}$, then $\alpha_\mathrm{L}$ can in general be a vector [97]. Also, if $\nabla_\mathrm{L}$ is constructed in an upwind fashion, it has been shown that larger linearly-stable time steps can be used [36,144]. Multiple TS expansions to a given cell face are possible, e.g., one from each neighboring cell centroid, but the two chosen are those from the two nearest centroids (i.e., cells $i$ and $i + 1$). With two expansions performed for $\phi$, a single face value is determined by choosing the TS expansion for $\phi$ that originates from the upwind centroid.

This basic principles of this advection method are similar to the formulation of Colella [36], and have been well established in earlier works [12, 195]. The advection scheme embodied in (29) and (30) is considered to be a multi-dimensional *unsplit* algorithm, i.e., a full multi-dimensional solution is updated in a single time step. It is multi-dimensional because *all* control volume faces in (29) are summed over and *all* $\nabla$ components in (30) are taken into account in one step. *Split* schemes, on the other hand, construct the multi-dimensional solution as a series of sequential, one-dimensional (1-D) solutions to (29) and (30). A 1-D solution is arrived at for a given direction (e.g., $x$) by considering only one $\nabla$ component (e.g., $\partial_x$) in (30) and only those faces in (29) whose unit normals are (anti)parallel to the direction of concern. For example, in a split scheme an $x$-direction solution is first obtained, evolving $\phi^n$ to an intermediate value $\phi^*$, then a $y$-direction solution follows, which brings the intermediate $\phi^*$ to the final $\phi^{n+1}$. This directional "sweeping" process is not necessarily less accurate than a single-step (unsplit) scheme, but it does induce numerical asymmetries that can often compromise solution quality [157]. The sweeping process is also virtually impossible on nonorthogonal or GU meshes that are not generally aligned with any particular coordinate direction.

### 3.6.1  Face Flux Velocities

Unfortunately all of the fluid variable/control volume positioning options detailed in Section 3.3.2 suffer from an identical problem: the velocity $\mathbf{u}$ is not always positioned on the mesh where it is needed. For advection schemes, the flux velocities needed in (29) reside at (or near) control volume faces. When momentum is advected, the control volume faces surround a centroid where the velocity $\mathbf{u}$ resides, hence face flux velocities $\mathbf{u}_f$ are not known, and therefore must be constructed. In most MAC schemes, $\mathbf{u}_f$ is usually constructed from simple averages of nearby $\mathbf{u}$ data [95], yet this procedure can be first-order accurate, and, worse yet, it is not guaranteed to satisfy the $\nabla \cdot \mathbf{u}_f = 0$ condition. We present below an elegant procedure for constructing face flux velocities first introduced by Bell and coworkers for schemes based on colocated positioning [10,

13, 144]. This approach has since been adopted successfully for a scheme based on MAC (staggered) positioning [185].

Here we will assume a scheme using colocated positioning, with velocities $\mathbf{u}_i$ residing at each cell centroid. In a manner similar to the $\phi$ TS expansion in (30), face flux velocities $\mathbf{u}_f$ are derived from a TS expansion in space (to cell face $i + \frac{1}{2}$) and time (to level $n + \frac{1}{2}$):

$$\mathbf{u}(i)_{i+\frac{1}{2}}^{n+\frac{1}{2}} = \mathbf{u}_i^n + \frac{\delta t}{2}\left(\frac{\partial \mathbf{u}}{\partial t}\right)_i^n + (\mathbf{x}_{i+\frac{1}{2}} - \mathbf{x}_i) \cdot (\boldsymbol{\nabla}_{\mathrm{L}}\mathbf{u})_i^n , \tag{31}$$

For velocities, however, the appropriate PDE to substitute for the $\mathbf{u}$ time derivative is the *full momentum equation* (9) at time level $n$. So, in effect, we are "paying a price" here for a time-centered flux velocity: solutions to an additional momentum equation must be obtained at each face. (See references [10, 13, 144, 152] for further details.) Since TS expansions are performed from the two closest cell centroids (cells $i$ and $i+1$) to cell face $i + \frac{1}{2}$, the two candidate values for the flux velocity must be resolved with the solution of a Riemann problem:

$$\mathbf{u}_{i+\frac{1}{2}}^{n+\frac{1}{2}} = R\left[\mathbf{u}(i+1)_{i+\frac{1}{2}}^{n+\frac{1}{2}}, \mathbf{u}(i)_{i+\frac{1}{2}}^{n+\frac{1}{2}}\right] , \tag{32}$$

where $\mathcal{R}$ is our general Riemann function. Following Bell and coworkers [10, 13, 144], the functional form of $\mathcal{R}$ is given from solutions to an inviscid Burger's equation. The function $\mathcal{R}$ is usually quite simple and intuitive, returning the TS expansion for $\mathbf{u}$ that originates from the upwind centroid.

As an interesting aside, why is this TS expansion approach for face velocities so invaluable? It obviously provides a rigorous, formal procedure for obtaining the time-centered, second-order fluxing velocities needed for advection, but is that it? No; it also provides the time-advanced *pressure-velocity coupling* so crucially needed for colocated schemes to be reliable and robust. As in the original Rhie-Chow prescription [148] for estimating face velocities from local centroid velocities, the important pressure gradient coupling terms can be found in the $\mathbf{u}$ temporal derivative term of the TS expansion. In fact, one could easily argue that the Rhie-Chow face velocity prescription represents a simplification of the temporal and spatial TS expansion in (31), i.e., the Rhie-Chow prescription simplifies the spatial derivative term (to an average) and only the uses the pressure gradient portion of the temporal derivative (the momentum equation).

An important point must be made about finding face flux velocities $\mathbf{u}_f$ with the aforementioned TS expansion: the solenoidal condition $\nabla \cdot \mathbf{u}_f = 0$ will *not* necessarily be satisfied, even if the initial cell centroid velocities $\mathbf{u}$ were solenoidal. Solenoidality can (and in general will) be broken while obtaining $\mathbf{u}_f$ from (31) and (32) because of the nonlinearities inherent in the gradient limiting procedure and the Riemann solution. One must therefore additionally apply a "MAC" projection [11,100,144] to $\mathbf{u}_f$, whereby a potential field that projects out the solenoidal part of $\mathbf{u}_f$ is found via a solution to (19). For example, consider the 2-D vortical (yet solenoidal) velocity field given in Figure 1(a). When (31) and (32) are used to construct $\mathbf{u}_f$, nonlinearities induced by

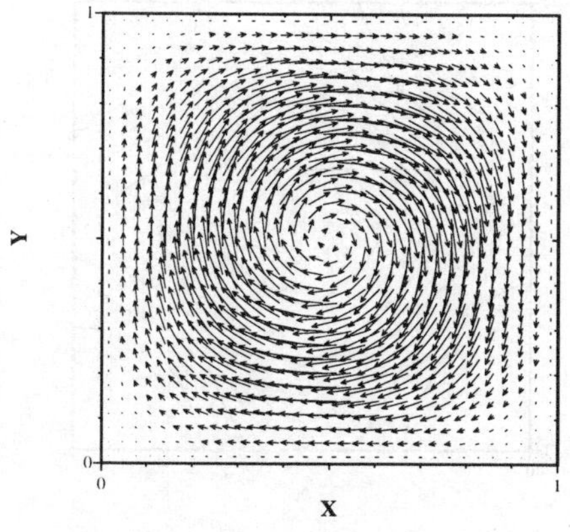

(a) Vortex velocity field, as prescribed in [10].

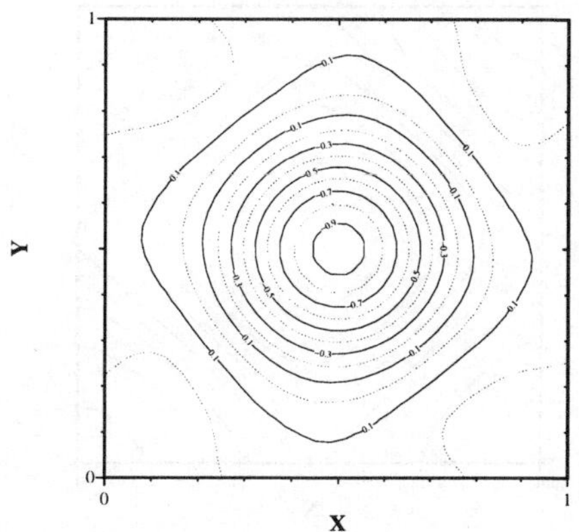

(b) Pressure field corresponding to the vortex.

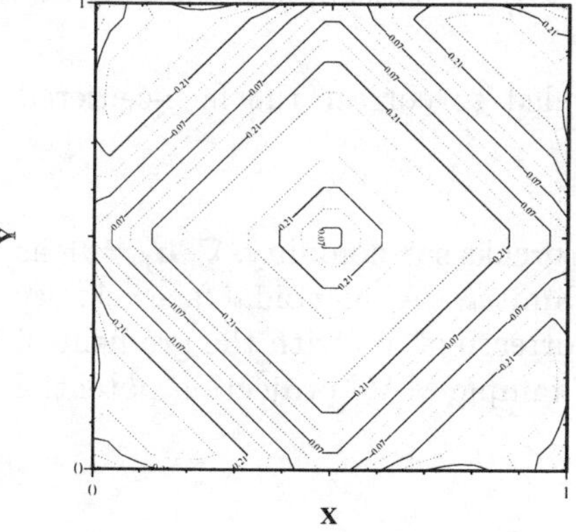

(c) Divergence of face-centered velocities resulting from application of the Barth limiter [9] for $\nabla_L$ in (31).

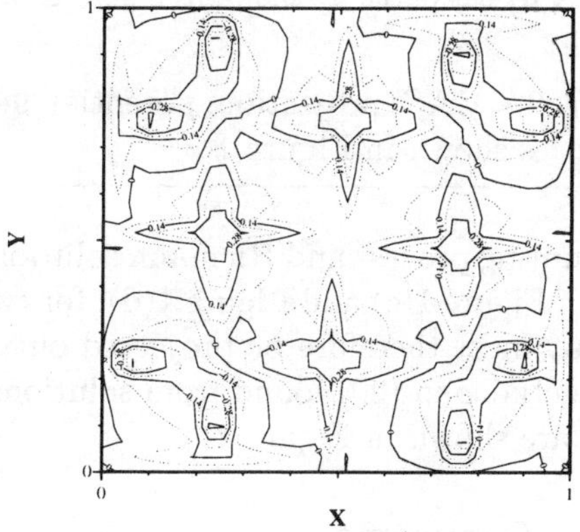

(d) Divergence of face-centered velocities resulting from application of the Venkat limiter [197] for $\nabla_L$ in (31).

Figure 1: Face-centered velocities constructed from the solenoidal cell-centered velocity field in (a) using equations (31) and (32) are not necessarily solenoidal, as shown in (c) and (d) for two different types of slope limiters.

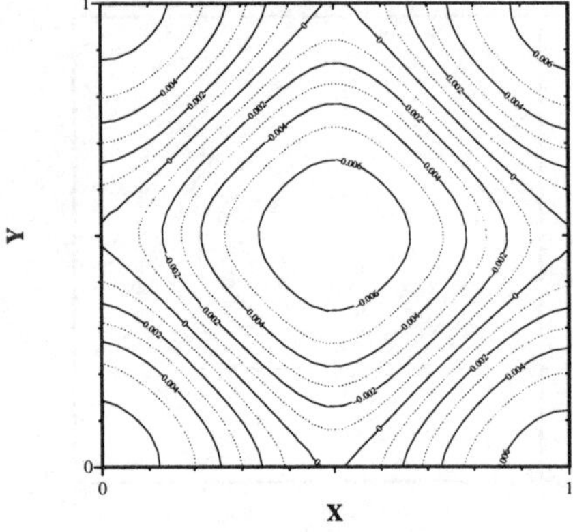

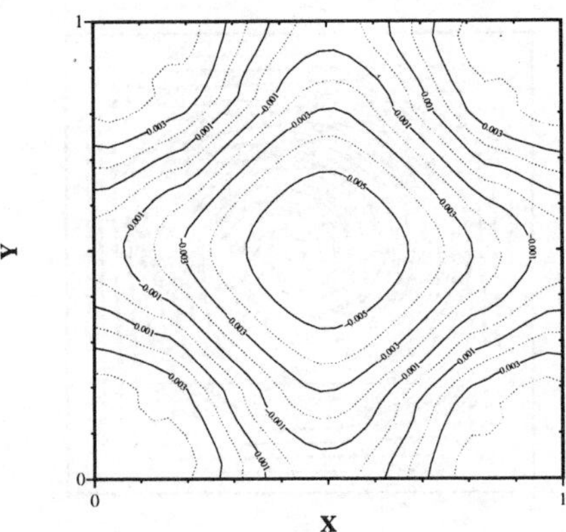

(a) MAC projection potential field $\varphi$ needed to project $\mathbf{u}_f^d$ out of the $\mathbf{u}_f$ shown in Figure 1(c).

(b) MAC projection potential field $\varphi$ needed to project $\mathbf{u}_f^d$ out of the $\mathbf{u}_f$ shown in Figure 1(d).

Figure 2: MAC projection potential fields are needed to correct the face-centered velocities shown in Figure 1.

the limiting process and Riemann solution procedure breaks solenoidality, $\nabla \cdot \mathbf{u}_f \neq 0$, as seen in Figure 1(c) and Figure 1(d), for two different limiters. A solenoidal face velocity field $\mathbf{u}_f^d$ must therefore be projected out of $\mathbf{u}_f$ by "correcting" $\mathbf{u}_f$ with the gradient of a potential $\varphi$ in (20) found from solutions to (19). Example MAC projection potential fields are shown in Figure 2.

### 3.6.2  Comments

The application of unsplit high-order Godunov advection schemes has been common-place for high speed flow algorithms over the last two decades, yet it was not until 1989 that such schemes were first applied to unsteady incompressible flows [10]. The dividends have in many cases been quite evident; see the examples available in references [3,10,11,13,36,152]. The central theme of the method is the reliance upon second order, multi-dimensional TS expansions in space *and* time for the dependent variables. These variables are constructed at cell faces, where they will be multi-valued, and then the upwind values are propagated in the solution. The complete, governing PDE is used to approximate the time derivatives. This approach exhibits excellent phase error properties and compares favorably with higher order Runge-Kutta methods.

The upwinding protects this method from stability restrictions related to the Reynolds number, thus it can be used for both viscous and inviscid flows. Results show that the method can resolve flows up to a grid Reynolds number $(U\rho h/\mu)$ of forty, beyond which solutions will degrade in a graceful fashion and produce physical, but not necessarily accurate, results [26]. The advection method is also robust in cases where the flow contains discontinuities and shear. For further information about projection method solutions of the type discussed here, see the recent excellent articles by Minion and coworkers, where the nature of under-resolved projection solutions is discussed [26] and stability-enhancing improvements are suggested [116]. Interesting under-resolved performance comparisons with many other incompressible NS solution methods are also presented [117].

## 3.7   Viscous Stresses

Assuming viscous stresses in the flow can be approximated as Newtonian, a finite volume discretization of viscous forces follows directly from the divergence of the viscous stress tensor:

$$\boldsymbol{\nabla} \cdot \boldsymbol{\tau} = \boldsymbol{\nabla} \cdot \mu \left( \boldsymbol{\nabla} \mathbf{u} + \boldsymbol{\nabla} \mathbf{u}^T \right) = \frac{1}{V} \oint \mu \left( \boldsymbol{\nabla} \mathbf{u} + \boldsymbol{\nabla} \mathbf{u}^T \right) \cdot d\mathbf{A} , \tag{33}$$

where the surface integral is approximated as a discrete sum over faces:

$$\boldsymbol{\nabla} \cdot \boldsymbol{\tau} \approx \frac{1}{V} \sum_f \mu_f \left( \boldsymbol{\nabla} \mathbf{u} + \boldsymbol{\nabla} \mathbf{u}^T \right)_f \cdot \mathbf{A}_f . \tag{34}$$

With this discretization, a viscous force estimate follows directly from approximations of velocity gradients and dynamic viscosities at control volume faces. These approximations are straightforward except in the proximity of fluid interfaces possessing abrupt property variations. In this case, face-centered operators for $\boldsymbol{\nabla}\mathbf{u}$ remain easy in a one-field model (where all $\mathbf{u}_k = \mathbf{u}$). The difficulty, however, arises in approximating the dynamic viscosity $\mu$ along those control volume faces that intersect or are coincidental with a fluid interface. In these situations, most (including myself) have "homogenized" the faces bounding these multi-material cells by invoking a simple geometric average for the viscosity:

$$\mu^s = \sum_k f_k \mu_k , \tag{35}$$

where the superscript "s" denotes a *serial* average, a term common in circuit analysis. Conversely, one might also use a *parallel* average:

$$\mu^p = \left( \sum_k \frac{f_k}{\mu_k} \right)^{-1} . \tag{36}$$

The parallel (harmonic) mean places more importance on the fluid with the smaller viscosity, whereas the serial (geometric) mean places more importance on the fluid with the higher viscosity. This choice of averaging is not significant when relative differences between fluid $\mu_k$ are small (e.g., $< 100\%$), but this is generally not true for free surface flows. Consider, for example, a simple two-fluid example: if $\mu_1 = 1$, $\mu_2 = 100$, and $f_1 = f_2 = 0.5$, then $\mu^s = 50.50$ and $\mu^p = 1.98$. Such large differences between $\mu^s$ and $\mu^p$ are to be expected in free surface flows, where viscosity ratios across the interface being two (or more) orders of magnitude are commonplace. Which one, then, is the correct expression? Since viscous fluxes at the interface must be continuous (as viewed from either side of the interface), one can use this condition to derive the correct value for $\mu_i$, the interface viscosity, which *is* $\mu_p$. This is completely analogous to the derivation of an appropriate interface thermal conductivity by invoking continuity of heat fluxes at the interface [135]. In effect, one cannot replace the average of a product $\mu(\nabla\mathbf{u}+\nabla\mathbf{u}^T)$ by a product of averages.

The importance of using the correct interface viscosity was recently recognized in [38] and validated in [190] for simple shear flows. In particular, it was pointed out in [38] that the stress tensor $\boldsymbol{\tau}$ will be continuous at the interface, even when some of its components will *not* be continuous. To see this, in the proximity of an interface with normal $\hat{\mathbf{n}}_k$, the velocity $\mathbf{u}$ can be decomposed into normal $\mathbf{u}_n = (\mathbf{u} \cdot \hat{\mathbf{n}}_k)\hat{\mathbf{n}}_k$ and tangential $\mathbf{u}_\tau = \mathbf{u} - \mathbf{u}_n$ components, and likewise for the gradient, $\nabla = \nabla_n + \nabla_\tau$. With these definitions it is easy to isolate the potentially discontinuous "interface shear" component of $\boldsymbol{\tau}$, namely $\nabla_n\mathbf{u}_\tau$. With this decomposition of $\boldsymbol{\tau}$, one might use the parallel viscosity $\mu^p$ for the shear component $\nabla_n\mathbf{u}_\tau$ and $\mu^s$ for all other components. This selective averaging was used in [38], where it generated improved results.

The difficulty lies in formulating an appropriate average viscosity expression that is suitable for arbitrary orientations of the flow field relative to the interface, and for the interface relative to control volume faces. First, it is important to realize that average viscosity values are needed at control volume faces, hence the volume fractions $f_k$ in (35) and (36) are those associated with *flux volumes*, not control volumes. These fractions are therefore the fractional volume of a material $k$ passing through a given control volume face over a given time step. Faces that do not intersect an interface, for example, should have a pure material value for the viscosity $\mu_f$. Second, if an interface does intersect a given face, then the relative interface/face orientation, expressed as a fraction $\eta$,

$$\eta_f = |\hat{\mathbf{n}}_k \cdot \hat{\mathbf{n}}_f|, \tag{37}$$

can be used to define a face viscosity that varies smoothly between the serial and parallel averages:

$$\mu_f = \eta_f\mu^s + (1 - \eta_f)\mu^p. \tag{38}$$

Use of this average viscosity expression, coupled with isolation of the interface shear component of $\boldsymbol{\tau}$, might provide an avenue for improved viscous stress modeling at fluid interfaces, but further investigations are needed. As a final note, Rudman [163] also recently called attention to the problem of estimating an average $\mu$ at the interface, noting that $\mu^p$ was found to give better results in practice. The larger serial average $\mu^s$ actually caused *increased*, unphysical acceleration of fluid elements in the lighter fluid near the interface. High velocities were generated in the lighter fluid because the viscous acceleration term $(\boldsymbol{\nabla} \cdot \boldsymbol{\tau})/\rho$ was amplified by a larger $\boldsymbol{\tau}$ (with $\mu^s$) in regions of small $\rho$. These effects need to be further investigated and quantified.

## 3.8 Numerical Linear Algebra

An incompressible flow algorithm constructed with the fractional-step projection method discussed in Section 3.4 requires solutions to as many as three linear systems of equations per computational time step. First, if the advection scheme in Section 3.6 is used, then solutions to a MAC projection equation, given by (19) where $\varphi$ is a potential field, must be found. Second, if an implicit treatment of viscous terms is desired, then solutions to the predictor velocity equation (25) must be obtained. Third, solutions to a pressure projection equation, also given by (19) where $\varphi$ is the pressure, are required. All three of these equations are basically elliptic in nature, and can be expressed in matrix notation as

$$A\mathbf{x} = \mathbf{b}, \tag{39}$$

where $A$ is a matrix resulting from the discretization, $\mathbf{x}$ is the solution vector, and $\mathbf{b}$ is a vector source term. Since for our equations the matrix $A$ arises from finite volume discretizations of the Laplacian, we expect $A$ to be sparse, positive definite ($\mathbf{x}^T A \mathbf{x} > 0$), and in general symmetric, hence our solution methods should take advantage of this structure. The total computational effort of our fractional-step scheme will be dominated by the effort required to find solutions to (39), therefore designing an efficient and scalable method for solving these systems of linear equations is of paramount importance.

Which method for the solution of (39) is recommended? There are several metrics one must take into account when considering a solution method for linear systems of equations: robustness, or ability to converge; efficiency, or convergence rate (if the method is iterative); scalability of the computational effort (relative to the number of unknowns $N$) required to find a solution; and complexity of implementation. Ideally, we desire a method that *always* converges (provided our equations are well-posed), that requires computational effort scaling linearly with $N$, and that exhibits *grid-independent iterations to convergence*. This last requirement may be restated as requiring our method to converge to a solution for a given physical domain in an iteration count that does not change with the number of grid points used to partition the domain. Of the possible solution methods briefly mentioned here, including direct and stationary iterative methods, Krylov subspace methods, multigrid methods, and hybrid methods, only the multigrid and hybrid methods have shown promise in meeting all of our requirements.

### 3.8.1 Direct and Stationary Iterative Methods

Since $A$ for our equations is not dense, but rather sparse and usually diagonally-dominant, direct solution methods such as Gaussian elimination and Cholesky or LU factorization are not recommended. Because of the non-constant coefficients of the operators, FFT or cyclic reduction methods are also not effective. These methods require computational effort that scale like $N^3$, which can improve to $N^2$ if the solution method takes advantage of the sparsity of $A$, but this scaling is not the linear scaling we desire. The primary attractiveness of direct solvers is their robustness, or ability to find a solution for any nonsingular $A$, e.g., especially when $A$ has a high condition number (ratio of maximum to minimum eigenvalues) arising from small mesh spacings or high density ratio flows. Stationary iterative methods, such as Jacobi, Gauss-Seidel, and symmetric successive over-relaxation (SSOR) [196], are attractive because of their ease of implementation, but they exhibit poor scaling, requiring computational effort that scales like $N^2$, and they can frequently exhibit a lack of robustness (inability or slowness to converge). While these stationary iterative methods are not recommended for finding solutions to (39), they do remain quite useful as preconditioners in Krylov subspace methods or as smoothers in the multigrid method. Further information about direct and stationary iterative methods can be found in a host of classic textbooks. I have found references [57, 177, 178] to be particularly useful.

### 3.8.2   Krylov Subspace Methods

Krylov subspace methods [57, 164] are iterative methods in which solutions to (39) are extracted from a subspace by imposing constraints on the residual vector $\mathbf{b} - A\mathbf{x}$, typically that it be orthogonal to $m$ linearly independent vectors in the subspace. The most popular and widely used Krylov subspace methods are the conjugate gradient (CG) algorithm if our positive definite matrix $A$ is symmetric or generalized minimal residual (GMRES) algorithm if $A$ is not symmetric. These methods are in general robust, with the CG method theoretically guaranteed to converge in $N$ iterations, and GMRES usually able to converge if $A$ does not have an excessive condition number and is diagonally-dominant. Krylov subspace methods are also relatively easy to implement.

The problem, however, is that Krylov subspace methods can be slow to converge unless the linear system given by (39) is first "preconditioned" with a preconditioning matrix $M$ that either multiplies the system from the left side (left preconditioning),

$$(M^{-1}A)\mathbf{x} = M^{-1}\mathbf{b}, \tag{40}$$

or from the right side (right preconditioning),

$$(AM^{-1})\mathbf{y} = \mathbf{b}; \quad \mathbf{y} = M\mathbf{x}. \tag{41}$$

The net effect of preconditioning is a linear operator (in parentheses above) that is closer to the identity matrix, which accelerates the convergence of the Krylov subspace method. The new linear operator, $AM^{-1}$ or $M^{-1}A$, will have a smaller condition number and eigenvalues that are more clustered than with $A$ alone. Choosing the preconditioning matrix $M$ is often *not* easy, as it must be an approximation to $A$ that is easily inverted. Unfortunately, preconditioning the Krylov subspace method is almost always necessary, as without it these methods converge too slowly. With preconditioning, one must also solve an additional preconditioning equation, given by $M\mathbf{z} = \mathbf{r}$, where $\mathbf{z}$ is a Krylov vector and $\mathbf{r}$ is a residual. Fortunately, approximate solutions for $\mathbf{z}$ are usually good enough, hence using simple iterative methods like SSOR or Jacobi to find $\mathbf{z}$ is often adequate.

If a good preconditioning matrix $M$ can be chosen, preconditioned Krylov subspace methods can be quite powerful and efficient linear solution methods. They have been perhaps the most popular choice for the past two decades, primarily because of their robustness and ease of implementation. The problem, however, is that preconditioned Krylov subspace methods do not exhibit the scaling we desire, requiring computational work that scales like $N^{5/4}$ (at best) and iterations to convergence somewhere between $N^{1/4}$ and $N^{1/2}$. Can this scaling problem of Krylov subspace methods be overcome? Perhaps, if one is willing to focus efforts on the preconditioner, as discussed in Section 3.8.4 below.

Additional information on Krylov space methods can be found in [45], where detailed algorithm templates sufficient for implementation are provided. See also reference [170] for an insightful introductory overview and [114, 166] for performance comparisons on linear systems arising from the NS equations.

### 3.8.3 Multigrid Methods

As stated previously, we seek a linear solution algorithm that requires computational work scaling like $N$, and we furthermore require that the method can find solutions in an iteration count that does not change with $N$. These requirements are usually met for linear elliptic solutions with a multigrid (MG) method [24], hence the scalability of the MG method is a powerful attraction. The basic premise of the MG method is the identification and suppression of long wavelength (low frequency) error modes in the residual via solutions of an equivalent linear system on a series of grids coarser than the base (finest) grid. This is in contrast to traditional iterative (Jacobi, SSOR, Gauss-Seidel) and Krylov subspace methods, which quickly eliminate only those high frequency error modes indicative of coupled nearest neighbor cells. Low frequency error modes, however, tend to persist without elimination until enough iterations have taken place for the long wavelength modes to be "seen". MG methods, on the other hand, immediately "see" these long wavelength modes on the coarser grids, which, once identified, can be suppressed after transfer back to the series of finer grids.

On each grid in the MG method, iterative approximate solutions to the linear system are obtained; rigorous solutions are not obtained on any one grid, but rather obtained on the base (finest) grid after many fine-to-coarse-to-fine (V) cycles. Space does not permit a detailed discussion of this powerful technique, but overviews can be found in the excellent monograph of Briggs [25] and the introductory textbook by Wesseling [201]. The reader is also encouraged to consult references [163, 174] for examples of the MG method applied to linear systems of equations arising specifically from incompressible interfacial flows.

One of the most important and difficult tasks in formulating an MG algorithm is the approximation of the intergrid transfer functions, namely the restriction (fine-to-coarse) and prolongation (coarse-to-fine) operators. Performance of the MG method, measured as scalability and convergence robustness, depends crucially upon the choice of these operators. One approach is using the intergrid transfer functions to define variational or Galerkin coarse grid operators. This task can be complicated and expensive, however, especially if the restriction and prolongation operators are anything but piecewise constant (e.g., stencil growth can occur). This fact has motivated others [152,159] to adopt the simpler approaches like those suggested in [112].

Experience has shown that the MG method at times lacks robustness, having a propensity to fail and/or exhibit slow convergence on the types of linear systems arising in incompressible interfacial flows. Here "tough" systems are those resulting from flows possessing large, abrupt, and localized changes in density and/or surface tension along an interface that is topologically complex. This lack of robustness can usually be traced to restriction and prolongation operators that are misrepresenting important interfacial physics because of inaccurate or inappropriate interpolation and/or smoothing functions. Robustness and convergence can be enhanced in many cases with more intelligent restriction and prolongation operators that do not incorrectly smooth across interfaces. Formulation and implementation of such operators in the presence of arbitrarily complex interface topologies, however, can be very expensive and cumbersome.

### 3.8.4  Hybrid Methods

Until the MG method can be made more robust and easily implemented, and Krylov subspace methods more scalable, more and more researchers are devising unified, *hybrid* methods aimed at combining the strengths of both methods while eliminating their weaknesses. Two basic types of hybrid methods have appeared in the literature to date. In the first approach, MG is the principal solution method, but a Krylov subspace method is used for solutions on one or more of the (coarser) grids rather than a simple iterative method. This approach helps to alleviate the MG robustness problem by relying on a robust Krylov method. In the second approach, a preconditioned Krylov subspace method is retained as the principal solution algorithm, but an MG-like (*multilevel*) method is used to obtain solutions to the preconditioning equation. Both approaches have merit and have exhibited improved performance, but it is not clear at this time which hybrid method exhibits the desired scalable performance without loss of robustness. One issue has become clear, however: scalable performance absolutely requires a multilevel algorithm, i.e., ideas inherent in the MG method *must* pervade.

The idea of using a symmetric MG algorithm to precondition a standard Krylov subspace (CG) method was first proposed and demonstrated by Kettler [85]. This idea, however, did not gain acceptance and popularity until the recent work of Tatebe [184]. Its use has since exploded, having shown utility in modeling incompressible flows [144,159], semiconductor performance [115], and groundwater flow [6]. It possesses the robustness lacking in many MG algorithms, able to find solutions on the most challenging of interfacial flow problems. A hybrid "MGCG" method, while usually scaling like a MG algorithm, occasionally exhibits CG-like scaling, hence additional research is needed to understand which aspects of the algorithm hinder consistent MG-like scaling. For most of the flow problems tested, the hybrid MGCG method requires less computational work than the MG method to seek a solution. This same basic approach was recently shown to be effective for the parallel solution of linear systems of equations on 3-D unstructured meshes [103]. In this case, a combined additive/multiplicative Schwarz [29, 164] technique was ideal for providing a means by which multilevel solutions to the preconditioning equation on parallel architectures could be obtained.

# 4   INTERFACE KINEMATICS

Formidable challenges must be met in the construction of methods designed to track interfaces of arbitrary topology on fixed grids. Features characterizing a given interface tracking method often result from design compromises driven by applications for which the method is intended. Accuracy and robustness, for example, are often possible only at the expense of decreased efficiency and increased complexity. If mass conservation is a design constraint, geometrically-based algorithms tend to result rather than simpler algebraically-based algorithms. Geometrically-based algorithms, on the other hand, tend to exhibit "numerical surface tension" when interface features are not resolved [156].

After introducing design requirements for an effective interface tracking algorithm in Section 4.1, most of the currently-used methods are reviewed briefly in Sections 4.2–4.9. How does one assess the merits of one method relative to another with controlled and fair quantitative comparisons? One approach is through the use of tougher (hence more representative) performance metrics and benchmarks, as discussed briefly in Section 4.10. Such performance metrics have been quite useful, serving to highlight and magnify algorithmic strengths and weaknesses in each method, as discussed in Section 4.11. In particular, a couple of excellent examples are given where "spotlighting" a particular weakness has motivated researchers to actually *fix* the problem.

## 4.1   Introduction

What specific capabilities might one design into an interface tracking algorithm targeted for interfacial flow simulation? For high fidelity interfacial flow simulations, we desire an interface tracking algorithm that:

- is globally *and* locally mass conservative;
- maintains (at a minimum) second order temporal and spatial accuracy;
- maintains compact interface discontinuity width;
- is topologically robust;

- is amenable to three-dimensions on meshes of arbitrary element type/connectivity;
- can accommodate additional interfacial physics models;
- can track interfaces bounding more than two materials;
- is computationally efficient;
- can be implemented by novices (given ample documentation); and
- can be readily maintained, improved, and extended.

We do not require that the algorithm be *implicit* in time, as this is unnecessary in most interfacial flow situations because resolution of the dynamics is desired. To date, no one algorithm discussed in the following adequately meets each design requirement itemized above. Interface tracking algorithms generally fall into one of two methods categories, with each category containing several schemes:

*tracking* methods: moving-mesh, front tracking, boundary integral, particle schemes;

*capturing* methods: continuum advection, volume tracking, level set, and phase field schemes.

Each of these schemes is discussed briefly in the sections that follow.

A *tracking* method is Lagrangian in nature, whereby the position history of discrete points $\mathbf{x}_i$ lying on the interface are tracked for all time by integrating the evolution equation

$$\frac{d\mathbf{x}_i}{dt} = \mathbf{u}_i , \tag{42}$$

where $\mathbf{u}_i$ is the velocity with which interface point $x_i$ moves. For moving-mesh, front tracking, and boundary integral schemes, the points $i$ correspond to the discrete points on the grid line representing the interface. For particle-based schemes, the points $i$ correspond to individual particles (with known identity) lying along the interface. In *capturing* methods, the interface is not explicitly tracked, but rather "captured" using a characteristic function $C$ that is the discontinuous Heaviside function in the limit of zero mesh spacing. For example, in the two-fluid case, $C$ can be expressed by:

$$C = \begin{cases} C_1 & \text{in fluid 1;} \\ C_2 & \text{in fluid 2;} \\ > C_1, \, < C_2 & \text{at the interface;} \end{cases} \tag{43}$$

where we assume $C_2 > C_1$. For finite mesh spacing, $C$ is *not* perfectly discontinuous, hence the region with $C_1 < C < C_2$ has finite width, on the order of the mesh spacing. Since a point on the interface must remain there, the evolution equation for $C$ is simply a statement of Lagrangian invariance:

$$\frac{DC}{Dt} = \frac{\partial C}{\partial t} + (\mathbf{u} \cdot \boldsymbol{\nabla}) C = 0 , \tag{44}$$

where $\mathbf{u}$ is the velocity at the interface. Exact Lagrangian information is not retained in capturing methods (discarded instead for $C$), hence the interface location is not known exactly. Its position is defined as the transition region where $C_1 < C < C_2$. In capturing methods, knowledge of $C$ allows one set of model equations to apply everywhere in the domain. Away from the interface, the equations reduce to the correct pure fluid equations, and within the interface, they contain appropriate discrete delta functions (formed from $C$) for interfacial terms [110, 139].

Although differences between each scheme can be inferred from the discussions in the next section, some categorical tracking/capturing differences pervade. In capturing methods, the grid is usually fixed, hence topological robustness is inherent. Compact interface width, on the other hand, is difficult to insure because of the potential for numerical diffusion in discretization of the $(\mathbf{u} \cdot \nabla) C$ term in regions where $C$ abruptly varies. In tracking methods, the interface is maintained as a discontinuity, yet 3-D topological robustness can be elusive.

## 4.2 Moving-Mesh Methods

Moving-mesh methods encompass all techniques that move the physical position of discrete grid points in the computational domain by integrating (42) forward in time. A moving-mesh method is Lagrangian if *every* point is moved, and mixed (Lagrangian-Eulerian) if grid points in a subset of the domain are moved. Mold filling simulations provide an excellent reason for using mixed methods [129], where the mold computational domain can be held stationary and the molten liquid is followed with a Lagrangian mesh [111, 127]. With the mesh boundary-fitted to the physical domain, the system of equations may be transformed into logical space [48, 175, 188] or expressed as integrals over discrete control volumes [198]. Useful overviews can be found in classical [151] and more recent textbooks [132]. If the interfacial flow problem to be modeled has regular interface topologies, than a moving-mesh method can yield very accurate solutions. "Regular" topologies here refer to single-valued interface topologies that do not undergo tearing, stretching, or merging. Unfortunately regular topologies are not characteristic of the free surface flows of interest to most researchers.

Discrete control volumes (elements) in the computational domain encompass the same parcel of fluid in Lagrangian moving-mesh methods. If the velocity field has appreciable shear or vorticity, then elements must distort and freely deform with the flow, yet they cannot because of their geometric limitations. Herein lies the principal drawback of these methods: element distortion leads to deterioration of solution accuracy and eventual termination of the simulation if element connectivity rules are violated. A solution to this problem is to remesh the domain, either by keeping the original mesh connectivity constant or allowing it to change. Lagrangian mesh methods that allow connectivity changes (including removal or creation of elements) are known as free Lagrange methods [50]. Both conventional Lagrangian [8, 70] and free Lagrange [49] methods have been used for incompressible interfacial flows. The principal problem with remeshing is the potential for global numerical diffusion and difficulty in demonstrating convergence.

## 4.3   Front Tracking Methods

Explicit front tracking has its roots in the original MAC method [68] and its extensions by Daly [39]. The interface is represented discretely by Lagrangian markers connected to form a front which lies within and moves through a stationary Eulerian mesh. As the front moves and deforms, interface points are added, deleted and reconnected as necessary. The interface can in principle undergo arbitrarily complex topology changes, provided the algorithm is capable of making logical topology decisions. Topological changes do not result from a set of localized interface physics models, instead resulting from algorithm intervention, via decisions such as whether or not to "cut" a front into pieces or to join two fronts into one. This algorithm as designed is *not* explicitly mass conservative, although conservation tends to be reasonably adhered to for topologically regular interfaces and high densities of marker particles (per unit interface length). Further details on the front tracking method can be found in [32, 56, 173, 175, 190].

Interface information is passed between the moving Lagrangian interface and the stationary Eulerian mesh using ideas borrowed from the immersed interface method [110, 139]. With this technique, the sharp interface is approximated by a smooth distribution function that is used to distribute the sources at the interface over mesh points nearest the interface. The front is therefore given a finite (mesh size) thickness to provide stability and smoothness. Numerical diffusion is absent since this thickness remains constant for all time. The front tracking distribution function is the previously-mentioned function $C$, hence this portion of the algorithm is akin to capturing methods. Mass conservation is violated in front tracking methods because no single $C$ contour will in general coincide with the interface formed by connecting the Lagrangian markers.

Front tracking methods have been successfully applied to a variety of interfacial flow problems: gas dynamics [32,56], incompressible flow with and without phase change [83, 84, 191, 192], and microstructure evolution in alloy solidification [81, 82]. It has also been recently demonstrated that front tracking methods can reasonably withstand 3-D topological challenges [191]. It remains to be seen, however, if future improvements to the front tracking conservation properties and implementation complexities are enough to attract widespread acceptance and use.

## 4.4   Particle-Based Methods

Particle-based methods are characterized by the use of discrete "particles" to represent macroscopic fluid parcels [118]. The Lagrangian NS equations are integrated on globs of fluid (the "particles") having properties such as mass, momentum, and energy. Using a particle-based method for modeling interfacial flows is attractive because difficult nonlinear advection terms in the NS equations are simply modeled as particle motion, and, by knowing the identity and position of each particle, material interfaces are automatically tracked. By using particle motion to approximate the advection terms, numerical diffusion across interfaces (where particles change identity) is virtually zero, hence interface widths remain compact. Particle-based methods can be roughly broken down into two principal categories: those that use particles in conjunction with a grid, namely the particle-in-cell (PIC) methods [61], and those that are "meshless" [14], such as the so-called smoothed particle hydrodynamics (SPH) methods [55, 119].

F. Harlow and coworkers invented the first particle-based method over forty years ago, the ingenious PIC method [60, 63]. The PIC method (and its countless variations) then enjoyed explosive use and development over the next twenty years or so [61], where it became *the* method for modeling highly distorted interfacial flows. It has enjoyed a slow and steady resurgence since the mid 1980s, in large part due to innovative new developments and improvements [21, 23, 92]. Relative to Harlow's *classical* PIC method, modern *full* PIC methods force particles to carry all relevant fluid information (rather than only identity, position, and mass). This formulation was first adopted in the novel FLIP algorithm [23] for interfacial flows. The SPH method, first invented by J. Monaghan and coworkers over twenty years ago, differs from PIC methods in that an accompanying grid is *not* used. Like PIC methods, SPH methods are particularly well suited (and are generally designed) for high-speed compressible flows such as those found in astrophysical applications and shock dynamics.

Why have particle-based methods not been used more often (or at all) for simulating free surface flows? Because particle-based methods can be prohibitively CPU and memory intensive, they have not had the ability to model incompressible flows until only recently [91, 120, 122, 123], they tend to be susceptible to subtle numerical instabilities [19], and they do not have adequate awareness and notoriety among many researchers interested in free surface flows. If sustained attention is paid to these outstanding issues (e.g., the memory use issue is addressed in [156]), then many of these shortcomings might be alleviated, whereby particle-based methods could play a vital role in free surface flow simulations. One powerful attraction to these methods is ease of implementation: they are typically no more complex to implement on 3-D unstructured meshes than on 2-D structured meshes.

## 4.5   Boundary Integral Methods

Methods of the boundary integral type [53, 73, 141, 146] can be highly accurate for modeling free surface flows, especially 2-D flows with relatively regular interface topologies. In this approach, the interface is explicitly tracked, as in moving-mesh or front tracking schemes, but the flow solution in the entire domain is deduced *solely* from information possessed by discrete points along the interface. For incompressible flows, the interface is characterized by a velocity potential, which is represented as a distribution of point dipoles. Boundary integral methods (BIM) were first used by Rosenhead [160] some sixty years ago to study vortex sheet roll-up, but it took another thirty years before extensions allowed the modeling of more general fluid interface problems [15]. Much evolution and enhancement has since occurred in the past two decades; see Yeung [209] and Hou [73] for reviews of earlier and more recent works, respectively. We include "vortex methods" in the class of boundary integral methods; see the excellent reviews by Leonard [105, 106]. These methods express the velocity of each discrete interface point as an integral relation that depends upon the vortex strength at that point.

The principal advantages gained by using boundary integral methods are the reduction of the flow problem by one dimension involving quantities of the interface only, and the potential for highly accurate solutions if the flow has topologically regular interfaces. Disadvantages include difficulties in extending the method to 3-D (although it has been done [106]), sensitivity to numerical instabilities because the underlying problems are very singular, and the need for local "surgery" of the interface in the event of topological changes, much like the front tracking method. BIM-based free surface flow simulations, however, continue to generate impressive solutions for a wide variety of free surface flow applications [180].

## 4.6  Continuum Advection Schemes

Simply discretizing (44) with a high resolution continuum advection scheme is quite appealing [13, 183], since such an algorithm is already an integral part of any flow solver. Continuum advection schemes refer to traditional methods for obtaining discrete numerical solutions to (44). This equation is perhaps the simplest form of a hyperbolic equation, hence the countless references and textbooks available on hyperbolic equation solution techniques can be drawn upon [47, 108, 205, 210]. We classify these schemes as continuum advection schemes since they are usually designed upon the premise that $C$ in (44) is smoothly varying. A simple yet highly diffusive example is a first-order upwind "donor cell" scheme. More accurate examples are the higher-order monotonic van Leer [194], PPM [37], and TVD [182] schemes. Techniques similar to these schemes should be used in free surface flow simulations for solutions to (44) if $C$ represents a passive scalar, energy, momentum, or density away from interfaces.

Continuum advection schemes have been employed as interface tracking methods in modeling the free surface flows found in mold filling scenarios [31, 137]. Asking such schemes to track interfaces, however, forces them to generate solutions to (44) when $C$ is discontinuous. This is problematic because these schemes are not designed for this purpose. The source of this problem lies in the algebraic treatment of the $(\mathbf{u} \cdot \nabla) C$ term; even with higher-order approximations, unacceptable broadening of the region where $C$ varies occurs. For example, a compressive, fourth-order PPM method was still found to unacceptably diffuse the interface [156]. Why? Because discontinuities in $C$ can only remain so after solutions to (44) result from *geometric* approximations to $(\mathbf{u} \cdot \nabla) C$; continuum advection schemes approximate this term *algebraically*, i.e, by taking spatial differences in $C$ directly across the interface. Higher-order approximations to these differences can obviously mitigate this numerical diffusion, yet not nearly well enough. Studies in addition to our own [156] found that interface widths can broaden to many cells (4–8) even when higher-order methods are used [51]. This is not satisfactory for an interface tracking method tasked to track discontinuous free surfaces.

One interesting approach to the interface diffusion problem is to transform the discontinuous $C$ function into another *smooth* function $\tilde{C}$, solve (44) with $\tilde{C}$, then transform $\tilde{C}$ back to $C$. In this way, the continuum advection scheme can generate solutions for which it was designed. This idea, proposed in [206, 207] and elsewhere, is also the basic premise of the level set method [167].

## 4.7 Volume Tracking Methods

Volume tracking methods remain today as perhaps the most widely used tracking method for modeling free surface flows, having been used for several decades at the United States National Laboratories (Livermore [41] and Los Alamos [72] and later Sandia [138]). The earlier work is typified by the SLIC [130] algorithm and the original method with the moniker VOF [71]. Its widespread use could be due to reasons other than performance: it is relatively easy to implement (at least in its crude forms), it is an older, well-documented algorithm, it is topologically robust, and its basis in volume fractions lends itself well to incorporation of other physics. Volume tracking methods differ from continuum advection schemes in one principal way: the $(\mathbf{u} \cdot \boldsymbol{\nabla}) C$ in (44) is approximated geometrically, based upon knowledge of a "reconstructed" interface position that is not unique.

Volume tracking methods originated in the early 1970s, when three methods were introduced within a short period of time: DeBar's method [41], Hirt and Hichols' VOF method [71, 128], and Noh and Woodward's SLIC method [130]. See [98, 157] for historical perspectives and accounts of the chronological developments that have taken place over the last three decades. In short, *substantial* evolutionary development and improvement has since taken place, rendering most of these original methods virtually obsolete [96]. Their spatial and temporal first order accuracy is just not competitive nor adequate for modern free surface flow simulations.

Without certain key developments occurring over the last decade, volume tracking methods would not have remained competitive with newer, impressive methods such as front tracking and level set techniques. Such developments include spatially second order, linearity-preserving interface geometry reconstructions, multi-dimensional time integration schemes [98, 157, 161], and extensions to 3-D unstructured meshes [147]. Perhaps the most important trend is the movement of volume tracking algorithms away from heuristic "case-by-case" logic [87] toward a mathematically formal algorithm based upon well-established geometric primitives [133, 157].

To illustrate the differences between first order (piecewise constant) and second order (piecewise linear) volume tracking schemes, consider the following simple 3-D

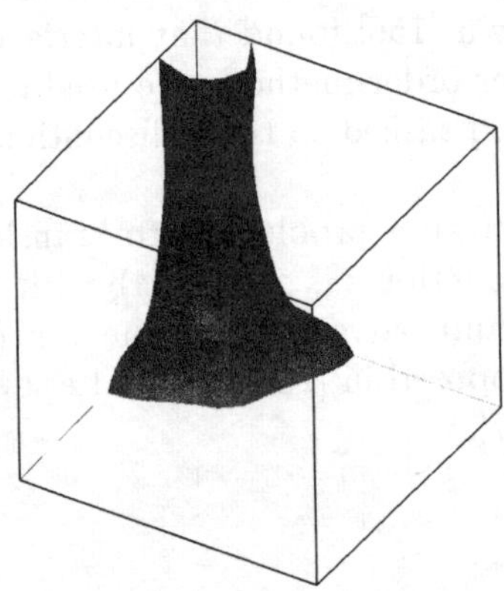

(a) Free surface as tracked by a piecewise constant volume tracking scheme at a time of 0.02 seconds.

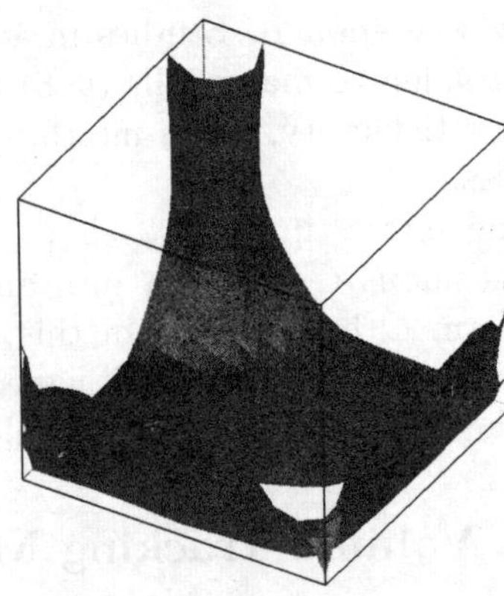

(b) Free surface as tracked by a piecewise constant volume tracking scheme at a time of 0.04 seconds.

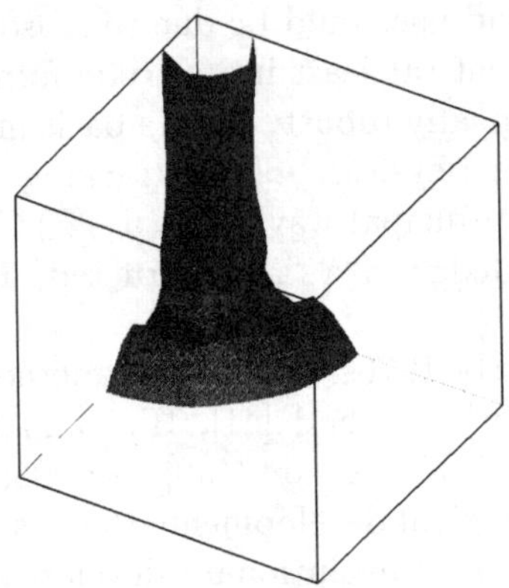

(c) Free surface as tracked by a piecewise linear volume tracking scheme at a time of 0.02 seconds.

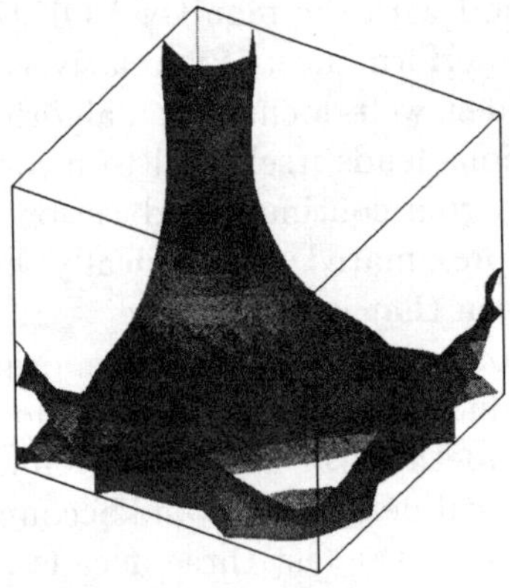

(d) Free surface as tracked by a piecewise linear volume tracking scheme at a time of 0.04 seconds.

Figure 3: Comparison of piecewise constant (top) and piecewise linear (bottom) volume tracking schemes for a simulation of a dense fluid being injected (under gravity) into the top corner of box partitioned with $16 \times 16 \times 16$ cubical cells. Shown are one-half volume fraction isosurfaces of the free surface.

free surface flow simulation. An $8 \times 8 \times 8$ centimeter (cm) box, initially occupied by a light fluid, is filled (under gravity) at the top corner with a heavier fluid (ten times as dense) injected downward with a 100 cm/s velocity. The heavier fluid fills the box at a rate of 300 cm$^3$/s, and the lighter fluid is allowed to exit the box from the opposite top corner. Both fluids are assumed inviscid and incompressible. The simulations, shown in Figure 3, were executed in parallel on four processors of an SGI Power Challenge. Although the piecewise constant and piecewise linear results in Figure 3 are perhaps more similar than one might expect, it is evident that the first order volume tracking results are less energetic, tending to impart excessive numerical surface tension to the free surface.

The recent evolutionary volume tracking developments have facilitated propagation of improvements to older implementations, reduced the likelihood of any two volume tracking algorithms generating vastly different solutions, reduced implementation difficulties, and helped new researchers to more easily identify weaknesses and devise improvements. In short, volume tracking algorithmic details (necessary for implementation) and improvements have finally become prevalent in the literature, which has enabled many interested researchers to use and take advantage of these improvements for higher fidelity simulations. A good example is the now widespread use of piecewise linear (spatially second order) schemes rather than the original piecewise constant (spatially first order) schemes.

Volume tracking methods are not without the need for improvement and further attention. Numerical surface tension is brought about when interfaces are approximated as piecewise linear, and this effect needs to be understood and quantified, especially for those interfacial flows where physical surface tension is important. A 3-D (piecewise planar) extension of this method that maintains formal second order spatial accuracy has also been elusive, with the algorithm for plane normal estimation being the culprit. Robust, multi-dimensional unsplit time-integration methods must also continue to evolve and improve, especially in 3-D. Spatially third order (or higher) schemes are needed in order to capture subgrid curvature; this is again more critical in 3-D. Finally, "thin filament" models are also needed for those situations where two (or more) interfaces pass through one computational cell.

Countering this list of weaknesses are many promising developments by researchers motivated to tackle these problems. For example, several 2-D unsplit time-integration schemes have been introduced recently by Pilliod [140], Rider and Kothe [157], Mosso [125, 126], Garrioch [52], and Price [142]. While the time-integration schemes of Mosso and Price appear the most accurate (assuming nodal velocities can be "reconstructed" [169]), none of these recent schemes appear easily extensible to 3-D. Price has also introduced a thin filament model [143] that enhances the ability of volume tracking methods to resolve fluid filaments having a width less than a mesh spacing.

Numerical surface tension is unfortunately present in any spatially second order volume tracking method because of the assumptions inherent in the reconstruction, namely a piecewise linear interface approximation constrained by mass conservation. A good example of the application of this surface tension is the breakup along interfaces that are resolve inadequately in the single vortex test problem [157]. High curvature regions are those regions having interfaces with a radii of curvature less than roughly a mesh spacing. The piecewise linear interface approximation immediately flattens these regions, effectively applying numerical surface tension. Thin filament regions can also be the recipients of numerical surface tension, because poor linear reconstructions occur in these regions from inaccurate interface normal estimations. Numerical surface tension in these high curvature and thin filament regions can be reduced (well below physical levels) with increased refinement, but Price has recently shown that a third order piecewise parabolic scheme is also successful in alleviating numerical surface tension [142].

Given these recent developments, modern volume tracking methods are certainly competitive with other interface tracking methods [156, 158]. Volume tracking will remain viable in the future, especially if these outstanding issues can be resolved.

## 4.8　Level Set Methods

Since introduction of the level set method by Osher and Sethian [134] only a decade ago, its use has exploded, evidenced by the recent textbook and references therein [167]. To date, the level set method has been used to model interfacial phenomena in the fields of material science, fluid mechanics, image enhancement, computer vision, and grid generation, to name a few. The mathematical formalism and rigor grounded in the level set method has helped to attract leading numerical analysts and mathematicians, resulting in its evolution, widespread promotion and use, and increasing range of applicability.

The basic premise of the level set method is to embed the propagating interface $\Gamma(t)$ as the zero level set of a higher dimensional function $\phi$, defined as $\phi(\mathbf{x}, t = 0) = \pm d$, where $d$ is the distance from $\mathbf{x}$ to $\Gamma(t = 0)$, chosen to be positive(negative) if $\mathbf{x}$ is outside(inside) the initial $\Gamma(t = 0)$. If the zero level set coincides with the initial interface, i.e., $\Gamma(t = 0) = \phi(\mathbf{x}, t = 0) = 0$, then a dynamical equation for $\phi(\mathbf{x}, t)$ that contains the embedded motion for $\Gamma(t)$ as the level set $\phi = 0$ can be derived [134]:

$$\frac{\partial \phi}{\partial t} + F|\boldsymbol{\nabla}\phi| = 0\,, \tag{45}$$

where $F$ is the speed of the interface $\Gamma$ in the outward normal direction. In general, $F$ is the sum of any applied interface propagation speed, a curvature-dependent speed, and the flow velocity normal to interface, i.e., $\mathbf{u} \cdot \hat{\mathbf{n}}$, where $\hat{\mathbf{n}} = \boldsymbol{\nabla}\phi/|\boldsymbol{\nabla}\phi|$ is the unit interface normal. For certain forms of $F$ this equation takes on a standard Hamilton-Jacobi form, but for most free surface flows, $F$ is only $\mathbf{u} \cdot \hat{\mathbf{n}}$, hence (45) is equivalent to (44).

Level set methods, then, propagate interfaces by integrating the same basic scalar evolution equation as other capturing methods. The *difference*, however, is that the scalar function $\phi$ in level set methods is *not* some discrete representation of a Heaviside ($H$) function, but rather a smoothly varying *distance* function. Herein lies a key advantage: highly accurate numerical solutions to (44) are possible (e.g., using the continuum advection schemes previously mentioned) since $\phi$ is smoothly varying. Many such schemes are readily available and easily implemented [108]. Herein also lies a key disadvantage: since $H$ is not directly integrated forward in time, it must be reconstructed each time step from the distance function $\phi$, $\tilde{H}(\phi)$ [180]. This procedure will *not* be mass conservative [156] since densities are defined from $\tilde{H}(\phi)$ rather than an $H$ that is directly integrated forward in time according to strict mass conservation.

Rigorous mass conservation is elusive in the level set method because $\phi$ does not necessarily remain a distance function after solutions to (45) are obtained. This is especially true if the interface has undergone large topological changes. So-called "reinitialization" algorithms have therefore been devised [180, 181] to insure $\phi$ remains a distance function, i.e., satisfying $|\nabla\phi| = 1$. Reinitialization can (and in general will) move the zero level set position; this violates mass conservation. More accurate information is needed about the front position so that movement of the zero level set does not occur during reinitialization. A global mass conservation constraint [180], acting like a Lagrange multiplier, has improved conservation properties of the reinitialization, but *local* constraints are still needed. Without local constraints, the zero level set might still move as much as a cell width during reinitialization, which artificially creates mass locally in some cells and destroys it in others. If local, *geometric* information is used to fix the zero level set position during reinitialization, the algorithm is likely to have many similarities to volume tracking algorithms [17].

## 4.9  Phase Field Formulations

Phase field formulations have been applied to crystal growth problems and Hele-Shaw flows over the past decade [27, 88, 199, 202], but only recently have they shown promise for NS flows [5, 77, 78]. Phase field models, like other Eulerian capturing methods, model interfacial forces as continuum forces by smoothing interface discontinuities and forces over thin but numerically resolvable layers. This smoothing allows conventional numerical approximations of interface kinematics on fixed grids. The phase field method also provides a continuum surface tension model [22] that is energetically and thermodynamically consistent [78]. Just recently a phase field model for dendritic solidification in the presence of melt convection (incompressible NS flow) was developed [189]. Phase field formulations are indeed showing promise and the ability to provide a powerful vehicle for the direct numerical simulation of interfacial phenomena.

In the case of free surface flows, the starting point is the van der Waal hypothesis, in which the interfacial energy density depends upon both $\phi$ (the phase field) and gradients in $\phi$. Cahn and Hilliard [28] extended this hypothesis to dynamical situations by approximating interfacial diffusion fluxes as being proportional to chemical potential gradients. Jacqmin [78] recently extended the Cahn-Hilliard equation to allow for the presence of flow. Equation (44) gives Jacqmin's evolution equation for $\phi$ (where $C = \phi$), except that the RHS side is the Laplacian of the chemical potential rather than zero. This term is quite interesting; it can be diffusive (positive) or anti-diffusive (negative), helping to regularize the interface width (not too diffuse or compact). Rather than relying on special numerical techniques in tracking algorithms to keep interface widths regular, physical mechanisms in the phase field method do the work. Simple central difference expressions for $(\mathbf{u} \cdot \nabla) C$ were found to be adequate in most cases [78].

The presence of the (anti)diffusive term on the RHS of (44) is problematic, however: at least three cells are needed through the interface so that the Laplacian can be properly discretized, otherwise the interface will "stick" to the mesh [78]. Current approaches aimed to alleviate this problem are to adaptively refine the interface region; this (or some other solution) will be required for phase field methods to be viable in 3-D. Perhaps the physical principles embodied in these formulations can be combined with the numerical techniques in tracking methods to yield an improved, unified approach. Phase field formulations for free surface flows are new and exciting, and deserve further attention and exploration.

## 4.10 Wanted: Tougher Performance Metrics!

A literature survey indicates that uniform translation is the customary barometer for interface tracking methods. Solid body rotation tests accompany translation tests in more complete analyses. Typical examples of such tests can be found in [7, 99]. An acceptable tracking method must translate and rotate fluid bodies without significant distortion or degradation of fluid interfaces. Mass should also be conserved rigorously in these cases. Translation and solid body rotation, however, enable only a *minimal* assessment of interface tracking algorithm integrity and capability because topology change is absent. Translation and rotation serve as useful debugging tests, but they are not sufficient for definitive analysis, in-depth understanding, or final judgment. Difficult (yet controlled) test problems having flows that bring about topology change elucidate algorithm strengths and weaknesses relevant to modeling interfacial flows. Also, with controlled test problems, assessment of the interface tracking method is not obscured by subtleties of physics and algorithms in the flow solver.

In addition to standard translation and solid body rotation, two new 2-D test problems were recently proposed by Kothe and Rider [156]. The problems have since become de facto interface tracking benchmark problems, witnessed by others using them to assess and draw conclusions about their own algorithms [52, 143, 161]. The test problems possess vortical flows that stretch and potentially tear any interfaces carried within the flow. The first problem, dubbed the *vortex problem*, contains a single vortex that will spin fluid elements, stretching them into a filament that spirals toward the vortex center. The flow field is taken from the "vortex-in-a-box" problem introduced in [10, 42]. The second problem, the *deformation problem*, has a flow field characterized by sixteen vortices as introduced in [176]. This flow field causes fluid elements to undergo large topological change. In the converged limit, fluid elements will not tear, instead forming thin filaments. The flow field in both problems is solenoidal, and is given cosinusoidal time-dependence following Leveque [109]. The temporal cosine term gives the flow the nice property of returning any fluid configuration to its initial state after one period. By forcing the flow to return to its initial state, quantitative comparison and evaluation can be simply computed with the help of error norms and convergence tests.

These test problems were used to provide a basis for comparison of four of the interface tracking methods discussed in this section: a fourth-order (monotone slope-limited) continuum advection scheme, a piecewise linear volume tracking method, a level set method, and a particle-based method [155]. The test problem results indicated that the particle-based method was the most accurate. A modern piecewise linear volume tracking method, while less powerful than the particle method, proved to be reasonably accurate, physically-based, robust, and relatively easy to implement. Its cost, however, was comparatively high, surpassed only by the particle scheme. Performance of the level set method fell short of the volume tracking method, but was an improvement upon conventional continuum advection methods. While topology changes (tearing or merging of interfaces) are treated naturally by level set methods, mass loss becomes troublesome under those conditions. Advection methods, on the other hand, do not have these difficulties, but instead suffer from excessive smearing of interfaces.

What motivated the inception of these test problems? Contrary to some opinion, it was not to "sell" one chosen method over all others, but instead to expose the weaknesses of each method, thereby motivating researchers to devise evolutionary algorithmic improvements needed by the interfacial flow modeling community. In this regard, these test problems have more than served their purpose, e.g., having motivated level set researchers to devise more conservative algorithms [180] and volume tracking researchers to devise spatially higher order algorithms [125, 126, 142, 143, 157]. Still needed are similar test problems for measuring the performance of these methods when they are tasked track interfaces (surfaces) in 3-D.

## 4.11  Random Observations

Using even well-designed continuum advection methods to track interfaces results in a minimum interface thickness of two to three cells. This introduces a limitation on the resolution of the method that may be unacceptable, as was shown for idealized test problems in [156] and for incompressible multiphase flows in [158]. Continuum advection schemes do not provide compact interfaces and result in excessive diffusion when bodies that are tracked become thin.

Using a level set method provides a crisp interface of constant thickness, but this thickness is commonly chosen to be greater than one mesh cell wide. Additionally, if the problem has a flow field with appreciable vorticity, the level set method has a propensity to lose or gain mass. In the case where the underlying distance function is reinitialized, level set methods have the tendency to gain mass. Without reinitialization, the method tends to lose mass.

Modern piecewise linear volume tracking methods do not suffer from these problems, but manifest other difficulties. The actual interface is completely contained in one mesh cell, therefore the interface width remains compact for all times. Volume tracking methods, by construction, rigorously conserve mass. Piecewise linear schemes, on the other hand, exhibit strong numerical surface tension when a body becomes thinner than the underlying mesh, e.g., filaments tend to clump or coagulate into strands of blobs. This is a consequence of mass conservation. In contrast, level set methods which do not exhibit this behavior, but rather lose or gain mass when a body becomes too thin to be resolvable by the mesh spacing.

Particle-based methods offer their own set of strengths and pitfalls. Chief among the strengths is their ability to faithfully advect bodies even when the body is not resolvable on the grid. For strong vortical flows, particle methods are unsurpassed. Marker-particle methods, while not rigorously conserving mass, do provide acceptable (even excellent) mass conservation. One reason for this is the minimal numerical diffusion possessed by this method. The major negative aspect of particle methods is their cost in terms of CPU time and storage.

It is useful to compare the cost (measured in CPU time) of each method. Experience indicates (not surprisingly) that additional cost is incurred when using a sophisticated interface tracking method [156]. The higher cost, however, is often easily justified by the results. The more expensive methods tend to be more robust and provide quality results regardless of the severity of the deformation. Particle methods, as expected, are the most expensive, followed by volume tracking methods. Choosing an inexpensive method (such as an upwind continuum advection scheme) to gain computational efficiency is an alternative only at the expense of decreased fidelity. The general conclusions regarding these methods can be seen vividly when these methods are integrated with the incompressible NS equations [144, 158].

# 5  SURFACE TENSION

In this section we describe methods for modeling interfacial surface tension. For complex topology interfacial flows, surface tension is commonly reformulated as a *localized volume force* as prescribed by the popular continuum surface force (CSF) model [22]. This basic approach, in fact, has since demonstrated the ability to model other interfacial physics such as phase change [84]. In the following the basic idea of the CSF model is reviewed, followed by a discussion of some important issues pertaining to the accurate estimation of interface normals, curvatures, and delta functions. We conclude with brief comments on implicit formulations, triple points, and energy consistency. In short, progress in the reliable modeling of surface tension this past decade has been remarkable, but much work remains.

## 5.1  Continuum Models for Surface Tension

The central theme of continuum surface tension models is formulation of interface dynamics as a localized volumetric force, which is quite different from earlier numerical models of interfacial phenomena. The basic premise is to model physical processes specific to and localized at fluid interfaces (e.g, surface tension) by applying the process to fluid elements everywhere within interface transition regions. Surface processes are replaced with volume processes whose integral effect properly reproduces the desired interface physics. This approach falls under the general class of immersed interface methods [110] whose origin dates back to the pioneering work of Peskin [139]. These methods lift all topological restrictions without sacrificing much in the way of accuracy and robustness. They have been verified extensively in 2-D flows via implementation in a classical algorithm for free surface flows [94,95], where complex interface phenomena such as breakup and coalescence have been modeled.

In the CSF model, interfacial surface phenomena (normally applied via a discrete boundary condition) are replaced as smoothly varying volumetric forces derived from a product of the appropriate interfacial physics per unit area and an approximation to surface (interface) integrals. The CSF formulation makes use of the fact that numerical models of discontinuities in finite volume schemes are really continuous transitions within which the fluid properties vary smoothly. It is not appropriate, therefore, to apply in these schemes a boundary condition at an interface "discontinuity", which in the case of surface tension is a pressure jump across the interface. Surface tension should instead act on fluid elements everywhere within the transition region. The relevant surface physics is a force per unit area arising from local interface curvature and local (tangential) variations in the surface tension coefficient. Application of interfacial physics then reduces to application of a localized force in the momentum equation, regardless of interface topology. The resulting simplicity, accuracy, and robustness has led to its widespread and popular use [59,94,95,113,121,149,150,165,171] in modeling complex interfacial flows.

Surface tension is reformulated as a volume force $\mathbf{F}_s$ satisfying

$$\mathbf{F}_s(\mathbf{x}_s) = \int \mathbf{f}_s(\mathbf{x})\delta(\mathbf{x} - \mathbf{x}_s)dS \,, \tag{46}$$

where $\mathbf{x}_s$ are points on the interface, the interface integral is over an arc length in 2-D or a surface in 3-D, and the delta function (having units of inverse length) is a product of one-dimensional delta functions:

$$\delta(\mathbf{x} - \mathbf{x}_s) = \Pi_i \delta(x_i - x_{is}) \,. \tag{47}$$

The surface tension force per unit interfacial area, $\mathbf{f}_s$, is given by [22]:

$$\mathbf{f}_s = \sigma\kappa\hat{\mathbf{n}} + \boldsymbol{\nabla}_s\sigma \,, \tag{48}$$

where $\sigma$ is the surface tension coefficient, $\boldsymbol{\nabla}_s$ is the surface gradient [22], $\hat{\mathbf{n}}$ is the interface unit normal, and $\kappa$ is the mean interfacial curvature, given by [200]:

$$\kappa = -\boldsymbol{\nabla} \cdot \hat{\mathbf{n}} \,. \tag{49}$$

The first term in (48) is a force acting normal to the interface, proportional to the curvature $\kappa$. The second term is a force acting tangential to the interface toward regions of higher surface tension coefficient ($\sigma$). The normal force tends to smooth and propagate regions of high curvature, whereas the tangential force tends to force fluid along the interface toward regions of higher $\sigma$.

Ideally the surface tension force $\mathbf{F}_s(\mathbf{x}_s)$ is discontinuous since the interface along $\mathbf{x}_s$ is discontinuous. But, in an Eulerian method, interfaces are instead represented in the computational domain as *finite thickness transition regions* within which fluid volume fractions smoothly vary from zero to one over a distance of $\mathcal{O}(h)$. The surface tension force $\mathbf{F}_s(\mathbf{x}_s)$ must therefore also vary smoothly, being nonzero only within these transition regions. In the original CSF model, (46) was approximated as [22]:

$$\mathbf{F}_s(\mathbf{x}) = \mathbf{f}_s(\mathbf{x}) \frac{|\nabla C(\mathbf{x})|}{[C]}, \tag{50}$$

where $C$ is the characteristic ("color") function and $[C]$ is the jump in $C$, which is unity if $C$ is taken to be a volume fraction. In effect, the integral in (46) is approximated as the gradient magnitude of the color function. This simple approximation, although proven effective in many instances, is not without its weaknesses and need for improvement, as pointed out in the next section.

In most studies, only the normal force in (48) is modeled for the volume force in (50), i.e., the surface tension coefficient $\sigma$ is taken to be constant. Details on modeling the tangential force can be found in reference [165]. The interested reader should also consult reference [99], where the volume force in (50) (for constant $\sigma$) is approximated as the divergence of a "stress" tensor. The advantages (if any) of approximating this force as the divergence of a tensor are not yet clear.

Despite the success of the CSF model and related immersed interface methods, outstanding issues remain [98], the most prominent of which are formulations for $\hat{\mathbf{n}}$ and $\kappa$ that exhibit higher accuracy and convergence properties. Another important issue is the discrete numerical form used for the interface delta function. If these issues can be resolved adequately, a wider range of surface tension-driven flows can be modeled reliably.

## 5.2 Estimating Interface Normals and Curvatures

The color function $C$, while smoothly varying, nevertheless transitions abruptly through the interface, hence estimating the first and second order spatial derivatives necessary for computing interface normals $\hat{\mathbf{n}}$ and curvatures $\kappa$ is prone to noise and inaccuracies [22]. Briefly outlined below are approaches for the difficult task of estimating interface normals and curvatures from an abruptly varying color function.

### 5.2.1  The CSF Approach

One solution is to first smooth the color function until its abrupt transition widens enough to support standard first and second order discretization stencils. In the original CSF model, the color function $C$ is convolved with a kernel $K$ to give a mollified color function $\tilde{C}$:

$$\tilde{C}(\mathbf{x}) = \int C(\mathbf{x}')K(|\mathbf{x} - \mathbf{x}'|)dV',\qquad(51)$$

where $K$ has compact support and satisfies the normalization

$$\int K(|\mathbf{x} - \mathbf{x}'|)dV' = 1.\qquad(52)$$

A quadratic B-spline was chosen for $K$ in the CSF model [22], but this choice is by no means unique. In discrete form, the convolution integral becomes a sum over all cell neighbors $n$ surrounding the reference cell $i$ within the support of the kernel $K$:

$$\tilde{C}_i = \sum_n C_n K(|\mathbf{x}_i - \mathbf{x}_n|) \quad \text{for} \quad |\mathbf{x}_i - \mathbf{x}_n| \le r,\qquad(53)$$

where $r$ is the radius of support of the kernel. Given the mollified color function $\tilde{C}$, interface normals are computed in a straightforward manner,

$$\hat{\mathbf{n}} = \frac{\boldsymbol{\nabla}\tilde{C}}{|\boldsymbol{\nabla}\tilde{C}|},\qquad(54)$$

using standard forms for the discrete gradient operator. Once interface normals are obtained, curvatures $\kappa$ follow from (49), using a conservative finite volume discretization:

$$\kappa = -\frac{1}{V}\int \boldsymbol{\nabla}\cdot\hat{\mathbf{n}}\,dV \approx -\frac{1}{V}\sum_{\mathrm{f}} \hat{\mathbf{n}}_{\mathrm{f}}\cdot\mathbf{A}_{\mathrm{f}},\qquad(55)$$

where $V$ is the control (cell) volume, $\mathbf{A}_{\mathrm{f}}$ the area of face f (pointing outward) on $V$, and $\hat{\mathbf{n}}_{\mathrm{f}}$ the unit interface normal on face $f$ of $V$. This discretization helps to preserve an important physical property of surface tension, namely that the net normal surface tension force (and also the curvature $\kappa$) should vanish over any closed surface.

### 5.2.2  Improved Approaches

One way to improve upon the accuracy of estimations for $\hat{\mathbf{n}}$ and $\kappa$ is to use what Rudman calls "higher order kernels" (HOK) for $K$ [162]. Examples are Peskin's kernel [139], the Nordmark [131] kernel used by Aleinov and Puckett [1], and the cubic and quintic B-splines (popular in the SPH field [119]) used by Rudman [163], Morton [124], and Morris [122]. These kernels do yield a more accurate and convergent curvature $\kappa$ that, like the CSF model, follows from (55), except that the interface normals used in (55) are *convolved* normals $\tilde{\mathbf{n}}$ given by

$$\tilde{\mathbf{n}}(\mathbf{x}) = \boldsymbol{\nabla}\tilde{C}(\mathbf{x}) = \boldsymbol{\nabla}\int C(\mathbf{x}')K(|\mathbf{x} - \mathbf{x}'|)dV' = \int C(\mathbf{x}')\boldsymbol{\nabla}K(|\mathbf{x} - \mathbf{x}'|)dV'.\qquad(56)$$

The convolved normal $\tilde{\mathbf{n}}$, then, results, from a convolution of the kernel gradient $\boldsymbol{\nabla} K$. This convolution places important requirements on $\boldsymbol{\nabla} K$, namely that it be bounded and at least piecewise continuous. Other kernel properties, e.g., continuity of second (and higher) derivatives, relation of support radius $r$ to mesh spacing and local radius of curvature, and monotonicity, all impact the accuracy of convolution operations. It is tempting to estimate curvatures by convolving kernel *second* derivatives (this was done in [1]), but experience has shown that this approach is not as reliable as convolving normals, then using a standard discretization to find $\kappa$. Accurately convolving kernel second derivatives places requirements on $K$ that are just too severe and restrictive. Remember that $K$ can be properly viewed as a discrete, finite approximation to a delta function, hence second derivatives are very likely to be ill-behaved (discontinuous and/or singular).

The HOK approach for estimating $\hat{\mathbf{n}}$ and $\kappa$ is a definite improvement over the original CSF prescription, witnessed by recent results that demonstrate both accuracy and convergence [1,122,124,163]. The HOK method, however, is not without problems, the most notable being a troubling sensitivity to grid orientation and the type of kernel used. Accurate and convergent $\hat{\mathbf{n}}$ and $\kappa$ estimates, therefore, continue to be an outstanding issue.

## 5.3   Interface Delta Function

The interface delta function is simply a discrete approximation to the integral in (46), which was proposed in the CSF model to be [22]

$$\delta = \frac{|\boldsymbol{\nabla} C(\mathbf{x})|}{[C]}. \tag{57}$$

This form is not unique; see, for example, reference [181] for other forms. One can think of the discrete $\delta$ as being a smoother, i.e., smoothing the surface tension force within and nearby the interface transition region. From this point of view, then, the forms chosen for $\delta$ could be the same as those chosen for the convolution kernels in the previous section. In fact, Jacqmin argues [79] that the form for $\delta$ needs to be consistent with the chosen kernel (based upon energy conservation constraints). This important issue needs to be further quantified.

## 5.4   Implicit Formulations

As pointed out in [22], a heuristic linear stability condition associated with an explicit (time level $n$) approximation of surface tension forces can be derived. With an explicit treatment, we require the time step $\delta t_{\mathrm{s}}$ to be small enough to resolve capillary waves traveling with phase speed $c_\phi$ over a distance not exceeding one-half of a mesh spacing $h$:

$$\frac{c_\phi \delta t_{\mathrm{s}}}{h} < \frac{1}{2}. \tag{58}$$

The capillary phase speed is given by

$$c_\phi = \sqrt{\frac{\sigma k}{2\langle \rho \rangle}},\tag{59}$$

where $\langle \rho \rangle$ is an average fluid density at the interface and $k = \pi/h$ is the largest possible wavenumber supportable on the mesh. Substituting (59) into (58), a conservative maximum upper bound is obtained for a linearly stable time step:

$$\delta t_\mathrm{s} < \sqrt{\frac{\langle \rho \rangle h^3}{2\pi\sigma}}.\tag{60}$$

For time steps $\delta t > \delta t_\mathrm{s}$, then, an explicit treatment of surface tension will be numerically unstable, i.e., capillary waves may grow in an unbounded fashion. Numerical experiments have proven this to be the case [204]. Physically, this instability can occur because interfacial curvature is likely to change magnitude (and even sign) over the course of a large time step, hence application of a surface tension force using an old-time curvature can induce large forcing errors that will be incorrect in magnitude and perhaps even direction! A closer look at (60) shows that an explicit surface tension time step constraint becomes restrictive for small mesh spacings (scaling like $h^{3/2}$) and large values of the surface tension coefficient (scaling like $\sigma^{-1/2}$). "Restrictive" here means that the overall integration time step $\delta t$ is held smaller than $\delta t_\mathrm{s}$, yet the dynamics of interest are occurring at much longer time scales, hence the solution algorithm cannot march forward to the desired time without consuming an excessive number of integration time steps.

Integrating forward in time with time steps larger than $\delta t_\mathrm{s}$ is allowable only if the surface tension force is approximated *implicitly*, i.e., using interface topology information at an advanced time $n+1$ level rather than the current time $n$ level. Even though interface information is known only at time level $n$, advanced-time interface normals and curvatures can be deduced with a Taylor series extrapolation from time level $n$ information:

$$\hat{\mathbf{n}}^{n+1} \approx \hat{\mathbf{n}}^n + \delta t \left(\frac{\partial \hat{\mathbf{n}}}{\partial t}\right)^n ; \qquad \kappa^{n+1} \approx \kappa^n + \delta t \left(\frac{\partial \kappa}{\partial t}\right)^n \approx \kappa^n - \delta t \boldsymbol{\nabla} \cdot \left(\frac{\partial \hat{\mathbf{n}}}{\partial t}\right)^n,\tag{61}$$

where (49) has been used and the $\partial/\partial t$ and $\boldsymbol{\nabla}$ operators have been assumed to commute. Using (6) and (54), an evolution equation for $\hat{\mathbf{n}}$ can be derived that depends on the normal component of the velocity field along the interface [18, 204]. Substitution of the $\hat{\mathbf{n}}$ evolution equation into the Taylor series expansion for $\kappa$ above gives a parabolic equation that describes the curvature as evolving according to a "surface diffusion" process by which regions of high curvature are transported to regions of low curvature [18, 204]. Estimates of the advanced-time curvature $\kappa^{n+1}$ result from implicit solutions to this parabolic equation, hence an additional linear solution is required. Initial studies by Brackbill [18] and Williams [204] indicate that the explicit time step constraint can be alleviated with this approach, and that only the curvature $\kappa$ need be implicit (instead of both $\kappa$ *and* $\hat{\mathbf{n}}$). An approach similar in concept to this work has also been recently published by Hou and coworkers [74].

Implicit surface tension algorithms are in their infancy, as many issues warrant further investigation. Chief among these is whether or not the extra computational effort incurred in estimating an advanced-time curvature offsets being able to march forward with larger time steps. Another is whether or not dynamical, topologically-complex interfacial processes (such as merging or breakup) can be accurately modeled with integration time steps much larger than capillary wave time scales (dictated by an explicit algorithm). In other words, how does dynamical accuracy degrade with the larger time steps one can use with an implicit surface tension model?

## 5.5 Triple Points

*Triple points* refer to regions in the flow where three (or more!) immiscible fluids meet. The interface formed at this junction is actually a curve in 3-D, hence the popular term "point" is correct only for 2-D idealizations. One would like resolution sufficient to insure no more than two materials residing in a computational cell at any time. But for arbitrarily complex interface topologies, these triple points can occur regularly. The most common situation occurs in *two*-fluid simulations, where the third "fluid" is actually a rigid wall. The correct surface force $\mathbf{F}_\mathrm{s}$ acting at this fluid junction is the *net* force $\sum_{k \neq l} \mathbf{F}_\mathrm{s}^{kl}$, where $\mathbf{F}_\mathrm{s}^{kl}$ is the surface force acting along the interface between fluids $k$ and $l$. Approximating $\mathbf{F}_\mathrm{s}^{kl}$ presents special topological challenges. Since the $kl$ interface "ends" at the multiple junction point, accurate estimations of $\kappa^{kl}$, $\hat{\mathbf{n}}^{kl}$, and $\delta^{kl}$ at the junction are difficult because of the sparsity of relevant color function data. In effect, special "one-sided" or "interior" difference stencils must be used to insure that interface geometry estimations are derived from the appropriate discrete color function data (only fluids $k$ and $l$ color functions). This is a task that is perhaps perfectly suited for data-dependent weighted least squares techniques [16]. Improved and more general triple point models have yet to be devised, hence this is an area that deserves further attention.

Wall adhesion (or wetting) is a term common for a three-fluid junction where the third fluid is a rigid wall. For this case, a very simple model presented in [22] has been surprisingly useful. The surface forces near the rigid wall are calculated exactly the same as along those interfaces separating two fluids, except that the boundary condition $\hat{\mathbf{n}} \cdot \hat{\mathbf{n}}_\mathrm{w} = \cos \theta_\mathrm{eq}$, where $\hat{\mathbf{n}}_\mathrm{w}$ is the wall normal and $\cos \theta_\mathrm{eq}$ is the equilibrium contact angle, is applied to $\hat{\mathbf{n}}$ prior to evaluating (54). The condition is applied only to those normals lying on a rigid boundary [94, 95]. This model insures that the surface forces will be qualitatively correct, tending to drive the interface toward an equilibrium contact angle configuration, but no assurances can be made about quantitative accuracy. This is largely because if the fluid is not quiescent near the triple point, the correct contact angle is not the static equilibrium one, but rather a dynamic angle $\theta_\mathrm{d} \neq \theta_\mathrm{eq}$ that is in for generating high resolution solutions for the incompressible NS equations, the tracking of interfaces, and the modeling of surface tension. Cited are those key algorithmic improvements and evolutionary enhancements occurring within the last decade that have brought these methods closer to rigorous validation and verification. Space did not permit discussion of other issues such as the application of boundary conditions, models for turbulence and phase change, and computer science issues such as parallelization [46, 93].

Remarkable progress has taken place: 2-D simulations of complex topology interfacial flows are now commonplace. *High fidelity* simulations, however, remain elusive, even in 2-D. The quest for improved accuracy must therefore continue before 3-D interfacial flow simulations can be reliable and quantitatively accurate. High fidelity is more important in 3-D, where relevant length scales are likely to remain under-resolved with the computing platforms of today and tomorrow. This field must continue to benefit from the evolutionary algorithm developments brought about when bright, driven researchers are challenged to model complex interfacial flows.

# 7   ACKNOWLEDGMENTS

I have had the good fortune over the last decade or so to develop friendships and collaborations with countless colleagues and mentors who have had the patience to provide helpful assistance and guidance during the course of my research endeavors. A list of these good folks would unfortunately be too long, but I do call special attention and thanks to my long-time friend and collaborator, Bill Rider, whose influence and guidance is pervasive throughout this manuscript. I also thank my wife (AdriAnne), who endured the annoyance of my pecking on the computer late at night, and my three sons (Brett, Brock, and Matt) who probably enjoyed the coach (Dad) missing a few basketball and baseball practices. And, finally, I extend my gratitude to the people at the International Centre for Mechanical Sciences (CISM) in Udine, Italy, whose overwhelming hospitality and kindness guaranteed that my experience as an invited lecturer at the CISM Free Surface Flow Workshop was a memorable one.

# References

[1] I. Aleinov and E. G. Puckett. Computing surface tension with high-order kernels. In H. A. Dwyer, editor, *Proceedings of the Sixth International Symposium on Computational Fluid Dynamics*, pages 13–18, Lake Tahoe, NV, 1995.

[2] A. S. Almgren, J. B. Bell, P. Colella, and L. H. Howell. An adaptive projection method for the incompressible Euler equations. In J. L. Thomas, editor, *Proceedings of the AIAA Eleventh Computational Fluid Dynamics Conference*, pages 530–539, 1993. See also AIAA Paper 93-3345.

[3] A. S. Almgren, J. B. Bell, and W. G. Szymczak. A numerical method for the incompressible Navier-Stokes equations based on an approximate projection. *SIAM Journal on Scientific Computing*, 17:358–369, 1996.

[4] A. A. Amsden and F. H. Harlow. The SMAC method: A numerical technique for calculating incompressible fluid flows. Technical Report LA–4370, Los Alamos National Laboratory, 1970.

[5] L. K. Antanovskii. A phase field model of capillarity. *Physics of Fluids*, 7:747–753, 1995.

[6] S. F. Ashby and R. D. Falgout. A parallel multigrid preconditioned conjugate gradient algorithm for groundwater flow simulations. *Nuclear Science and Engineering*, 124:145–159, 1996.

[7] N. Ashgriz and J. Y. Poo. FLAIR - flux line-segment model for advection and interface reconstruction. *Journal of Computational Physics*, 93:449–468, 1991.

[8] P. Bach and O. Hassager. An algorithm for the use of the Lagrangian specification in Newtonian fluid mechanics and applications to free surface flow. *Journal of Fluid Mechanics*, 152:173–210, 1985.

[9] T. J. Barth and D. C. Jesperson. The design and application of upwind schemes on unstructured meshes. Technical Report AIAA–89–0366, AIAA, 1989. Presented at the 27th Aerospace Sciences Meeting and Exhibit.

[10] J. B. Bell, P. Colella, and H. M. Glaz. A second-order projection method of the incompressible Navier-Stokes equations. *Journal of Computational Physics*, 85:257–283, 1989.

[11] J. B. Bell, P. Colella, and L. Howell. An efficient second-order projection method for viscous incompressible flow. In D. Kwak, editor, *Proceedings of the AIAA Tenth Computational Fluid Dynamics Conference*, pages 360–367, 1991. AIAA Paper 91-1560.

[12] J. B. Bell, C. N. Dawson, and G. R. Shubin. An unsplit, higher order Godunov method for scalar conservation laws in multiple dimensions. *Journal of Computational Physics*, 74:1–24, 1988.

[13] J. B. Bell and D. L. Marcus. A second-order projection method for variable-density flows. *Journal of Computational Physics*, 101:334–348, 1992.

[14] T. Belytschko, Y. Krongauz, D. Organ, M. Fleming, and P. Krysl. Meshless methods: An overview and recent developments. *Computer Methods in Applied Mechanics and Engineering*, 139:3–47, 1996.

[15] G. Birkhoff. Helmholtz and Taylor instability. In *Symposium on Applied Mathematics*, volume 13, pages 55–76, Providence, RI, 1962. American Mathematical Society. Volume XIII.

[16] Å. Björk. *Numerical Methods for Least Squares Problems*. SIAM, 1996.

[17] A. A. Bourlioux. A coupled level set volume of fluid algorithm for tracking material interfaces. In H. A. Dwyer, editor, *Proceedings of the Sixth International Symposium on Computational Fluid Dynamics*, pages 15–22, Lake Tahoe, NV, 1995.

[18] J. Brackbill, 1995. Personal Communication.

[19] J. U. Brackbill. The ringing instability in particle-in-cell calculations of low-speed flow. *Journal of Computational Physics*, 75:469–484, 1988.

[20] J. U. Brackbill and D. B. Kothe. Dynamical modeling of surface tension. Technical Report LA-UR-96-1706, Los Alamos National Laboratory, 1996.[2]

[21] J. U. Brackbill, D. B. Kothe, and H. M. Ruppel. FLIP: A low-dissipation, particle-in-cell method for fluid flow. *Computer Physics Communications*, 48:25–38, 1988.

[22] J. U. Brackbill, D. B. Kothe, and C. Zemach. A continuum method for modeling surface tension. *Journal of Computational Physics*, 100:335–354, 1992.

[23] J. U. Brackbill and H. M. Ruppel. FLIP: A method for adaptively zoned, particle-in-cell calculations of fluid flows in two dimensions. *Journal of Computational Physics*, 65:314–343, 1985.

[24] A. Brandt. Multi-level adaptive solutions to boundary-value problems. *Mathematics of Computation*, 31:333–390, 1977.

[25] W. L. Briggs. *A Multigrid Tutorial*. SIAM, Philadelphia, PA, 1987.

[26] D. L. Brown and M. L. Minion. Performance of underresolved two-dimensional incompressible flow simulations. *Journal of Computational Physics*, 122:165–183, 1995.

[27] G. Caginalp. Stefan and Hele-Shaw models as asymptotic limits of the phase-field equations. *Physical Review A*, 39:5887–5896, 1989.

[28] J. W. Cahn and J. E. Hilliard. Free energy of a nonuniform system. III. Nucleation in a two-component incompressible fluid. *Journal of Chemical Physics*, 31:688–699, 1959.

---

[2] Available at `http://lune.mst.lanl.gov/Telluride/Text/publications.html`.

[29] X.-C. Cai. A family of overlapping Schwarz algorithms for nonsymmetric and indefinite elliptic problems. In D. E. Keyes, Y. Saad, and D. G. Truhlar, editors, *Domain-Based Parallelism and Problem Decomposition Methods in Computational Science and Engineering*, pages 1–19, Philadelphia, PA, 1995. SIAM.

[30] L. S. Caretto, A. D. Gosman, S. V. Patankar, and D. B. Spalding. Two calculation procedures for steady, three-dimensional with recirculation. In *Proceedings of the Third International Conference on Numerical Methods for Fluid Dynamics*, Paris, France, 1972.

[31] K. S. Chan, K. A. Pericleous, and M. Cross. Numerical simulation of flows encountered during mold-filling. *Applied Mathematical Modelling*, 15:624–631, 1991.

[32] I-L. Chern, J. Glimm, O. McBryan, B. Plohr, and S. Yaniv. Front tracking for gas dynamics. *Journal of Computational Physics*, 62:83–110, 1986.

[33] A. J. Chorin. A numerical method for solving incompressible viscous flow problems. *Journal of Computational Physics*, 2:12–26, 1967.

[34] A. J. Chorin. Numerical solution of the Navier-Stokes equations. *Mathematics of Computation*, 22:745–762, 1968.

[35] A. J. Chorin and G. Marsden. *A Mathematical Introduction to Fluid Mechanics*. Springer-Verlag, 1993.

[36] P. Colella. Multidimensional upwind methods for hyperbolic conservation laws. *Journal of Computational Physics*, 87:171–200, 1990.

[37] P. Colella and P. Woodward. The piecewise parabolic method (PPM) for gasdynamical simulations. *Journal of Computational Physics*, 54:174–201, 1984.

[38] A. V. Coward, Y. Y. Renardy, M. Renardy, and J. R. Richards. Temporal evolution of periodic disturbances in two-layer couette flow. *Journal of Computational Physics*, 132:346–361, 1997.

[39] B. J. Daly. Numerical study of two fluid Rayleigh-Taylor instability. *Physics of Fluids*, 10:297–307, 1967.

[40] B. J. Daly and W. E. Pracht. Numerical study of density-current surges. *Physics of Fluids*, 11:15–30, 1968.

[41] R. DeBar. Fundamentals of the KRAKEN code. Technical Report UCIR–760, Lawrence Livermore National Laboratory, 1974.

[42] J. K. Dukowicz. New methods for conservative rezoning (remapping) for general quadrilateral meshes. Technical Report LA–10112–C, Los Alamos National Laboratory, 1984.

[43] A. S. Dvinsky and J. K. Dukowicz. Null-space-free methods for the incompressible Navier-Stokes equations on non-staggered curvilinear grids. *Computers and Fluids*, 22:685–696, 1993.

[44] J. Eggers. Nonlinear dynamics and breakup of free-surface flows. *Reviews of Modern Physics*, 69:865–929, 1997.

[45] R. Barrett et al. *Templates for the Solution of Linear Systems: Building Blocks for Iterative Methods*. SIAM, Philadelphia, PA, 1993.

[46] R. C. Ferrell, D. B. Kothe, and J. A. Turner. PGSLib: A library for portable, parallel, unstructured mesh simulations. In *Proceedings of the Eighth SIAM Conference on Parallel Processing for Scientific Computing*, Minneapolis, MN, March 14–17, 1997.[3]

[47] J. H. Ferziger and M. Peric. *Computational Methods for Fluid Dynamics*. Springer-Verlag, 1996.

[48] C. A. J. Fletcher. *Computational Techniques for Fluid Dynamics: Volume II*. Springer-Verlag, 1988.

[49] M. J. Fritts and J. P. Boris. The Lagrangian solution of transient problems in hydrodynamics using a triangular mesh. *Journal of Computational Physics*, 31:173–215, 1979.

[50] M. J. Fritts, W. P. Crowley, and H. E. Trease. *The Free Lagrange Method*. Springer-Verlag, New York, 1985. Lecture Notes in Physics, Volume 238.

[51] B. Fryxell, E. Müller, and D. Arnett. Instabilities and clumping in SN 1987a. I. Early evolution in two dimensions. *Astrophysical Journal*, 367:619–634, 1991.

[52] S. H. Garrioch and B. R. Baliga. A multidimensional advection technique for PLIC volume-of-fluid methods. In *Proceedings of the Fifth Annual Conference of the Computational Fluid Dynamics Society of Canada*, pages 11(3)–11(8), University of Victoria, Victoria, British Columbia, 1997.

[53] A. S. Geller, S. H. Lee, and L. G. Leal. Motion of a particle normal to a deformable surface. *Journal of Fluid Mechanics*, 169:27–69, 1986.

[54] D. Gidaspow. *Multiphase Flow and Fluidization*. Academic Press, New York, 1994.

---

[3]Available at `http://lune.mst.lanl.gov/Telluride/Text/publications.html`.

[55] R. A. Gingold and J. J. Monaghan. Smoothed particle hydrodynamics: Theory and application to non-spherical stars. *Monthly Notices of the Royal Astronomical Society*, 181:375–389, 1977.

[56] J. Glimm and O. A. McBryan. A computational model for interfaces. *Advances in Applied Mathematics*, 6:422–435, 1985.

[57] G. H. Golub and C. F. Van Loan. *Matrix Computations*. Johns Hopkins University Press, 1989.

[58] D. Gottlieb and S. A. Orzag. *Numerical Analysis of Spectral Methods: Theory and Applications*. SIAM, Philadelphia, PA, 1977.

[59] H. Haj-Hariri, Q. Shi, and A. Borhan. Effect of local property smearing on global variables: Implication for numerical simulations of multiphase flows. *Physics of Fluids, Part A*, 6:2555–2557, 1994.

[60] F. H. Harlow. Hydrodynamic problems involving large fluid distortions. *Journal of the Association for Computing Machinery*, 4:137–142, 1957.

[61] F. H. Harlow. PIC and its progeny. *Computer Physics Communications*, 48:1–11, 1988.

[62] F. H. Harlow and A. A. Amsden. A simplified MAC technique for incompressible fluid flow calculations. *Journal of Computational Physics*, 6:322–325, 1970.

[63] F. H. Harlow and M. W. Evans. A machine calculation method for hydrodynamics problems. Technical Report LAMS–1956, Los Alamos National Laboratory, 1955.

[64] F. H. Harlow and J. P. Shannon. Distortion of a splashing liquid drop. *Science*, 157:547–550, 1967.

[65] F. H. Harlow and J. P. Shannon. The splash of a liquid drop. *Journal of Applied Physics*, 38:3855–3866, 1967.

[66] F. H. Harlow, J. P. Shannon, and J. E. Welch. Liquid waves by computer. *Science*, 149:1092–1093, 1965.

[67] F. H. Harlow and J. E. Welch. Numerical calculation of time-dependent viscous incompressible flow of fluid with a free surface. *Physics of Fluids*, 8:2182–2189, 1965.

[68] F. H. Harlow and J. E. Welch. Numerical study of large-amplitude free-surface motions. *Physics of Fluids*, 9:842–851, 1966.

[69] C. W. Hirt, A. A. Amsden, and J. L. Cook. An arbitrary Lagrangian-Eulerian computing method for all flow speeds. *Journal of Computational Physics*, 14:227–253, 1974.

[70] C. W. Hirt, J. L. Cook, and T. D. Butler. A Lagrangian method for calculating the dynamics of an incompressible fluid with a free surface. *Journal of Computational Physics*, 5:103–124, 1970.

[71] C. W. Hirt and B. D. Nichols. Volume of Fluid (VOF) method for the dynamics of free boundaries. *Journal of Computational Physics*, 39:201–225, 1981.

[72] K. S. Holian, S. J. Mosso, D. A. Mandell, and R. Henninger. MESA: A 3-D computer code for armor/anti-armor applications. Technical Report LA–UR–91–569, Los Alamos National Laboratory, 1991.

[73] T.Y. Hou. Numerical solutions to free boundary problems. *Acta Numerica*, pages 335–415, 1995.

[74] T.Y. Hou, J.S. Lowengrub, and M.J. Shelley. Removing the stiffness from interfacial flows with surface tension. *Journal of Computational Physics*, 114:312–338, 1994.

[75] L. H. Howell. A multilevel adaptive projection method for unsteady incompressible flow. In N. D. Melson, T. A. Manteuffel, and S. F. McCormick, editors, *Proceedings of the Sixth Copper Mountain Conference on Multigrid Methods*, pages 243–257, Copper Mountain, CO, 1993.

[76] R. I. Issa. Solution of implicitly discretized fluid flow equations by operator splitting. *Journal of Computational Physics*, 62:40–65, 1986.

[77] D. Jacqmin. An energy approach to the continuum surface method. Technical Report 96–0858, AIAA, 1996. Presented at the 34th Aerospace Sciences Meeting.

[78] D. Jacqmin. Phase-field numerics of two-phase Navier-Stokes flows. Technical report, NASA Lewis Research Center, 1997. Submitted to the *Journal of Computational Physics*.

[79] D. Jacqmin. A variational approach to deriving smeared interface surface tension models. In V. Venkatakrishnan, M. D. Salas, and S. R. Chakravarthy, editors, *Workshop on Barriers and Challenges in Computational Fluid Dynamics*, pages 231–240, Boston, MA, 1998. Kluwer Academic Publishers.

[80] Y. N. Jeng and J. L. Chen. Truncation error analysis of the finite volume method for a model steady convective equation. *Journal of Computational Physics*, 100:64–76, 1992.

[81] D. Juric. Direct numerical simulation of solidification microstructures affected by fluid flow. In *Modeling of Casting, Welding, and Advanced Solidification Processes VIII*, New York, 1998.[4] TMS Publishers.

---

[4] Available at `http://lune.mst.lanl.gov/Telluride/Text/publications.html`.

[82] D. Juric and G. Tryggvason. A front-tracking method for dendritic solidification. *Journal of Computational Physics*, 123:127–148, 1996.

[83] D. Juric and G. Tryggvason. Numerical simulations of phase change in microgravity. *Heat Transfer in Microgravity Systems*, 332:33–44, 1996.

[84] D. Juric and G. Tryggvason. Computations of boiling flows. Technical Report LA–UR–97–1145, Los Alamos National Laboratory, 1997. Accepted for publication in the *International Journal of Multiphase Flow*.

[85] R. Kettler and J. A. Meijerink. A multigrid method and a combined multigrid-conjugate gradient method for elliptic problems with strongly discontinuous coefficients. Technical Report 604, Shell Corporation, Rijswijk, The Netherlands, 1981.

[86] J. Kim and P. Moin. Application of a fractional step method to incompressible Navier-Stokes equations. *Journal of Computational Physics*, 59:308–323, 1985.

[87] S.-O. Kim and H. C. No. Second-order model for free surface convection and interface recontruction. *International Journal for Numerical Methods in Fluids*, 26:79–100, 1998.

[88] R. Kobayashi. Modeling and numerical simulations of dendritic crystal growth. *Physica D*, 63:410–423, 1993.

[89] D. B. Kothe. PAGOSA: A massively-parallel, multi-material hydrodynamics model for three-dimensional high-speed flow and high-rate deformation. Technical Report LA–UR–92–4306, Los Alamos National Laboratory, 1992.

[90] D. B. Kothe. Computer simulation of metal casting processes: A new approach. Technical Report LALP–95–197, Los Alamos National Laboratory, 1995.[5]

[91] D. B. Kothe and J. U. Brackbill. FLIP-INC: A Particle-in-Cell method for incompressible flows, 1992. Unpublished manuscript.

[92] D. B. Kothe, J. U. Brackbill, and C. K. Choi. Implosion symmetry of heavy-ion-driven inertial confinement fusion targets. *Physics of Fluids B*, 2:1898–1906, 1990.

[93] D. B. Kothe, R. C. Ferrell, J. A. Turner, and S. J. Mosso. A high resolution finite volume method for efficient parallel simulation of casting processes on unstructured meshes. In *Proceedings of the Eighth SIAM Conference on Parallel Processing for Scientific Computing*, Minneapolis, MN, March 14–17, 1997.[6]

---

[5] Available at http://lune.mst.lanl.gov/Telluride/Text/publications.html.
[6] Available at http://lune.mst.lanl.gov/Telluride/Text/publications.html.

[94] D. B. Kothe and R. C. Mjolsness. RIPPLE: A new model for incompressible flows with free surfaces. *AIAA Journal*, 30:2694–2700, 1992.

[95] D. B. Kothe, R. C. Mjolsness, and M. D. Torrey. RIPPLE: A computer program for incompressible flows with free surfaces. Technical Report LA–12007–MS, Los Alamos National Laboratory, 1991.

[96] D. B. Kothe and W. J. Rider. Comments on modelling interfacial flows with volume-of-fluid methods. Technical Report LA–UR–3384, Los Alamos National Laboratory, 1994.[7]

[97] D. B. Kothe and W. J. Rider. Constrained minimization for monotonic reconstruction. In D. Kwak, editor, *Proceedings of the Thirteenth AIAA Computational Fluid Dynamics Conference*, pages 955–964, 1997.[8] AIAA Paper 97–2036.

[98] D. B. Kothe, W. J. Rider, S. J. Mosso, J. S. Brock, and J. I. Hochstein. Volume tracking of interfaces having surface tension in two and three dimensions. Technical Report AIAA 96–0859, AIAA, 1996.[9] Presented at the 34rd Aerospace Sciences Meeting and Exhibit.

[99] B. Lafaurie, C. Nardone, R. Scardovelli, S. Zaleski, and G. Zanetti. Modelling merging and fragmentation in multiphase flows with SURFER. *Journal of Computational Physics*, 113:134–147, 1994.

[100] M. Lai, J. B. Bell, and P. Colella. A projection method for combustion in the zero Mach number limit. In J. L. Thomas, editor, *Proceedings of the AIAA Eleventh Computational Fluid Dynamics Conference*, pages 776–783, 1993. See also AIAA Paper 93–3369.

[101] M. Lai and P. Colella. An approximate projection method for incompressible flow. Submitted to the *Journal of Computational Physics*, 1995.

[102] M. F. Lai. *A Projection Method for Reacting Flow in the Zero Mach Number Limit*. PhD thesis, University of California at Berkeley, 1993.

[103] B. Lally, R. Ferrell, D. Knoll, D. Kothe, and J. Turner. Parallel two-level additive-Schwarz preconditioning on 3D unstructured meshes for solution of solidification problems. In T. Manteuffel and S. McCormick, editors, *Proceedings of the Eighth Copper Mountain Conference on Iterative Methods*, Copper Mountain, CO, 1998.[10]

---

[7]Available at `http://lune.mst.lanl.gov/Telluride/Text/publications.html`.
[8]Available at `http://lune.mst.lanl.gov/Telluride/Text/publications.html`.
[9]Available at `http://lune.mst.lanl.gov/Telluride/Text/publications.html`.
[10]Available at `http://lune.mst.lanl.gov/Telluride/Text/publications.html`.

[104] B. E. Launder and D. B. Spalding. The numerical computation of turbulent flows. *Computer Methods in Applied Mechanics and Engineering*, 3:269–289, 1974.

[105] A. Leonard. Vortex methods for flow simulation. *Journal of Computational Physics*, 37:289–335, 1980.

[106] A. Leonard. Computing three-dimensional incompressible flows with vortex elements. *Annual Reviews in Fluid Mechanics*, 17:523–559, 1985.

[107] M. Lesieur. *Turbulence in Fluids*. Kluwer Academic Publishers, 1997.

[108] R. J. Leveque. *Numerical Methods for Conservation Laws*. Birkhäuser, 1990.

[109] R. J. Leveque. High-resolution conservative algorithms for advection in incompressible flow. *SIAM Journal on Numerical Analysis*, 33:627–665, 1996.

[110] R. J. Leveque and Z. Li. The immersed interface method for elliptic equations with discontinuous coefficients and singular sources. *SIAM Journal on Numerical Analysis*, 31:1019–1044, 1994.

[111] R. W. Lewis, S. E. Navti, and C. Taylor. A mixed Lagrangian-Eulerian approach to modelling fluid flow during mould filling. *International Journal for Numerical Methods in Fluids*, 25:931–952, 1997.

[112] C. Liu, Z. Liu, and S. McCormick. An efficient multigrid scheme for elliptic equations with discontinuous coefficients. *Communications in Applied Numerical Methods*, 8:621–631, 1992.

[113] H. Liu, E. J. Lavernia, and R. H. Rangel. Numerical investigation of micropore formation during substrate impact of molten droplets in plasma spray processes. *Atomization and Sprays*, 4:369–384, 1994.

[114] A. Meister. Comparison of different Krylov subspace methods embedded in an implicit finite volume scheme for the computation of viscous and inviscid flow fields on unstructured grids. *Journal of Computational Physics*, 140:311–345, 1998.

[115] J. C. Meza and R. S. Tuminaro. A multigrid preconditioner for the semiconductor equations. *SIAM Journal on Scientific Computing*, 17:118–132, 1996.

[116] M. L. Minion. On the stability of Godunov projection methods for incompressible flow. *Journal of Computational Physics*, 123:435–449, 1996.

[117] M. L. Minion and D. L. Brown. Performance of under-resolved two-dimensional incompressible flow simulations, II. *Journal of Computational Physics*, 138:734–765, 1997.

[118] J. J. Monaghan. Particle methods for hydrodynamics. *Computer Physics Reports*, 3:71–124, 1985.

[119] J. J. Monaghan. Smoothed particle hydrodynamics. *Annual Reviews in Astronomy and Astrophysics*, 30:543–574, 1992.

[120] J. J. Monaghan. Simulating free surface flows with SPH. *Journal of Computational Physics*, 110:399–406, 1994.

[121] N. B. Morley, A. A. Gaizer, and M. A. Abdou. Estimates of the effect of a plasma momentum flux on the free surface of a thin film of liquid metal. In *ISFNT-3: Third International Symposium on Fusion Nuclear Technology*, 1994.

[122] J. P. Morris. Simulating surface tension with smoothed particle hydrodynamics. *International Journal for Numerical Methods in Fluids*, 1997. Submitted.

[123] J. P. Morris, P. J. Fox, and Y. Zhu. Modeling low Reynolds number incompressible flows using SPH. *Journal of Computational Physics*, 136:214–226, 1997.

[124] D. E. Morton. *Numerical Simulation of an Impacting Drop*. PhD thesis, The University of Melbourne, Australia, 1997.

[125] S. J. Mosso, B. K. Swartz, D. B. Kothe, and S. P. Clancy. Recent enhancements of volume-tracking algorithms for irregular grids. Technical Report LA-CP-96-227, Los Alamos National Laboratory, 1996.[11]

[126] S. J. Mosso, B. K. Swartz, D. B. Kothe, and R. C. Ferrell. A parallel, volume-tracking algorithm for unstructured meshes. In P. Schiano, A. Ecer, J. Periaux, and N. Satofuka, editors, *Parallel Computational Fluid Dynamics: Algorithms and Results Using Advanced Computers*, pages 368–375, Capri, Italy, 1997.[12] Elsevier Science.

[127] F. Muttin, T. Coupez, M. Bellet, and J. L. C. Chenot. Lagrangian finite element analysis of time dependent viscous free surface flow using an automatic remeshing technique. *International Journal for Numerical Methods in Engineering*, 36:2001–2015, 1993.

[128] B. D. Nichols and C. W. Hirt. Methods for calculating multi-dimensional, transient free surface flows past bodies. Technical Report LA–UR–75–1932, Los Alamos National Laboratory, 1975.

[129] W. F. Noh. CEL: A time-dependent, two-space-dimensional, coupled Euler-Lagrange code. *Methods in Computational Physics*, 3:117–179, 1964.

---

[11] Available at http://lune.mst.lanl.gov/Telluride/Text/publications.html.
[12] Available at http://lune.mst.lanl.gov/Telluride/Text/publications.html.

[130] W. F. Noh and P. R. Woodward. SLIC (simple line interface method). In A. I. van de Vooren and P. J. Zandbergen, editors, *Lecture Notes in Physics 59*, pages 330–340, 1976.

[131] H. O. Nordmark. Rezoning for higher order vortex methods. *Journal of Computational Physics*, 97:366–397, 1991.

[132] E. S. Oran and J. P. Boris. *Numerical Simulation of Reactive Flow*. Elsivier, 1987.

[133] J. O'Rourke. *Computational Geometry in C*. Cambridge, 1993.

[134] S. Osher and J. A. Sethian. Fronts propagating with curvature-dependent speed: Algorithms based on Hamilton-Jacobi formulation. *Journal of Computational Physics*, 79:12–49, 1988.

[135] S. V. Patankar. *Numerical Heat Transfer and Fluid Flow*. Hemisphere, 1980.

[136] M. Peric, R. Kessler, and G. Scheuerer. Comparison of finite-volume numerical methods with staggered and colocated grids. *Computers and Fluids*, 16:389–403, 1988.

[137] K. A. Pericleous, K. S. Chan, and M. Cross. Free surface flow and heat transfer in cavities: The SEA algorithm. *Numerical Heat Transfer, Part B*, 27:487–507, 1995.

[138] J. S. Perry, K. G. Budge, M. K. W. Wong, and T. G. Trucano. RHALE: A 3-D MMALE code for unstructured grids. In ASME, editor, *Advanced Computational Methods for Material Modeling, AMD-Vol. 180/PVP-Vol. 268*, pages 159–174, 1993.

[139] C. S. Peskin. Numerical analysis of blood flow in the heart. *Journal of Computational Physics*, 25:220–252, 1977.

[140] J. E. Pilliod, Jr. An analysis of piecewise linear interface reconstruction algorithms for volume-of-fluid methods. Master's thesis, University of California at Davis, 1992.

[141] C. Pozrikidis. *Boundary Integral and Singularity Methods for Linearized Viscous Flow*. Cambridge University Press, 1992.

[142] G. R. Price, G. T. Reader, R. D. Rowe, and J. D. Bugg. A piecewise parabolic interface calculation for volume tracking. In *Proceedings of the Sixth Annual Conference of the Computational Fluid Dynamics Society of Canada*, University of Victoria, Victoria, British Columbia, 1998.

[143] G. R. Price and R. D. Rowe. An extended piecewise linear volume-of-fluid algorithm for reconstruction of thin interfaces. In *Proceedings of the Fifth Annual Conference of the Computational Fluid Dynamics Society of Canada*, pages 11(9)–11(14), University of Victoria, Victoria, British Columbia, 1997.

[144] E. G. Puckett, A. S. Almgren, J. B. Bell, D. L. Marcus, and W. J. Rider. A second-order projection method for tracking fluid interfaces in variable density incompressible flows. *Journal of Computational Physics*, 130:269–282, 1997.

[145] G. D. Raithby and G. E. Schneider. Numerical solution of problems in incompressible fluid flow: Treatment of the velocity-pressure coupling. *Numerical Heat Transfer*, 2:417–440, 1979.

[146] J. M. Rallison and A. Acrivos. A numerical study of the deformation and burst of a viscous drop in an extensional flow. *Journal of Fluid Mechanics*, 89:191–200, 1978.

[147] A. V. Reddy, D. B. Kothe, C. Beckermann, R. C. Ferrell, and K. L. Lam. High resolution finite volume parallel simulations of mould filling and binary solidification on unstructured 3D meshes. In *Solidification Processing 1997: Proceedings of the Fourth Decennial International Conference on Solidification Processing*, pages 83–87, The University of Sheffield, Sheffield, UK, July 7–10, 1997.[13]

[148] C. M. Rhie and W. L. Chow. A numerical study of the turbulent flow past an isolated airfoil with trailing edge separation. *AIAA Journal*, 21:1525–1532, 1983.

[149] J. R. Richards, A. N. Beris, and A. M. Lenhoff. Steady laminar flow of liquid-liquid jets at high Reynolds numbers. *Physics of Fluids, Part A*, 5:1703–1717, 1993.

[150] J. R. Richards, A. M. Lenhoff, and A. N. Beris. Dynamic breakup of liquid-liquid jets. *Physics of Fluids, Part A*, 6:2640–2655, 1994.

[151] R. D. Richtmyer and K. W. Morton. *Difference Methods for Initial Value Problems*. Wiley-Interscience, 1967.

[152] W. J. Rider. Approximate projection methods for incompressible flow: Implementation, variants and robustness. Technical Report LA–UR–94–2000, Los Alamos National Laboratory, 1994.

[153] W. J. Rider. The robust formulation of approximate projection methods for incompressible flows. Technical Report LA–UR–94–3015, Los Alamos National Laboratory, 1994.

---

[13]Available at `http://lune.mst.lanl.gov/Telluride/Text/publications.html`.

[154] W. J. Rider. Filtering nonsolenoidal modes in numerical solutions of incompressible flows. Technical Report LA–UR–94–3014, Los Alamos National Laboratory, 1998. Accepted for publication in the *International Journal for Numerical Methods in Fluids*.

[155] W. J. Rider and D. B. Kothe. A marker particle method for interface tracking. In H. A. Dwyer, editor, *Proceedings of the Sixth International Symposium on Computational Fluid Dynamics*, pages 976–981, Lake Tahoe, NV, 1995.

[156] W. J. Rider and D. B. Kothe. Stretching and tearing interface tracking methods. Technical Report AIAA 95–1717, AIAA, 1995.[14] Presented at the 12th AIAA CFD Conference.

[157] W. J. Rider and D. B. Kothe. Reconstructing volume tracking. *Journal of Computational Physics*, 141:112–152, 1998.

[158] W. J. Rider, D. B. Kothe, S. J. Mosso, J. H. Cerruti, and J. I. Hochstein. Accurate solution algorithms for incompressible multiphase fluid flows. Technical Report AIAA 95–0699, AIAA, 1995.[15] Presented at the 33rd Aerospace Sciences Meeting and Exhibit.

[159] W. J. Rider, D. B. Kothe, E. G. Puckett, and I. D. Aleinov. Accurate and robust methods for variable density incompressible flows with discontinuities. In V. Venkatakrishnan, M. D. Salas, and S. R. Chakravarthy, editors, *Workshop on Barriers and Challenges in Computational Fluid Dynamics*, pages 213–230, Boston, MA, 1998.[16] Kluwer Academic Publishers.

[160] L. Rosenhead. The point vortex approximation of a vortex sheet. *Proceedings of the Royal Society of London, Series A*, 134:170–192, 1932.

[161] M. Rudman. Volume tracking methods for interfacial flow calculations. *International Journal for Numerical Methods in Fluids*, 24:671–691, 1997.

[162] M. Rudman. A kernal-based surface tension algorithm. Technical Report (unpublished), Commonweath Scientific and Industrial Research Organization (CSIRO), Highett, Victoria, Australia, 1998.

[163] M. Rudman. A volume-tracking method for incompressible multifluid flows with large density variations. *International Journal for Numerical Methods in Fluids*, 1998. In Press.

[164] Y. Saad. *Iterative Methods for Sparse Linear Systems*. PWS Publishing Company, 1996.

---

[14]Available at `http://lune.mst.lanl.gov/Telluride/Text/publications.html`.
[15]Available at `http://lune.mst.lanl.gov/Telluride/Text/publications.html`.
[16]Available at `http://lune.mst.lanl.gov/Telluride/Text/publications.html`.

[165] G. P. Sasmal and J. I. Hochstein. Marangoni convection with a curved and deforming free surface in a cavity. *Journal of Fluids Engineering*, 116:577–582, 1994.

[166] W. Schönauer and R. Weiss. An engineering approach to generalized conjugate gradient methods and beyond. *Applied Numerical Mathematics*, 19:175–206, 1995.

[167] J. A. Sethian. *Level Set Methods*. Cambridge University Press, 1996.

[168] M. Shashkov. *Conservative Finite-Difference Methods on General Grids*. CRC Press, 1996.

[169] M. Shashkov, B. Swartz, and B. Wendroff. Local reconstruction of a vector field from its normal components on the faces of grid cells. *Journal of Computational Physics*, 139:406–409, 1998.

[170] J. R. Shewchuk. An introduction to the conjugate gradient method without the agonizing pain. Technical Report CMU-CS-94-125, Carnegie Mellon University, 1994.

[171] Q. Shi and H. Haj-Hariri. The effects of convection on the motion of deformable drops. Technical Report AIAA 94–0834, AIAA, 1994. Presented at the 32nd Aerospace Sciences Meeting and Exhibit.

[172] T. M. Shih, C. H. Tan, and B. C. Hwang. Effects of grid staggering on numerical schemes. *International Journal for Numerical Methods in Fluids*, 9:193–212, 1989.

[173] W. Shyy. *Computational Modeling for Fluid Flow and Interfacial Transport*. Elsevier Science, 1994.

[174] W. Shyy, S. S. Thakur, H. Ouyang, J. Liu, and E. Blosch. *Computational Techniques for Complex Transport Phenomena*. Cambridge University Press, 1997.

[175] W. Shyy, H. Udaykumar, M. Rao, and R. Smith. *Computational Fluid Dynamics with Moving Boundaries*. Taylor and Francis, 1996.

[176] P. K. Smolarkiewicz. The multi-dimensional Crowley advection scheme. *Monthly Weather Review*, 110:1968–1983, 1982.

[177] G. Strang. *Linear Algebra and Its Applications*. Harcourt Brace Jovanovich, 1976.

[178] G. Strang. *Introduction to Applied Mathematics*. Wellesley-Cambridge, 1986.

[179] J. C. Strikwerda. *Finite Difference Schemes and Partial Differential Equations.* Wadswork and Brooks/Cole, 1989.

[180] M. Sussman and P. Smereka. Axisymmetric free boundary problems. *Journal of Fluid Mechanics*, 341:269–294, 1997.

[181] M. Sussman, P. Smereka, and S. Osher. A level set approach for computing solutions to incompressible two-phase flow. *Journal of Computational Physics*, 114:146–159, 1994.

[182] P. K. Sweby. High resolution TVD schemes using flux limiters. In B. Engquist, editor, *Lectures in Applied Mathematics*, volume 22, pages 289–309, 1985.

[183] W. G. Szymczak, J. C. W. Rogers, J. M. Solomon, and A. E. Berger. A numerical algorithm for hydrodynamic free boundary problems. *Journal of Computational Physics*, 106:319–336, 1993.

[184] O. Tatebe. The multigrid preconditioned conjugate gradient method. In N. D. Melson, T. A. Manteuffel, and S. F. McCormick, editors, *Proceedings of the Sixth Copper Mountain Conference on Multigrid Methods*, pages 621–634, Copper Mountain, CO, 1993.

[185] E. Y. Tau. A second-order projection method for the incompressible Navier-Stokes equations in arbitary domains. *Journal of Computational Physics*, 115:147–152, 1994.

[186] R. Temam. Sur l'approximation de la solution des équations de Navier-Stokes par la méthod des pas fractionnaires (II). *Archives for Rational Mechanics and Analysis*, 33:377–385, 1969.

[187] R. Temam. Sur l'approximation de la solution des équations de Navier-Stokes par la méthod des pas fractionnaires (I). *Archives for Rational Mechanics and Analysis*, 32:135–153, 1969.

[188] J. F. Thompson, Z. U. A. Warsi, and C. W. Mastin. *Numerical Grid Generation.* Elsevier Science, New York, 1985.

[189] X. Tong, C. Beckermann, and A. Karma. Phase field simulation of dendritic growth with convection. In B. Thomas and C. Beckermann, editors, *Modeling of Casting, Welding, and Advanced Solidification Processes VIII*, New York, 1998. TMS Publishers.

[190] G. Tryggvason, B. Bunner, O. Ebrat, and W. Tauber. Computations of multi-phase flow by a finite difference/front tracking method. I. Multi-fluid flows. In *Lecture Notes for the 29th Computational Fluid Dynamics Lecture Series*, Karman Institute for Fluid Mechanics, Belgium, February 23–27, 1998.

[191] S. O. Unverdi and G. Tryggvason. A front-tracking method for viscous, incompressible, multi-fluid flows. *Journal of Computational Physics*, 100:25–37, 1992.

[192] S. O. Unverdi and G. Tryggvason. Computations of multi-fluid flows. *Physica D*, 60:70–83, 1992.

[193] J. P. van Doormal and G. D. Raithby. Enhancements of the simple method for predicting incompressible fluid flows. *Numerical Heat Transfer*, 7:147–163, 1984.

[194] B. van Leer. Towards the ultimate conservative difference scheme. IV. A new approach to numerical convection. *Journal of Computational Physics*, 23:276–299, 1977.

[195] B. van Leer. Multidimensional explicit difference schemes for hyperbolic conservation laws. In R. Glowinski and J.-L. Lions, editors, *Computing Methods in Applied Sciences and Engineering VI*, pages 493–497, 1984.

[196] R. Varga. *Matrix Iterative Analysis*. Prentice-Hall, Englewood Cliffs, NJ, 1962.

[197] V. Venkatakrishnan. On the accuracy of limiters and convergence to steady state solutions. Technical Report 93–0880, AIAA, 1993. Presented at the 31st Aerospace Sciences Meeting.

[198] M. Vinokur. An analysis of finite-difference and finite-volume formulations of conservation laws. *Journal of Computational Physics*, 81:1–52, 1989.

[199] S. L. Wang, R. F. Sekerka, A. A. Wheeler, B. T. Murray, S. R. Coriell, and G. B McFadden. Thermodynamically consistent phase-field models for solidification. *Physica D*, 69:189–200, 1993.

[200] C. E. Weatherburn. On differential invariants in geometry of surfaces, with some applications to mathematical physics. *Quarterly Journal of Mathematics*, 50:230–269, 1927.

[201] P. Wesseling. *An Introduction to Multigrid Methods*. Wiley, 1992.

[202] A. A. Wheeler, B. T. Murray, and R. J. Schaefer. Computation of dendrites using a phase-field model. *Physica D*, 66:243–262, 1993.

[203] D. C. Wilcox. *Turbulence Modeling for CFD*. DCW Industries, Inc., 1993.

[204] T. L. Williams. An implicit surface tension model. Master's thesis, Memphis State University, 1993.

[205] P. Woodward and P. Colella. The numerical simulation of two-dimensional fluid flow with strong shocks. *Journal of Computational Physics*, 54:115–173, 1984.

[206] T. Yabe. Unified solver CIP for solid, liquid, and gas. *Computational Fluid Dynamics Review*, 1998. To appear.

[207] T. Yabe and F. Xiao. Description of complex and sharp interface with fixed grids in incompressible and compressible fluids. *Computers in Mathematics Applications*, 29:15–25, 1995.

[208] N. N. Yanenko. *The Method of Fractional Steps*. Springer-Verlag, 1971.

[209] R. W. Yeung. Numerical methods in free-surface flows. *Annual Reviews in Fluid Mechanics*, 14:395–442, 1982.

[210] S. T. Zalesak. A preliminary comparison of modern shock-capturing schemes: Linear advection. In R. Vichnevetsky and R. S. Stepleman, editors, *Advances in Computer Methods for Partial Differential Equations*, volume 6, pages 15–22. IMACS, 1987.

[211] O. C. Zienkiewicz. *The Finite Element Method*. McGraw-Hill, New York, NY, 1977.